Sustainable Energy Systems: Emerging Technologies and Practices in Renewable Energy Storage

Contents

Editor
Muhammad Khalid
King Fahd University of Petroleum
and Minerals (KFUPM)
Dhahran, Saudi Arabia

Editorial Office
MDPI
St. Alban-Anlage 66
4052 Basel, Switzerland

This is a reprint of articles from the Special Issue published online in the open access journal *Sustainability* (ISSN 2071-1050) (available at: https://www.mdpi.com/journal/sustainability/special_issues/energy_systems_technologies_practices_storage).

For citation purposes, cite each article independently as indicated on the article page online and as indicated below:

Lastname, A.A.; Lastname, B.B. Article Title. *Journal Name* **Year**, *Volume Number*, Page Range.

ISBN 978-3-0365-8914-5 (Hbk)
ISBN 978-3-0365-8915-2 (PDF)
doi.org/10.3390/books978-3-0365-8915-2

Sustainable Energy Systems: Emerging Technologies and Practices in Renewable Energy Storage

Editor

Muhammad Khalid

Basel • Beijing • Wuhan • Barcelona • Belgrade • Novi Sad • Cluj • Manchester

About the Editor

Muhammad Khalid

Dr. Muhammad Khalid earned his Ph.D. in Electrical Engineering from the School of Electrical Engineering and Telecommunications at the University of New South Wales (UNSW), Sydney, Australia, in 2011. Currently, he serves as an associate professor in the Electrical Engineering Department at King Fahd University of Petroleum and Minerals (KFUPM), Dhahran, Saudi Arabia. Additionally, he is a research affiliate with the Interdisciplinary Research Center for Renewable Energy and Power Systems within KFUPM. His diverse research interests include renewable power systems, energy storage, hydrogen technologies, net-zero energy buildings, sustainability, electric vehicles, AI, machine learning, and smart grids. He has made significant contributions to these fields, authoring and co-authoring numerous research articles. He has received numerous academic accolades and prestigious research fellowships.

Article

A Comprehensive Review on Residential Demand Side Management Strategies in Smart Grid Environment

Sana Iqbal [1], Mohammad Sarfraz [1], Mohammad Ayyub [1], Mohd Tariq [1,*], Ripon K. Chakrabortty [2], Michael J. Ryan [2] and Basem Alamri [3]

[1] Department of Electrical Engineering, ZHCET, Aligarh Muslim University, Aligarh 202002, India; sanaiqbal339@yahoo.com (S.I.); msarfraz@zhcet.ac.in (M.S.); mayyub09@gmail.com (M.A.)

[2] Capability Systems Centre, School of Engineering and Information Technology, University of New South Wales, Canberra, ACT 2612, Australia; r.chakrabortty@adfa.edu.au (R.K.C.); mike.ryan@ieee.org (M.J.R.)

[3] Department of Electrical Engineering, College of Engineering, Taif University, Taif 21944, Saudi Arabia; b.alamri@tu.edu.sa

* Correspondence: tariq.ee@zhcet.ac.in

Abstract: The ever increasing demand for electricity and the rapid increase in the number of automatic electrical appliances have posed a critical energy management challenge for both utilities and consumers. Substantial work has been reported on the Home Energy Management System (HEMS) but to the best of our knowledge, there is no single review highlighting all recent and past developments on Demand Side Management (DSM) and HEMS altogether. The purpose of each study is to raise user comfort, load scheduling, energy minimization, or economic dispatch problem. Researchers have proposed different soft computing and optimization techniques to address the challenge, but still it seems to be a pressing issue. This paper presents a comprehensive review of research on DSM strategies to identify the challenging perspectives for future study. We have described DSM strategies, their deployment and communication technologies. The application of soft computing techniques such as Fuzzy Logic (FL), Artificial Neural Network (ANN), and Evolutionary Computation (EC) is discussed to deal with energy consumption minimization and scheduling problems. Different optimization-based DSM approaches are also reviewed. We have also reviewed the practical aspects of DSM implementation for smart energy management.

Keywords: demand response; demand-side management; energy consumption optimization; energy efficiency; load scheduling; smart grid; smart home

Citation: Iqbal, S.; Sarfraz, M.; Ayyub, M.; Tariq, M.; Chakrabortty, R.K.; Ryan, M.J.; Alamri, B. A Comprehensive Review on Residential Demand Side Management Strategies in Smart Grid Environment. *Sustainability* **2021**, *13*, 7170. https://doi.org/10.3390/su13137170

Academic Editor: Yateendra Mishra

Received: 2 May 2021
Accepted: 9 June 2021
Published: 25 June 2021

Publisher's Note: MDPI stays neutral with regard to jurisdictional claims in published maps and institutional affiliations.

1. Introduction

The emerging era of the smart grid not only assists utilities in conserving energy, reducing cost, increasing grid transparency, sustainability and efficiency, but also has captivated consumer attention via Demand Side Management (DSM), which is an important aspect of the smart grid. However, the exponentially increasing demand for electricity at consumer premises is still a pressing issue for both utilities and consumers. According to the forecast by the National Institution for Transforming India (NITI) Ayog, the electricity demand in India for the residential sector is predicted to grow 6–13 times by the year 2047 [1]. Smart energy management refers to planning, monitoring, controlling, and optimizing energy through smart solutions or intelligent means whose ultimate objective is to maximize productivity and comfort on the one hand, and to minimize the energy cost and pollution on the other hand [2]. To achieve these objectives effectively, there is a need for the electric grid to transition from the traditional centralized version to one that uses smart technologies and is known as the smart grid [3]. A smart grid is an electricity network based on digital technology that has the provision for full-duplex communication, as well as bidirectional power flow between utilities and customers [4]. To ensure grid sustainability, the residential customers, as a part of electricity demand, must have a better understanding and awareness of the evolving grid's worth. In India, the domestic sector accounts for 24.76% of the total electricity consumption in 2019 [5].

The smart home (utilizing home automation, or domotics), one of the key components of a smart grid, is a dwelling that serves the residents with security, healthcare, comfort, and remote control of the home appliances through smart technology [6,7]. Smart home energy management plays an important role in Demand Side Management (DSM), one of the aspects of the smart grid [8], which deals with controlling and optimizing the various smart home appliances according to the user needs and preferences to reduce the electricity consumption and therefore the cost, enhancing energy efficiency, and maintaining a clean and green environment [9]. Although various researchers have been working in this field for years in achieving said objectives, still there is a need for state-of-the-art technologies and developments to provide optimal solutions in maximizing user comfort levels and assisted living as well as energy consumption and wastage reduction. Figure 1 shows the block diagram of the energy management framework.

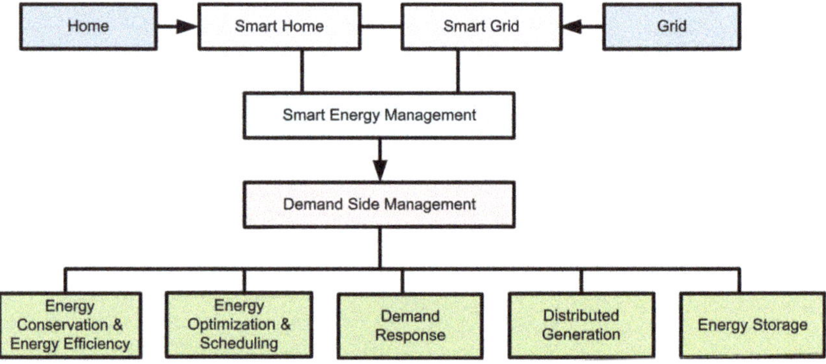

Figure 1. Energy management framework.

The main contributions of this paper are as follows:

1. Description of various DSM strategies.
2. Conduct of a comprehensive review of previous and current research works on DSM through soft computing and optimization techniques.
3. Proposal of new viewpoints and challenges for further research.

The rest of the paper is organized as follows: Section 2 describes DSM strategies. Section 3 addresses the hardware and communication technology in DSM. Section 4 provides the application of soft computing techniques for DSM. Section 5 discusses the optimization-based DSM approaches. Section 6 reviews DSM approaches and their hardware implementation. Section 7 discusses the challenges for future research. The paper is concluded in Section 8.

2. Demand Side Management

Demand Side Management is the planning, controlling, and execution that directly or indirectly influences the user-side demand of the electric meter. The DSM program reduces the energy costs of electricity, which in the long run will restrict the need for more capacity building transmission and distribution networks [10].

The significant objectives of demand-side management [11] are as follows:

1. Reduction in generation margin;
2. Improvement of the economic viability of the grid and its operating efficiency;
3. Improvement of the economic viability of the distribution network;
4. Maintenance of demand-supply balance with renewable;
5. Increasing the efficiency of the overall energy supply system.

Reducing the generation side peak demand is very expensive and according to a study done in [12], at least 10% of supply cost provides only 1% hours per year. To deal with such a challenge, DSM offers a cost-effective opportunity. DSM reduces the overall peak load demand by modifying the energy consumption pattern of the consumer that enhances the grid stability, which in turn reduces the energy consumption cost and carbon footprints [13,14]. Various DSM strategies (as shown in Figure 1) include—Energy Conservation and Energy Efficiency, Energy Consumption Optimization and Scheduling, Demand Response, Distributed Generation, and Energy Storage. Figure 2 illustrates the role of DSM strategies [15]. These roles include peak shaving, valley filling, strategic conservation, load shifting, and time-shifting. Peak shaving and valley filling are the direct load control techniques. Strategic conservation involves direct consumer-side demand reduction. Load shifting and time shifting shift the demand from peak hours to off-peak hours. Peak shaving is carried out through energy efficiency, incentive-based DR, and distributed generation, valley filling through price-based DR, strategic conservation through energy conservation and energy optimization, load shifting, and time-shifting through scheduling and energy storage, respectively.

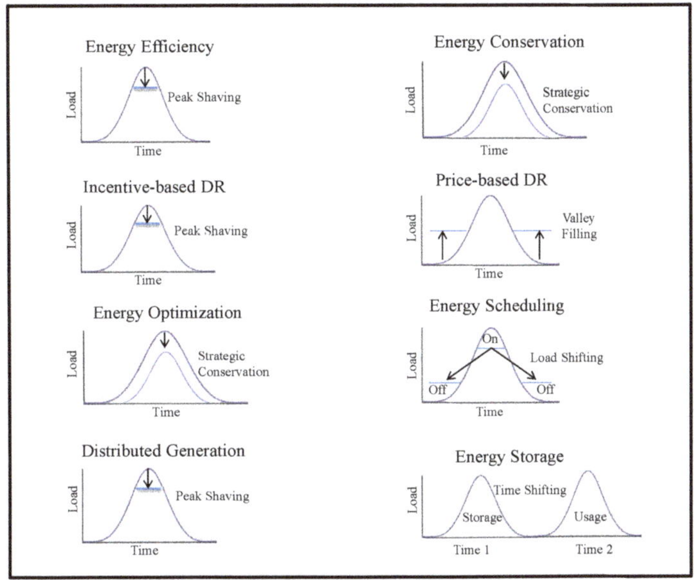

Figure 2. Role of DSM strategies.

2.1. Energy Conservation and Energy Efficiency

Energy conservation is at the heart of energy management that should be considered as a moral, religious, and societal duty. Both energy conservation and energy efficiency aim at saving energy and the environment, but with different methodologies. To clarify the subsequent confusion among consumers, we compared the two with examples in Table 1, which shows the basic differences between energy conservation and energy efficiency.

Table 1. Comparison between energy conservation and energy efficiency.

Attributes	Energy Conservation	Energy Efficiency
Meaning	Changing behavior or habits for using less energy	Using the technology that uses less energy
User-interaction	Yes	May or may not
Type of load	Traditional loads	Digital loads
User comfort	Compromise	Maximum
Examples	• Switching off lights or fans when leaving the room • Using natural day light • Walking instead of driving	• Replacing incandescent bulbs with LEDs • Using Solar panels • Using Electric vehicles

The government of India has adopted certain approaches to maintain the consumer demand with the view to minimize carbon dioxide growth rate to protect citizens and environment from its hazardous effects [16]. These approaches include:

- Greater use of renewable energy sources.
- Shifting towards super-critical technologies for conventional power plants.
- Energy efficient innovative measures under the overall realm of the Energy Conservation Act 2001.

The Ministry of Power has implemented many energy efficient programs through the Bureau of Energy Efficiency (BEE) in the fields of household lighting, commercial buildings, standards, and product marketing, and demand-side management.

2.1.1. Energy Conservation and Energy Efficiency Programs

Is bold necessary? If not, we'd like to change them to normal. The following highlighted parts are the same. You can convert them to normal. Bold is not necessary.

- **Standards and Labeling programs**—To provide consumers with a choice regarding the energy-saving potential and thus the cost-saving potential of the related product in the market. These programs aid the vision of energy surplus India with 24 * 7 power to all [1]. Please check if this should be multiplication sign. No this is not the multiplication sign.
- **Energy Conservation Buildings Code**—To set minimum energy standards for large commercial buildings having a connected load of 100 kW or contract demand of 120 KVA and above. For the residential sector, Eco-Niwas Samhita is launched to set various standards for limited heat gain and heat loss and for achieving natural ventilation and daylighting. Figure 3 shows the Eco-Niwas Samhita Scheme in the Residential sector [1].
- **Strengthening Institutional Capacity of States**—To set up State Designated Agencies for initiating the energy conservation activities at the state level.
- **School Education Program**—To promote energy efficiency in schools through the formation of Energy Clubs. BEE is realizing the Students Capacity Building Programme under the Energy Conservation awareness scheme for the XII five year plan.
- **Human Resource Development**—To implement energy-efficient technologies and practices in various sectors, a sound policy is required for the creation, retention, and up-gradation of skills of human resources.
- **National Mission for Enhanced Energy Efficiency**—One of the eight missions under the National Action Plan on Climate Change (NAPCC) is the National Mission for Enhanced Energy Efficiency (NMEEE). The goal of NMEEE is to improve energy efficiency by establishing a favorable regulatory and policy regime for encouraging innovative sustainability in energy efficiency.

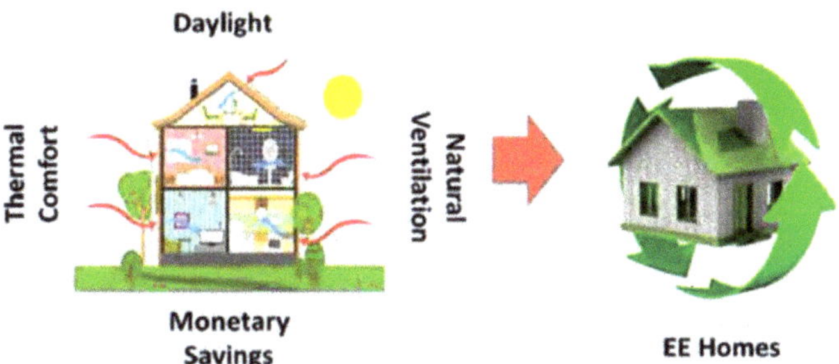

Figure 3. Eco-Niwas Samhita Scheme in the residential sector [1].

2.1.2. Energy Efficiency Projects in India

- **Energy Efficiency in light Bulb**: Domestic Efficient Lighting Program (DELP) scheme (now renamed as Unnat Jeevan by Affordable LEDs and Appliances for All (UJALA)) is designed to monetize energy consumption reduction in the household sector and to attract investments therein. Approximately 45,865 mn kWh of energy were saved per year according to the Ministry of Power, and carbon emissions were reduced by 3, 71, 50, 810 tonnes. For the fiscal year 2019-20, nearly 40 crores of LED bulbs were distributed under UJALA Yojana, resulting in cost savings of Rs 18,341 crores per year [17].
- **Energy Efficiency in Street Lighting**: The inefficient sodium and mercury vapor street lights were replaced by efficient LED street lights in many cities with a payback period of nearly two years. New technologies in LED-based street lights offer noise and pollution sensors, with remote control facilities.
- **Energy Efficiency in Water Pumping**: Five States in the Agricultural sector and 8 States in the Municipal sector replaced the traditional pump with its energy-efficient counterpart. The profound transition towards solar energy is making the water pumping system even smarter and efficient than the previous technologies.

Table 2 shows the international collaboration with India in energy efficiency.

Table 2. International collaboration under energy efficiency programs [1].

S. No.	International Collaboration	Programmes
1	Indo-US	Development of ECBC, Energy Efficient HVAC systems, Capacity Building for Institutional Financing
2	Indo-UK	Industrial Energy Efficiency, DSM Action Plans, Carbon Budgeting Approach
3	Indo-Japan	Energy Conservation Guidelines and Manuals, Waste Heat Recovery Projects, Joint Policy Researchers, Capacity Building and Industrial Energy Efficiency Programmes
4	Indo-German	Energy-Efficient Cooling, Energy Efficiency Standards for Multistorey Buildings, Perform, Achieve, and Trade (PAT) cycle
5	Indo-Switzerland	Smart GHAR Project, Energy Efficient Buildings via Integrated Design Method, Training Programmes

2.2. Demand Response

Demand response (DR) is a process in which the utility may curb the load at customer premises or remotely detach such customer appliances to avoid huge capital investments in generation capacity. DR acts as a resource to deal with a high spike in fuel prices, brownouts, blackouts, and other emergency conditions. DR engages customer participation through various incentives and penalties [10,18]. Scientists and researchers are now showing interest in residential DR programs, which enable a customer to decrease their electricity consumption and manage smart appliances [19–21]. The DR classification as given by the US Department of Energy [22] is shown in Figure 4, and its functional strategy is shown in Figure 5.

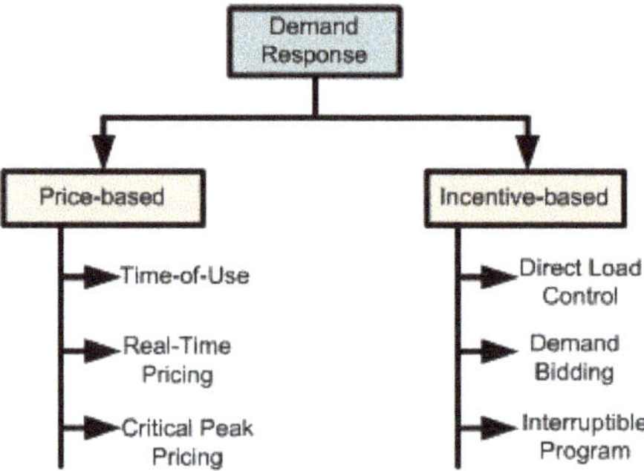

Figure 4. Demand response classification.

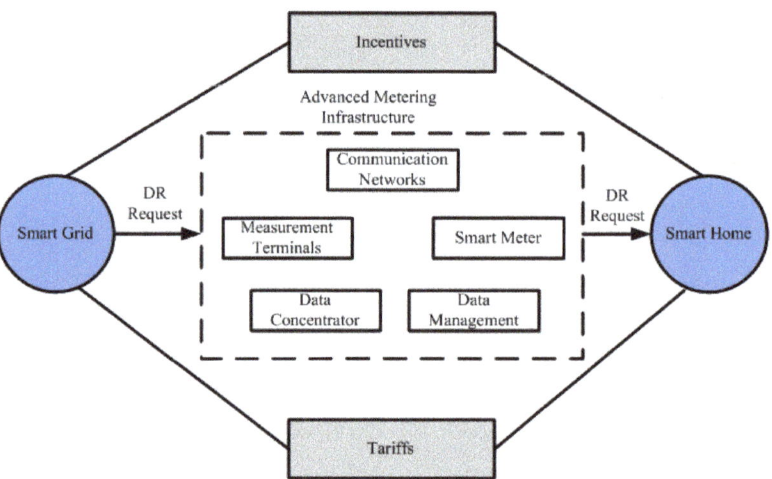

Figure 5. DR functional strategy.

Utility sends the DR request to the consumer via Advanced Metering Infrastructure (AMI), which is an integration of smart meters, communication networks, measurement terminals, data concentrators, and data management systems [23]. AMI replaces the conventional meters with smart meters to promote two-way communication for remote monitoring and control applications. Table 3 highlights the basic differences between the aforesaid DR programs. Price-based DR programs are time-dependent programs that require price design and involve voluntary participation. On the other hand Incentive-based DR programs are time-independent programs that require baseline estimation and involve voluntary, mandatory, or market-based participation.

Table 3. Difference between price-based and incentive-based DR programs.

Attributes	Price-Based DR	Incentive-Based DR
Price-variation Requirement	Time-dependent Price-design	Time-independent Baseline estimation
Discounts offered	Time-varying	Fixed or Time-varying
Consumer participation	Voluntary	Voluntary, mandatory, or market-based
Applicability	Mostly addressed to have a propensity for less electricity use during peak hours	Mostly addressed during overload periods or emergencies

2.2.1. Price-Based DR Program

In a price-based DR program or indirect load control, consumers modify their energy consumption patterns at peak demand times in response to different time-based pricing schemes called tariffs. This in turn offers financial benefits to users. The various pricing schemes include Time-Of-Use pricing (TOU), Critical Peak Pricing (CPP), and Real-Time Pricing (RTP). TOU is a widely used tariff in which usage charges are divided into different time slots for different seasons of the year or hours of the day [24]. Generally, prices are higher during peak hours and lower during off-peak hours, so consumers may respond through scheduling. CPP is quite similar to TOU, but here the prices change periodically often during the summer when the system is overloaded. The participants are notified of the new price a day ahead [25,26]. RTP or dynamic pricing is the one where the hourly prices fluctuate and participants are notified about the time beforehand. RTP implementation requires real-time communication between utilities and customers and an energy management controller for modifying the energy consumption pattern resulting in overall price reduction [21].

2.2.2. Incentive-Based DR Program

In incentive-based DR programs, participants reduce their energy consumption during overload periods, and they avail financial incentives in return. Incentive-based DR programs include Direct Load Control (DLC), demand bidding, and interruptible programs. In DLC, as the name signifies, the utility can directly switch on or off the customer's air conditioner or water heater based on the mutual contract [19,27,28]. Demand bidding is also known as negawatt or buyback program, is a market-based program where customers bid for the load they are willing to reduce [29]. Once the bid is accepted and customers commit according to the requirement, and is paid for that. Interruptible programs allow customers to shift their load to off-peak hours or shut down especially during emergencies. Enrolled customers may get penalties if they fail to respond during the event [30].

2.3. Energy Optimization and Scheduling

Optimization refers to the selection of the best possible element from several alternatives to achieve a target. Mathematically, it deals with finding the maxima or minima of a function that is subjected to some constraints [31]. Energy consumption optimization is used to find the optimal parameters needed for smart energy management [32]. Two important parameters include current indoor parameters and user-desired parameters [33]. The difference between the two produces the error, which is minimized using optimization to minimize energy consumption. Traditional energy management that is based on load forecasting and machine learning where the data is taken from traditional meters fails to predict hourly consumption [34]. The issue is overcome either by replacing these meters with digital ones or by using DR-based load forecasting [26]. Energy prediction is a prerequisite for energy consumption optimization [35]. The user comfort level is a prime factor in considering the optimization problem. Many researchers have proposed various optimization techniques for controlling energy consumption without jeopardizing user comfort.

As the conventional grid is transforming into a decentralized grid, load scheduling is now replacing load shedding. Demand-side management involves scheduling of home appliances by modifying their energy consumption pattern. Scheduling is a load management technique wherein the smart home appliances are shifted from on-peak hours to off-peak hours (during DR programs) [36], thereby shaving the peaks and filling the valleys resulting in load factor improvement.

2.4. Distributed Generation

Distributed Generation or Decentralized Generation (DG) is an electricity source directly connected to the distribution level or on the user-end side [37]. DG serves as a backup plan to the demand side, which not only mitigates the transmission losses, but also improves well-being, as no one wants the high transmission lines to pass over their residence. DG technologies provide economic benefits for cogeneration, peak-shaving, and standby power applications [38]. The DG technologies in smart homes include renewables (solar and wind), gas turbines, microturbines, and fuel cells [39].

With the rapid sustainable development and the need for emission-free generation, renewable energy penetration seems to the largest among the DG technologies. These are some of the sources of cost-effective, clean, and green energy. In the solar energy domain, photovoltaic technology is the most prevailing among smart homes. Another renewable technology is wind power, which is also growing worldwide after solar. Wind power is generated by wind turbines, which include a fan, generator, gearbox, tower, and safety mechanisms [40]. Biomass is also used in smart homes, especially for cooking and heating. Biomass technologies include combustion, gasification, and biogas [41,42]. According to the annual report of the Ministry of New and Renewable Energy (MNRE), nearly 86 GW of renewable power capacity has been set up by December 2019 in India, and the target is to extend it further to 175 GW by the year 2022 [43].

Now, in the era of renewable energy sources coupled with information and communication technology, the devices are becoming smart. The LED light bulb, LED street light, and energy-efficient water pump are enriched with smart technology and get power from solar photovoltaic systems and are now called smart home lights, smart street lights, and smart water pumps. In India, the present solar energy contribution includes 1,721,343 smart home lights, 679,772 smart street lights, and 246,074 smart water pumps, respectively, [43]. PLease confirm the number format. I confirm.

2.5. Energy Storage

Energy storage is the ultimate solution to overcome the intermittency challenges associated with renewable power [44]. Storing renewable power will abort the dependency on the grid power supply. With intelligent energy management, customers are engaged to buy and store electricity when it is available in plenty or when prices are low. With smart metering, customers can reduce their consumption and therefore cost during peak load hours (or high price periods). Energy storage technologies in the smart home include batteries, ultracapacitors, and electric vehicles [45,46]. Commonly used batteries are lead-acid, lithium-ion, zinc-bromine, zinc-iron, etc. [10]. Electric Vehicle (EV) or portable energy storage replacing the conventional vehicles are one of the promising green energy technologies that will turn the entire energy scenario shortly. EV technology comprises a battery, hybrid, plug-in hybrid, and fuel cell. In India, Faster Adoption and Manufacturing of Electric Vehicles (FAME-II) provides inputs on different aspects of electric mobility. Renewable energy-based charging infrastructure is in progress [43].

3. Hardware and Communication Technology

Home Energy Management System (HEMS) uses smart sensors to collect information and communicate with the smart appliances to perform the specific action. Various research projects have been carried out in the framework of intelligent HEMS, leveraging smart technologies to build HEMS hardware and control algorithms. In [47], authors presented a hardware demonstration for DR management comprising of home energy management unit, load controllers, PC, communication module, and a smart meter. A hardware HEMS is developed for controlling domestic loads in response to pricing signals in the department of Electrical Power Systems of Politehnica University of Bucharest [48]. The system employed a smart meter, bipolar fuses, relays, Raspberry PI, and wireless router. The work in [49] proposed an intelligent HEMS for DR management. The hardware setup consists of wireless modules, controllers, and smart plugs. The study in [50] provided the architecture and practical implementation of an IoT and cloud computing-based HEMS on a project circuit board. The hardware components used are the WeMos D1 Mini microcontroller with a built-in Wi-Fi module, current and voltage sensors, power module, multiplexer, and relay.

To apply appropriate appliance scheduling and energy management measures, smart HEMS needs communication technology. A smart home uses wireless sensor networks to connect home appliances with HEMS. Most widely used communication technologies include BACnet [51], Digital Addressable Lighting Interface (DALI) [52], Zigbee [53], Bluetooth [54], Wi-Fi [55], and Power Line Communication (PLC) [56]. BACnet was developed by Ashrae for controlling HVAC systems. The authors in [57] introduced the building automation and communication requirement using BACnet. The DALI protocol is used to provide communication between the fuzzy controller and LED luminaires [58]. Zigbee, Bluetooth, and Wi-Fi are wireless communication technologies. Zigbee is mostly preferred for communication in smart homes due to its low power requirement, simplicity, reasonable range, low cost, and support to a large number of network nodes [59]. An intelligent HEMS is designed for demand response and load management via Zigbee based on IEEE 802.15.4 standard [60]. A Zigbee-based protection system is constructed for building safety against fire [61]. An intelligent cloud home energy management system is proposed using the Zigbee protocol to overcome the intermittency challenges associated with renewable power [62]. HEMS based on Zigbee technology is developed that is capable of monitoring energy usage with accuracy, and thus is well suited to energy conservation and planning [63]. The hardware demonstrations for DR management using the Zigbee protocol are presented [47,64].

The use of Bluetooth is limited as it provides short-range communication (up to 10 m) and requires more power consumption than Zigbee. Researchers introduced a novel Bluetooth-based HEMS capable of reducing the peak load demand and electricity cost while maintaining user comfort [65]. Wi-Fi on the other hand provides a communication range of more than 100 m with high speed, but it requires more power consumption and additional components than Zigbee. The hardware demonstration of DSM for controlling air conditioners through Wi-Fi technology and DR programs is discussed [66]. A Wi-Fi smart plug is designed for monitoring and controlling smart home appliances. This inexpensive solution enables a user to remotely switch on/off the devices [67]. PLC provides high security at low cost, but it offers low speed and low data transmission quality. The study in [56] described HEMS that used power line communication to provide real-time information on energy consumption patterns.

4. Soft Computing Based DSM

Owing to the myriads of applications, soft computing techniques have been successfully applied to solve complex problems (imprecise or uncertain) of intelligent building control [68]. Based on the type of soft computing techniques, the DSM can be classified as Fuzzy Logic (FL) based DSM, Artificial Neural Network (ANN) based DSM, and Evolutionary Computation (EC) based DSM.

4.1. FL Based DSM

Fuzzy logic has been extensively used for controlling and monitoring home appliances for many years due to its simplicity, adaptability, flexibility, and outstanding capability in dealing with uncertainties and nonlinearities [69,70]. D. Kolokotsa et al. [71] designed fuzzy PD, fuzzy PID, and adaptive fuzzy PD controllers. They proved that the adaptive controller gives optimum performance and results in effective energy saving (25–30% more than ON-OFF controller) when user preferences are critical and suggested to use fuzzy PD for visual control and adaptive fuzzy PD for thermal and air quality control. A genetic algorithm (GA) tuned fuzzy controller is proposed in [72] for controlling the indoor building parameters and energy consumption minimization. The study in [73] showed that the fuzzy P controller yields an annual energy saving of 76% for electric lighting as compared to the fuzzy PD, fuzzy PI, fuzzy PID, and adaptive fuzzy PD controllers. These three works focused on providing thermal, visual, and air quality comfort via a smart card. An adaptive fuzzy controller is developed for ensuring the thermal comfort of the Heating Ventilation and Air-Conditioning (HVAC) system [74]. A fuzzy-based automatic roller blind [75] was designed for luminance control on account of the availability of solar radiation. The aim is to utilize the maximum daylight illumination effectively [75]. Improved adaptive fuzzy controllers are developed for controlling the air handling unit of the HVAC system in real-time where the GA is used for rule matrix and membership adjustment [76,77]. Authors in [78] presented an intelligent coordinator control with five fuzzy controllers for thermal, visual, and air quality control.

A scheduling problem for air conditioner temperature control based on day-ahead pricing is modeled using FL, and the temperature forecasting is done through immune clonal selection programming [79]. The concept of adaptive actively spheres integration with FL [80] is proposed, where the system learns and adapts to the changing human behavior and artifacts. As the era of the smart grid is emerging, various technologies like smart sensors, communication, and smart home appliance commitment are becoming a topic of interest to many researchers. The work in [81] applied fuzzy logic for scheduling smart home appliances based on the day-ahead pricing scheme and user comfort. Novel agent-based energy optimization of the HVAC system for higher education building is proposed [82]. Intelligent agents are used for prediction, control, sensing, and data processing. The experimental results showed 3% energy saving while maintaining user thermal comfort [82]. Authors used the synergy of wireless sensor networks, fuzzy logic, and smart grid incentives to design a smart thermostat for the HVAC system using a programmable communicating thermostat [83]. Then an adaptive model is developed to adjust the user's changing preferences [84]. The results are compared with the existing thermostat and it is observed that developed systems automatically respond to DR programs and resulted in a significant reduction in load demand without user discomfort [83,84]. Researchers in [85] proposed a fuzzy logic-based behavioral controller for HEMS.

A fuzzy logic-based smart LED lighting system is designed to provide visual comfort. The experimentation encompasses the DALI protocol for communication, daylight, user movement, and preferences [58]. The HVAC system is controlled using FL concepts and the performance is compared with the conventional on-off controller. The study implemented the simulation using the Building Control Virtual Test Bed platform [86]. The thermal comfort provided by the fuzzy controller is found superior to the on-off controller. A fuzzy logic-based smart HEMS for battery and load management was proposed in [87], which used Wi-Fi communication technology and IoT based monitoring. An additional humidity parameter is introduced in the fuzzy system and the rules are generated automatically using the combinatorial method [88]. Additionally, the study also utilized IoT-based sensors and a feedback loop. It is concluded that the proposed method can reduce energy consumption by up to 50%. The study in [8] classified the appliances based on their energy consumption pattern [89] and accordingly designed fuzzy controllers to control the HVAC and the illumination system.

4.2. ANN Based DSM

ANN is a machine learning approach that is flooded with numerous applications due to its simplicity, adaptability, real-time fast solution, and self-organization. An ANN-based predictive and adaptive control logic is developed for providing thermal comfort. The proposed logic used two predicted models and a hardware framework that resulted in accurate prediction and better thermal comfort than the conventional logic [90]. They also considered the humidity factor. Then a discrete model predictive approach is developed for an HVAC system. ANN is used for model prediction and branch and bound approach for optimization [91]. Simulations showed an energy saving of 50%.

An ANN-based HEMS is proposed with the DR program to maintain the energy consumption below the demand limit, and the system is trained using a Levenberg–Marquardt and feed-forward network [92]. An hourly energy consumption predictor [93] is developed using a multilayer perceptron. Recently, ANN was used for forecasting DR signals and energy consumption patterns for maintaining an energy-efficient smart home [94–96]. A hybrid Lightning Search algorithm (LSA)-ANN-based HEMS was developed [97]. For optimal scheduling, the LSA selects the appropriate neurons and learning rate. Deep Extreme Learning Machine (DELM) based energy consumption predictors were proposed and subsequently compared with other machine learning methods [98,99]. The DELM predictor outperformed the other methods. A hybrid Adaptive Neuro-Fuzzy Inference System (ANFIS) controller [100] was proposed to control the temperature and air quality concerning changing demands [101].

4.3. EC Based DSM

Evolutionary computation is well known for its highly optimized solutions, and therefore is widely used to solve complex nonlinear, nonconvex, and constrained optimization problems. An efficient energy management reset scheme using evolutionary programming was proposed [102] with 7% energy-saving potential. Authors used Binary Particle Swarm Optimization (BPSO) for scheduling interruptible loads for cost and interruption minimization [103]. They divided the swarms into subswarms for significant scheduling improvement. A day-ahead load scheduling method capable of handling a variety of loads was developed using a heuristic evolutionary algorithm [13].

The authors in [104] proposed an energy management system for micro-grids equipped with wind-turbines. The economic dispatch problem is solved by Ant Colony Optimization (ACO). The study in [105] used the dual pricing model RTP with Inclined Block Rate (IBR) for efficient load scheduling. The optimization of the operational time of the appliances is performed using GA [106]. GA was compared with ACO [107] and also with Particle Swarm Optimization (PSO) [108,109] for maximizing user comfort. Multi-objective optimization problems were solved using a non-dominated sorting GA [110,111] and PSO [112]. The works in [113,114] used Artificial Bee Colony for appliance scheduling for energy management considering renewables as well. The algorithm yields a cost reduction of about 47%. Different heuristic algorithms GA, ACO, BPSO, Wind-Driven Optimization, Bacterial Foraging Optimization, and Hybrid GA-PSO were compared, wherein the GA based controllers outperformed the other methods [115–117]. They also considered TOU and IBR dual models.

A multi-agent control system [118] with hybrid multi-objective GA is developed for energy-efficient buildings. The developed method resulted in 31.6% energy efficiency. A real-time appliance scheduling is performed by Binary Backtracking Search Algorithm for energy management [119]. For electricity cost and peak load reduction, HEMS comprising of GA, Cuckoo Search Algorithm, BPSO, and Crow Search Algorithm [120] were designed with RTP and TOU pricing models, respectively, [121,122]. The studies also considered energy storage and renewable energy options. An optimal energy scheduler for load reliability was investigated and the optimization problem was solved using PSO [123]. A real-time electricity scheduler was developed for smart home energy management, considering renewables and energy storage resources [124]. GA was used to solve the multiobjective optimization problem. A day-ahead load forecasting was assumed before scheduling, and a hybrid Harmony Search-PSO algorithm was used for optimal scheduling via a human–machine interface, central controller, and different loads [125].

The study in [126] introduced and implemented the Lightlearn controller based on reinforcement learning. Due to its adaptive nature, it learned the user's behavior and adapted to controlling actions accordingly. Recent research presented a bi-level deep reinforcement learning approach for appliance scheduling. Besides, it incorporated charge and discharge schedules of energy storage and EV [127]. In [128], a load scheduling problem was solved via the Dijkstra algorithm, and the simulation results are compared with GA, Optimal Pattern Recognition Algorithm, and BPSO. The results showed a cost reduction of about 51%. Renewable generation and storage systems were also considered. Table 4 summarizes the soft computing based DSM.

Table 4. Soft computing based DSM.

References	Method	Objective	Contribution
[71–73]	FL	thermal, visual, and air quality comfort	Fuzzy P, Fuzzy PD, Fuzzy PID, Adaptive fuzzy PD controller, and GA tuned Fuzzy controller
[74]	FL	thermal comfort	An adaptive fuzzy controller
[75]	FL	visual comfort	fuzzy-based automatic roller blind
[76,77]	FL	air quality comfort	Improved adaptive fuzzy controllers in real-time
[78]	FL	thermal, visual, and air quality comfort	intelligent coordinator control with five fuzzy controllers
[79,81]	FL	thermal comfort	A scheduling problem for air conditioner temperature control based on day-ahead pricing is modeled smart thermostat for the HVAC system using
[83,84]	FL	thermal comfort	a programmable communicating thermostat and an adaptive model to adjust the user's changing preferences
[58]	FL	visual comfort	A fuzzy logic-based smart LED lighting system
[86]	FL	thermal comfort	Fuzzy based controller for HVAC system using the Building Control Virtual Test Bed platform
[87]	FL	-	A fuzzy logic-based smart HEMS for battery and load management
[8]	FL	thermal, visual and air quality comfort	fuzzy controllers to control the HVAC and illumination system
[90]	ANN	thermal comfort	ANN-based predictive and adaptive control logic
[91]	ANN	thermal and air quality comfort	discrete model predictive approach is developed for an HVAC system
[94–96]	ANN	thermal comfort	forecasting DR signals and energy consumption patterns for maintaining an energy-efficient smart home
[97]	ANN	visual comfort	hybrid Lightning Search algorithm LSA-ANN-based HEMS
[100,101]	ANN	thermal and air quality comfort	hybrid Adaptive Neuro-Fuzzy Inference System (ANFIS) controller
[103]	EC	load scheduling	Binary Particle Swarm Optimization (BPSO) for scheduling interruptible loads for cost and interruption minimization
[13]	EC	load scheduling	A day-ahead scheduling method using a heuristic evolutionary algorithm
[104]	EC	economoc dispatch problem	an energy management system for micro-grids equipped with wind-turbines using ACO
[105]	EC	load scheduling	dual pricing model RTP with Inclined Block Rate (IBR)
[107–109]	EC	user comfort	GA is compared with ACO and PSO
[113,114]	EC	load scheduling	used Artificial Bee Colony for energy management considering renewables as well
[115–117]	EC	energy cost reduction	Different heuristic algorithms GA, ACO, BPSO, Wind-Driven Optimization, Bacterial Foraging Optimization, and Hybrid GA-PSO are compared
[119]	EC	energy management	A real-time appliance scheduling is performed by Binary Backtracking Search Algorithm
[120–122]	EC	electricity cost and peak load reduction	home energy management schemes comprising of GA, Cuckoo Search Algorithm, BPSO, and Crow Search Algorithm
[123]	EC	load scheduling	An optimal energy scheduler for load reliability using PSO
[124]	EC	load scheduling	A real-time electricity scheduler considering renewables and energy storage resources
[125]	EC	load scheduling	A hybrid Harmony Search-PSO algorithm
[126]	EC	visual comfort	Lightlearn controller based on reinforcement learning
[127]	EC	load scheduling	a bi-level deep reinforcement learning approach
[128]	EC	load scheduling	Dijkstra algorithm compared with GA, Optimal Pattern Recognition Algorithm, and BPSO

5. Optimization Based DSM

Game Theory is one of the most powerful and widely used optimization techniques. Autonomous Game-Theory-based DSM is presented in [129,130]. The players act as users and their strategies as daily home appliances schedules [131]. There is only interaction among participating users rather than utilities, and a single energy source is shared by the users. Their ultimate objective was to minimize energy costs and Peak to Average Ratio (PAR). Then, Game Theory was also used for load scheduling considering renewable sources [132], EV [133], and for realizing user-aware DSM considering user preferences [134].

Researchers in [135] used two optimization methods for RTP-based DSM. Stochastic optimization was used for price minimization and controlling associated financial risks. On the other hand, robust optimization deals with the price uncertainty intervals [136]. A mixed-integer programming optimization was used for smart home appliance scheduling [137–139], with EV and energy storage in [140]. Reference [141] transformed the Mixed-Integer Linear Programming (MILP) problem into a convex programming optimization one for flexible and efficient performance. To deal with the uncertainties such as price-elasticities of demand, [24] proposed a TOU tariff design using stochastic optimization based on quadratically constrained quadratic programming and RTP design in [142]. Simulated Annealing was used for DSM [143], which uses white tariff, an extension of TOU tariff. Researchers introduced a cost-efficient scheduling approach using fractional programming while considering service fees and renewables [144].

To implement an incentive-based DR program [145] proposed a practical load scheduling optimization algorithm for user satisfied energy management. A comparative study among Linear programming, PSO, Extended PSO, adaptive dynamic programming, and self-learning procedures was made for smart load scheduling while considering data uncertainties [146]. In a study, a multiobjective mixed-integer non-linear programming optimization was used for energy saving and maintaining thermal comfort [147]. The scheduling problem was solved by interval number optimization in [148]. At first, the uncertain parameters were transformed into interval numbers and then successively solved by BPSO coupled with Integer linear programming. A metaheuristic optimization method that is a hybrid bacterial foraging-GA is proposed to handle multiple constraints and improve search efficiency [149]. Dynamic programming is used for real-time appliance scheduling. A heuristic optimization based appliance scheduling and energy management system was developed, which considered both renewable sources as well as user preferences [150].

A recent study [151] used MILP with normalized weighted sum and compromise programming for solving scheduling problems considering the TOU pricing scheme. The work in [8] scheduled the appliances using the Bat algorithm [152], Flower pollination, and hybrid Bat Flower pollination optimization techniques, respectively. A novel appliance scheduling optimization for a flexible and comfortable environment contributed to peak load reduction while considering socio-technical factors [153]. Table 5 summarizes optimization-based DSM.

Table 5. Optimization based DSM.

References	Optimization Method	Objective	Contribution
[129,130]	Game Theory	minimize energy costs and Peak to Average Ratio (PAR)	Autonomous Game-Theory-based DSM
[132–134]	Game Theory	load scheduling	realizing user-aware DSM considering user preferences and renewable sources
[135]	Stochastic-Robust	price minimization	RTP-based DSM
[137–140]	Mixed-integer programming	load scheduling	smart home appliance scheduling
[143]	Simulated Annealing	energy optimization	DSM using white tariff
[146]	Linear programming	load scheduling	A comparative study among Linear programming, PSO, Extended PSO, Adaptive dynamic programming, and Self-learning procedures
[148]	Interval number optimization	load scheduling	BPSO coupled with Integer linear programming
[150]	Dynamic programming	load scheduling	A heuristic optimization based energy management system considering both renewable sources as well as user preferences
[151]	MILP	load scheduling	normalized weighted sum and compromise programming for solving scheduling problems considering the TOU pricing scheme
[152]	Bat algorithm, Flower pollination, and hybrid Bat Flower pollination	load scheduling	Energy management scheduler for smart home

Table 6 shows the comparison between the soft computing DSM and optimization DSM.

Table 6. Comparison between the soft computing DSM and optimization DSM.

S. No.	Soft Computing Based DSM	Optimization-Based DSM
1.	Set of computational techniques and algorithms that are used to deal with complex problems [154].	Selection of the best possible element from several alternatives to achieve a target [31]
2.	Does not require a mathematical model	Requires mathematical model
3.	Approximate solutions	Accurate solutions
4.	Fast	Time-consuming
5.	May use heuristics or learning methods	Require iterative methods
6.	Simplicity, adaptability, and flexibility	Robustness, stochastic, and optimality
7.	Best suited for real-world problems	It may be difficult to solve real-world problems
8.	Examples- Fuzzy logic [8], Artificial neural network [101], Genetic algorithm [124], Particle swarm optimization [125], Ant colony optimization [104], Cuckoo search algorithm [120], etc.	Examples- Game theory [129], Mixed-integer linear programming [151], Dynamic programming [150], Simulated annealing [143], Interval number optimization [148], Stochastic and Robust optimization [135], etc.

6. Miscellaneous

In addition to the soft computing and optimization based DSM, there are some other approaches, and this very section summarizes those works in the literature. Two of the major features of a smart grid are the integration of renewable energies [155] and storage resources and increased customer participation. For integration, [156] designed and tested an embedded system in which a microcontroller switches between the various power sources. The energy peaks are managed by the home gateway and utility server via the GSM modem. A net energy saving of about 33% is achieved. Since these integrations may also cause power supply uncertainty. To overcome the issue, [157] developed TOU-based DSM schemes for both prosumers and consumers. The latter is achieved by presenting a scheduling algorithm that takes into account the customer preferences and RTP using Analytical Hierarchy Process and Piecewise Cubic Hermite Interpolating Polynomial [158].

In DLC, it is quite difficult to decide which appliance to turn on or off while maintaining user comfort [159]. Researchers proposed a naïve control method for controlling an electric water heater without a temperature parameter. Instead, a time-varied weight matrix and heating durations are used to generate a customer satisfaction prediction index, which in turn selects the appropriate heater for DLC [160]. Asha Radhakrishnan and M.P. Selvan proposed a DR based off-line scheduling algorithm considering renewable sources [36]. The method comprises load classification, load prioritization, and application of tariff plans.

The hardware implementation of DR programs and cloud computing methods considering customer's preferences and load priority for energy management are presented [62,161]. A smart residential energy management system is designed for appliances and battery scheduling [162]. Graph theory and the Fast greedy approach [163] were used for efficient load scheduling implementation of thermostatic devices [164]. Then model predictive controllers [165] were used for scheduling both thermostatic [166–169] as well as non-thermostatic appliances [170]. An energy-saving smart LED lighting system is developed using sensors and microcontrollers. The experimental results achieved 55% and 65% energy saving in continuous and discrete pattern environment [171]. The work in [172] discussed DR management through practical implementation. The required algorithm is designed based on user indices and engagement plans. The authors proposed an energy management algorithm considering renewable power, battery state of charge level, grid availability, and different tariffs [173]. From the simulation, it is shown that energy-saving with the proposed algorithm with renewable energy is about 28% whereas it is 25% without renewable energy.

Recently researchers developed a residential load simulator using MATLAB-Simulink graphical user interface [174], which cannot only model the smart appliances, but also the local generation resources for extracting the power profiles. In [47] and [64], authors presented a hardware demonstration for DR management using the Zigbee protocol, considering load priority [175] and user preferences. The performance analysis of global model based anticipated building energy management system was developed for energy management [176]. A real-time rule-based DR controller with load shifting and curtailment mechanisms was proposed in [177]. The study in [178] conducted a quality of experience perception analysis and based on user profile proposed a smart HEMS considering the degree of annoyance and renewable energy resources. The hardware demonstration of DSM for controlling air conditioners through Wi-Fi technology and DR programs was discussed [66]. The control methodology in [64] used the combination of fuzzy controller, rolling optimization, and real-time control strategy for appliance scheduling in a DR environment. For efficient utilization of energy storage systems [179] developed a nonhomogenous hidden Markov model that formulates the energy storage management problem and used piecewise linear approximation for further solving. Table 7 summarizes the miscellaneous DSM approaches.

Table 7. Miscellaneous DSM approaches.

References	Contribution
[156]	Integration of renewable energies through microcontroller based embedded system
[157]	TOU-based DSM schemes for both prosumers and consumers and a scheduling algorithm that takes into account the customer preferences
[160]	A naïve control method for controlling an electric water heater without a temperature parameter
[36]	DR based off-line scheduling algorithm considering renewable sources
[62,161]	The hardware implementation of DR programs and cloud computing methods considering customer's preferences and load priority
[162]	A smart residential energy management system for appliances and battery scheduling
[163]	Efficient load scheduling implementation of thermostatic devices using Graph theory and the Fast greedy approach
[165–169]	Model predictive controllers for scheduling thermostatic appliances
[172]	DR management through practical implementation
[173]	Energy management algorithm considering renewable power, battery state of charge level, grid availability, and different tariffs
[174]	Residential load simulator using MATLAB-Simulink graphical user interface
[47,64]	Hardware demonstration for DR management using Zigbee protocol
[177]	Real-time rule-based DR controller with load shifting and curtailment mechanisms
[66]	Hardware demonstration of DSM for controlling air conditioners through Wi-Fi technology and DR programs
[64]	Appliance scheduling in a DR environment using the combination of fuzzy controller, rolling optimization, and real-time control strategy

7. Discussion and Future Works

From a technical point of view, the most challenging proposals are as follows:

1. As the number of HVAC systems is increasing, heat dissipation from the condensing coil is also increasing, thereby causing environmental issues indirectly affecting human comfort. To overcome the challenge there is a need for the development of a DSM scheme that can accommodate this heat which can either be used for space heating or in kitchen applications.
2. The majority of the research focused on thermal, visual, and air quality comfort, but did not consider humidity, social comfort, and assisted living in their experiments.
3. Design and real-time implementation of hybrid DR controllers considering both technical and economic aspects of the grid to provide enough knowledge of the system (experience) concerning decentralized control and to maintain the reliability of the grid (to control the peaks at off-peak hours).
4. Integration of Fuzzy Logic with metaheuristic algorithms capable of energy prediction, optimization, and scheduling in real-time could give the best results for energy consumption minimization without affecting the degree of comfort.
5. The system should also include renewable energy resources, energy storage devices, and an IoT based protocol to maintain the flexibility and security within the smart home.

8. Conclusions

This paper provides a review of the previous and ongoing research on DSM. DSM strategies are described and a comparison is made between energy conservation and energy efficiency, price-based and incentive-based DR programs, energy optimization and scheduling, and distributed generation and energy storage. We addressed soft computing techniques namely FL, ANN, EC, and different optimization techniques for energy management and scheduling using renewables and storage devices and finally compared them. From a sustainable point of view, DSM is economically viable, provides grid stability, improves the demand and supply-side efficiency, and is environmentally friendly. It is still a developing and promising area of the smart grid. We hope that this review can help new researchers and readers gain insights into various terminologies and methodologies adopted in DSM implementation.

Author Contributions: All authors contributed equally in the preparation of the manuscript. All authors have read and agreed to the published version of the manuscript.

Funding: This work was supported in part by the Taif University Researchers Supporting Project, Taif University, Taif, Saudi Arabia, under Grant TURSP-2020/278, and in part by the Project CRGS/Mohd Tariq/01 from the Hardware-In-the-Loop (HIL) Laboratory, Department of Electrical Engineering, Aligarh Muslim University, India.

Institutional Review Board Statement: Not applicable.

Informed Consent Statement: Not applicable.

Data Availability Statement: Not applicable.

Conflicts of Interest: The authors have no conflict of interest.

References

1. Bureau of Energy Efficiency. Marching towards Energy Efficient Economy. 2019. Available online: https://beeindia.gov.in/sites/default/files/English_BEE_0.pdf (accessed on 11 July 2020).
2. Ahmad, Z.; Abbasi, M.H.; Khan, A.; Mall, I.S.; Khan, M.F.N.; Sajjad, I.A. Design of IoT Embedded Smart Energy Management System. In Proceedings of the 2020 International Conference on Engineering and Emerging Technologies (ICEET), Lahore, Pakistan, 22–23 February 2020; pp. 1–5. [CrossRef]
3. Dileep, G. A survey on smart grid technologies and applications. *Renew. Energy* **2020**, *146*, 2589–2625. [CrossRef]
4. Sarker, E.; Halder, P.; Seyedmahmoudian, M.; Jamei, E.; Horan, B.; Mekhilef, S.; Stojcevski, A. Progress on the demand side management in smart grid and optimization approaches. *Int. J. Energy Res.* **2021**, *45*, 36–64. [CrossRef]
5. Central Electricity Authority. Growth of Electricity Sector in India from1947 to 2019. 2019. Available online: http://www.cea.nic.in/reports/others/planning/pdm/growth_2019.pdf (accessed on 7 August 2020).
6. Lobaccaro, G.; Carlucci, S.; Löfström, E. A review of systems and technologies for smart homes and smart grids. *Energies* **2016**, *9*, 348. [CrossRef]
7. Pau, G.; Collotta, M.; Ruano, A.; Qin, J. Smart home energy management. *Energies* **2017**, *10*, 382. [CrossRef]
8. Khalid, R.; Javaid, N.; Rahim, M.H.; Aslam, S.; Sher, A. Fuzzy energy management controller and scheduler for smart homes. *Sustain. Comput. Inf. Syst.* **2019**, *21*, 103–118. [CrossRef]
9. Makhadmeh, S.N.; Khader, A.T.; Al-Betar, M.A.; Naim, S.; Abasi, A.K.; Alyasseri, Z.A.A. Optimization methods for power scheduling problems in smart home: Survey. *Renew. Sustain. Energy Rev.* **2019**, *115*. [CrossRef]
10. *Smart Grid Handbook for Regulators and Policy Makers*; India Smart Grid Forum: New Delhi, India, 2017.
11. Strbac, G. Demand side management: Benefits and challenges. *Energy Policy* **2008**, *36*, 4419–4426. [CrossRef]
12. Hopkins, A.S. Best Practices in Utility Demand Response Programs. 2017. Available online: https://www.synapse-energy.com/sites/default/files/Utility-DR-Presentation-17-010.pdf (accessed on 19 July 2020).
13. Logenthiran, T.; Srinivasan, D.; Shun, T.Z. Demand side management in smart grid using heuristic optimization. *IEEE Trans. Smart Grid* **2012**, *3*, 1244–1252. [CrossRef]
14. Hussain, I.; Mohsin, S.; Basit, A.; Khan, Z.A.; Qasim, U.; Javaid, N. A review on demand response: Pricing, optimization, and appliance scheduling. *Procedia Comput. Sci.* **2015**, *52*, 843–850. [CrossRef]
15. Harish, V.; Kumar, A. Demand side management in India: Action plan, policies and regulations. *Renew. Sustain. Energy Rev.* **2014**, *33*, 613–624. [CrossRef]
16. Ministry of Power, Government of India. Energy Efficiency. 2019. Available online: https://powermin.nic.in/en/content/energy-efficiency (accessed on 2 February 2021).
17. Gupta, P. Budget 2019: Thousands of Crores Saved Annually by Using LED, Says Nirmala Sitharaman. 2019. Available online: https://www.financialexpress.com/budget/budget-2019-thousands-of-crores-saved-annually-by-using-led-says-nirmala-sitharaman/1631289/ (accessed on 11 July 2020).
18. Haider, H.T.; See, O.H.; Elmenreich, W. A review of residential demand response of smart grid. *Renew. Sustain. Energy Rev.* **2016**, *59*, 166–178. [CrossRef]
19. Shareef, H.; Ahmed, M.S.; Mohamed, A.; Al Hassan, E. Review on home energy management system considering demand responses, smart technologies, and intelligent controllers. *IEEE Access* **2018**, *6*, 24498–24509. [CrossRef]
20. Conejo, A.J.; Morales, J.M.; Baringo, L. Real-time demand response model. *IEEE Trans. Smart Grid* **2010**, *1*, 236–242. [CrossRef]
21. Vardakas, J.S.; Zorba, N.; Verikoukis, C.V. A survey on demand response programs in smart grids: Pricing methods and optimization algorithms. *IEEE Commun. Surv. Tutor.* **2015**, *17*, 152–178. [CrossRef]

22. US Department of Energy. *Benefits of Demand Response in Electricity Markets and Recommendations for Achieving Them*; A Report to the United States Congress Pursuany to Section 1252 of the Energy Policy Act of 2005; US Department of Energy: Washington, DC, USA, 2006.

23. Guerrero-Prado, J.S.; Alfonso-Morales, W.; Caicedo-Bravo, E.; Zayas-Pérez, B.; Espinosa-Reza, A. The Power of Big Data and Data Analytics for AMI Data: A Case Study. *Sensors* **2020**, *20*, 3289. [CrossRef] [PubMed]

24. De Sá Ferreira, R.; Barroso, L.A.; Lino, P.R.; Carvalho, M.M.; Valenzuela, P. Time-of-use tariff design under uncertainty in price-elasticities of electricity demand: A stochastic optimization approach. *IEEE Trans. Smart Grid* **2013**, *4*, 2285–2295. [CrossRef]

25. Herter, K. Residential implementation of critical-peak pricing of electricity. *Energy Policy* **2007**, *35*, 2121–2130. [CrossRef]

26. Zhou, Q.; Guan, W.; Sun, W. Impact of demand response contracts on load forecasting in a smart grid environment. In Proceedings of the 2012 IEEE Power and Energy Society General Meeting, San Diego, CA, USA, 22–26 July 2012; pp. 1–4.

27. Palensky, P.; Dietrich, D. Demand side management: Demand response, intelligent energy systems, and smart loads. *IEEE Trans. Ind. Inf.* **2011**, *7*, 381–388. [CrossRef]

28. Lang, C.; Okwelum, E. The mitigating effect of strategic behavior on the net benefits of a direct load control program. *Energy Econ.* **2015**, *49*, 141–148. [CrossRef]

29. Aalami, H.; Moghaddam, M.P.; Yousefi, G. Modeling and prioritizing demand response programs in power markets. *Electr. Power Syst. Res.* **2010**, *80*, 426–435. [CrossRef]

30. Faria, P.; Vale, Z. Demand response in electrical energy supply: An optimal real time pricing approach. *Energy* **2011**, *36*, 5374–5384. [CrossRef]

31. Rao, S.S. *Engineering Optimization: Theory and Practice*, 4th ed.; John Wiley & Sons: Hoboken, NJ, USA, 2009.

32. Shah, A.S.; Nasir, H.; Fayaz, M.; Lajis, A.; Shah, A. A review on energy consumption optimization techniques in IoT based smart building environments. *Information* **2019**, *10*, 108. [CrossRef]

33. Wang, Z.; Yang, R.; Wang, L. Multi-agent control system with intelligent optimization for smart and energy-efficient buildings. In Proceedings of the 36th Annual Conference on IEEE Industrial Electronics Society, IECON 2010, Glendale, AZ, USA, 7–10 November 2010; pp. 1144–1149.

34. Li, X.; Wen, J.; Bai, E.W. Developing a whole building cooling energy forecasting model for on-line operation optimization using proactive system identification. *Appl. Energy* **2016**, *164*, 69–88. [CrossRef]

35. Cai, H.; Shen, S.; Lin, Q.; Li, X.; Xiao, H. Predicting the energy consumption of residential buildings for regional electricity supply-side and demand-side management. *IEEE Access* **2019**, *7*, 30386–30397. [CrossRef]

36. Radhakrishnan, A.; Selvan, M. Load scheduling for smart energy management in residential buildings with renewable sources. In Proceedings of the 2014 Eighteenth National Power Systems Conference (NPSC), Guwahati, India, 18–20 December 2014; pp. 1–6.

37. Ackermann, T.; Andersson, G.; Soder, L. Distributed Generation: A Definition. *Electr. Power Syst. Res.* **2001**, *57*, 195–204. [CrossRef]

38. De Almeida, A.T.; Moura, P.S.; Gellings, C.W.; Parmenter, K.E. Distributed Generation and Demand Side Management. In *Handbook of Energy Efficiency and Renewable Energy*; Taylor & Francis Group, LLC: Abington, UK, 2007; Chapter 5, pp. 1–54.

39. U.S. Environmental Protection Agency. Distributed Generation of Electricity and Its Environmental Impacts. 2020. Available online: https://www.epa.gov/energy/distributed-generation-electricity-and-its-environmental-impacts (accessed on 21 July 2020).

40. Fthenakis, V.; Kim, H.C. Land use and electricity generation: A life-cycle analysis. *Renew. Sustain. Energy Rev.* **2009**, *13*, 1465–1474. [CrossRef]

41. Goh, C.S. Energy: Current Approach. In *Sustainable Construction Technologies*; Elsevier: Amsterdam, The Netherlands, 2019; Chapter 6, pp. 145–159.

42. Zhou, B.; Li, W.; Chan, K.W.; Cao, Y.; Kuang, Y.; Liu, X.; Wang, X. Smart home energy management systems: Concept, configurations, and scheduling strategies. *Renew. Sustain. Energy Rev.* **2016**, *61*, 30–40. [CrossRef]

43. Ministry of New and Renewable Energy. Annual Report 2019–20. 2020. Available online: https://mnre.gov.in/img/documents/uploads/file_f-1585710569965.pdf (accessed on 14 July 2020).

44. Revuelta, P.S.; Litrán, S.P.; Thomas, J.P. Distributed Generation. In *Active Power Line Conditioners*; Elsevier Inc.: Amsterdam, The Netherlands, 2016; Chapter 8, pp. 285–322. [CrossRef]

45. Longe, O.M.; Ouahada, K.; Rimer, S.; Harutyunyan, A.N.; Ferreira, H.C. Distributed demand side management with battery storage for smart home energy scheduling. *Sustainability* **2017**, *9*, 120. [CrossRef]

46. Mehrjerdi, H. Multilevel home energy management integrated with renewable energies and storage technologies considering contingency operation. *J. Renew. Sustain. Energy* **2019**, *11*, 025101. [CrossRef]

47. Kuzlu, M.; Pipattanasomporn, M.; Rahman, S. Hardware Demonstration of a Home Energy Management System for Demand Response Applications. *IEEE Trans. Smart Grid* **2012**, *3*, 1704–1711. [CrossRef]

48. Paunescu, C.I.; Zabava, T.; Toma, L.; Bulac, C.; Eremia, M. Hardware home energy management system for monitoring the quality of energy service at small consumers. In Proceedings of the 2014 16th International Conference on Harmonics and Quality of Power (ICHQP), Bucharest, Romania, 25–28 May 2014; pp. 24–28.

49. Shakeri, M.; Amin, N.; Pasupuleti, J.; Mehbodniya, A.; Asim, N.; Tiong, S.K.; Low, F.W.; Yaw, C.T.; Samsudin, N.A.; Rokonuzzaman, M.; et al. An autonomous home energy management system using dynamic priority strategy in conventional homes. *Energies* **2020**, *13*, 3312. [CrossRef]

50. Hashmi, S.A.; Ali, C.F.; Zafar, S. Internet of things and cloud computing-based energy management system for demand side management in smart grid. *Int. J. Energy Res.* **2021**, *45*, 1007–1022. [CrossRef]

51. Inoue, M.; Higuma, T.; Ito, Y.; Kushiro, N.; Kubota, H. Network architecture for home energy management system. *IEEE Trans. Consumer Electr.* **2003**, *49*, 606–613. [CrossRef]

52. Digital Illumination Interface Alliance. Standards. 2018. Available online: https://www.dali-alliance.org/dali/ (accessed on 25 May 2021).

53. Han, D.M.; Lim, J.H. Design and implementation of smart home energy management systems based on zigbee. *IEEE Trans. Consumer Electr.* **2010**, *56*, 1417–1425. [CrossRef]

54. Lilakiatsakun, W.; Seneviratne, A. Wireless home networks based on a hierarchical Bluetooth scatternet architecture. In Proceedings of the Ninth IEEE International Conference on Networks, ICON 2001, Bangkok, Thailand, 10–11 October 2001; pp. 481–485.

55. Tozlu, S.; Senel, M.; Mao, W.; Keshavarzian, A. Wi-Fi enabled sensors for internet of things: A practical approach. *IEEE Commun. Mag.* **2012**, *50*, 134–143. [CrossRef]

56. Son, Y.S.; Pulkkinen, T.; Moon, K.D.; Kim, C. Home energy management system based on power line communication. *IEEE Trans. Consumer Electr.* **2010**, *56*, 1380–1386. [CrossRef]

57. Kastner, W.; Neugschwandtner, G.; Soucek, S.; Newman, H.M. Communication systems for building automation and control. *Proc. IEEE* **2005**, *93*, 1178–1203. [CrossRef]

58. Liu, J.; Zhang, W.; Chu, X.; Liu, Y. Fuzzy logic controller for energy savings in a smart LED lighting system considering lighting comfort and daylight. *Energy Build.* **2016**, *127*, 95–104. [CrossRef]

59. Abubakar, I.; Khalid, S.; Mustafa, M.; Shareef, H.; Mustapha, M. Application of load monitoring in appliances' energy management—A review. *Renew. Sustain. Energy Rev.* **2017**, *67*, 235–245. [CrossRef]

60. Han, D.M.; Lim, J.H. Smart home energy management system using IEEE 802.15. 4 and zigbee. *IEEE Trans. Consumer Electr.* **2010**, *56*, 1403–1410. [CrossRef]

61. Huang, L.C.; Chang, H.C.; Chen, C.C.; Kuo, C.C. A ZigBee-based monitoring and protection system for building electrical safety. *Energy Build.* **2011**, *43*, 1418–1426. [CrossRef]

62. Byun, J.; Hong, I.; Park, S. Intelligent cloud home energy management system using household appliance priority based scheduling based on prediction of renewable energy capability. *IEEE Trans. Consumer Electr.* **2012**, *58*, 1194–1201. [CrossRef]

63. Peng, C.; Huang, J. A home energy monitoring and control system based on ZigBee technology. *Int. J. Green Energy* **2016**, *13*, 1615–1623. [CrossRef]

64. Zhou, S.; Wu, Z.; Li, J.; Zhang, X.P. Real-time energy control approach for smart home energy management system. *Electr. Power Compon. Syst.* **2014**, *42*, 315–326. [CrossRef]

65. Collotta, M.; Pau, G. A novel energy management approach for smart homes using bluetooth low energy. *IEEE J. Sel. Areas Commun.* **2015**, *33*, 2988–2996. [CrossRef]

66. Saha, A.; Kuzlu, M.; Pipattanasomporn, M. Demonstration of a home energy management system with smart thermostat control. In Proceedings of the 2013 IEEE PES Innovative Smart Grid Technologies Conference (ISGT), Washington, DC, USA, 24–27 February 2013; pp. 1–6.

67. Wang, L.; Peng, D.; Zhang, T. Design of smart home system based on WiFi smart plug. *Int. J. Smart Home* **2015**, *9*, 173–182. [CrossRef]

68. Vesselényi, T.; Moldovan, O.; Bungau, C.; Csokmai, L. A survey on soft computing techniques used in intelligent building control. *Recent Innov. Mechatron.* **2014**, *1*. [CrossRef]

69. Wu, Y.; Zhang, B.; Lu, J.; Du, K. Fuzzy logic and neuro-fuzzy systems: A systematic introduction. *Int. J. Artif. Intell. Expert Syst.* **2011**, *2*, 47–80.

70. Feng, G. A survey on analysis and design of model-based fuzzy control systems. *IEEE Trans. Fuzzy Syst.* **2006**, *14*, 676–697. [CrossRef]

71. Kolokotsa, D.; Tsiavos, D.; Stavrakakis, G.; Kalaitzakis, K.; Antonidakis, E. Advanced fuzzy logic controllers design and evaluation for buildings' occupants thermal–visual comfort and indoor air quality satisfaction. *Energy Build.* **2001**, *33*, 531–543. [CrossRef]

72. Kolokotsa, D.; Stavrakakis, G.; Kalaitzakis, K.; Agoris, D. Genetic algorithms optimized fuzzy controller for the indoor environmental management in buildings implemented using PLC and local operating networks. *Eng. Appl. Artif. Intell.* **2002**, *15*, 417–428. [CrossRef]

73. Kolokotsa, D. Comparison of the performance of fuzzy controllers for the management of the indoor environment. *Build. Environ.* **2003**, *38*, 1439–1450. [CrossRef]

74. Calvino, F.; La Gennusa, M.; Rizzo, G.; Scaccianoce, G. The control of indoor thermal comfort conditions: Introducing a fuzzy adaptive controller. *Energy Build.* **2004**, *36*, 97–102. [CrossRef]

75. Lah, M.T.; Zupančič, B.; Peternelj, J.; Krainer, A. Daylight illuminance control with fuzzy logic. *Solar Energy* **2006**, *80*, 307–321.

76. Navale, R.L.; Nelson, R.M. Use of genetic algorithms to develop an adaptive fuzzy logic controller for a cooling coil. *Energy Build.* **2010**, *42*, 708–716. [CrossRef]

77. Khan, M.W.; Choudhry, M.A.; Zeeshan, M. An efficient design of genetic algorithm based adaptive fuzzy logic controller for multivariable control of hvac systems. In Proceedings of the 2013 5th Computer Science and Electronic Engineering Conference (CEEC), Colchester, UK, 17–18 September 2013; pp. 1–6.

78. Dounis, A.I.; Tiropanis, P.; Argiriou, A.; Diamantis, A. Intelligent control system for reconciliation of the energy savings with comfort in buildings using soft computing techniques. *Energy Build.* **2011**, *43*, 66–74. [CrossRef]

79. Hong, Y.Y.; Lin, J.K.; Wu, C.P.; Chuang, C.C. Multi-objective air-conditioning control considering fuzzy parameters using immune clonal selection programming. *IEEE Trans. Smart Grid* **2012**, *3*, 1603–1610. [CrossRef]

80. Wagner, C.; Goumopoulos, C.; Hagras, H. Emerging and adaptive fuzzy logic based behaviours in activity sphere centred ambient ecologies. *Pervas. Mobile Comput.* **2012**, *8*, 500–521. [CrossRef]

81. Mohsenzadeh, A.; Shariatkhah, M.H.; Haghifam, M. Applying fuzzy techniques to model customer comfort in a smart home control system. In Proceedings of the 22nd International Conference and Exhibition on Electricity Distribution (CIRED 2013), Stockholm, Sweden, 10–13 June 2013; pp. 1–4.

82. Al-Daraiseh, A.; El-Qawasmeh, E.; Shah, N. Multi-agent system for energy consumption optimisation in higher education institutions. *J. Comput. Syst. Sci.* **2015**, *81*, 958–965. [CrossRef]

83. Keshtkar, A.; Arzanpour, S.; Keshtkar, F.; Ahmadi, P. Smart residential load reduction via fuzzy logic, wireless sensors, and smart grid incentives. *Energy Build.* **2015**, *104*, 165–180. [CrossRef]

84. Keshtkar, A.; Arzanpour, S. An adaptive fuzzy logic system for residential energy management in smart grid environments. *Appl. Energy* **2017**, *186*, 68–81. [CrossRef]

85. Ainsworth, N.; Johnson, B.; Lundstrom, B. A fuzzy-logic subsumption controller for Home Energy Management Systems. In Proceedings of the 2015 North American Power Symposium (NAPS), Charlotte, NC, USA, 4–6 October 2015; pp. 1–7.

86. Anastasiadi, C.; Dounis, A.I. Co-simulation of fuzzy control in buildings and the HVAC system using BCVTB. *Adv. Build. Energy Res.* **2018**, *12*, 195–216. [CrossRef]

87. Krishna, P.N.; Gupta, S.R.; Shankaranarayanan, P.; Sidharth, S.; Sirphi, M. Fuzzy Logic Based Smart Home Energy Management System. In Proceedings of the 2018 9th International Conference on Computing, Communication and Networking Technologies (ICCCNT), Bengaluru, India, 10–12 July 2018; pp. 1–5.

88. Ain, Q.U.; Iqbal, S.; Khan, S.A.; Malik, A.W.; Ahmad, I.; Javaid, N. IoT operating system based fuzzy inference system for home energy management system in smart buildings. *Sensors* **2018**, *18*, 2802. [CrossRef] [PubMed]

89. Lee, J.W.; Lee, D.H. Residential electricity load scheduling for multi-class appliances with time-of-use pricing. In Proceedings of the 2011 IEEE GLOBECOM Workshops (GC Wkshps), Houston, TX, USA, 5–9 December 2011; pp. 1194–1198.

90. Moon, J.W.; Kim, J.J. ANN-based thermal control models for residential buildings. *Build. Environ.* **2010**, *45*, 1612–1625. [CrossRef]

91. Ferreira, P.; Ruano, A.; Silva, S.; Conceicao, E. Neural networks based predictive control for thermal comfort and energy savings in public buildings. *Energy Build.* **2012**, *55*, 238–251. [CrossRef]

92. Ahmed, M.S.; Mohamed, A.; Shareef, H.; Homod, R.Z.; Ali, J.A. Artificial neural network based controller for home energy management considering demand response events. In Proceedings of the 2016 International Conference on Advances in Electrical, Electronic and Systems Engineering (ICAEES), Putrajaya, Malaysia, 14–16 November 2016; pp. 506–509.

93. Wahid, F.; Ghazali, R.; Fayaz, M.; Shah, A.S. Statistical Features Based Approach (SFBA) for Hourly Energy Consumption Prediction Using Neural Network. *Int. J. Inf. Technol. Comput. Sci.* **2017**, *9*, 23–30. [CrossRef]

94. Lu, R.; Hong, S.H.; Yu, M. Demand Response for Home Energy Management Using Reinforcement Learning and Artificial Neural Network. *IEEE Trans. Smart Grid* **2019**, *10*, 6629–6639. [CrossRef]

95. Hafeez, G.; Alimgeer, K.S.; Wadud, Z.; Khan, I.; Usman, M.; Qazi, A.B.; Khan, F.A. An Innovative Optimization Strategy for Efficient Energy Management With Day-Ahead Demand Response Signal and Energy Consumption Forecasting in Smart Grid Using Artificial Neural Network. *IEEE Access* **2020**, *8*, 84415–84433. [CrossRef]

96. Yousefi, M.; Hajizadeh, A.; Norbakhsh Soltani, M.; Hredzak, B. Predictive Home Energy Management System with Photovoltaic Array, Heat Pump and Plug-in Electric Vehicle. *IEEE Trans. Ind. Inf.* **2020**, *17*, 430–440. [CrossRef]

97. Ahmed, M.S.; Mohamed, A.; Homod, R.Z.; Shareef, H. Hybrid LSA-ANN based home energy management scheduling controller for residential demand response strategy. *Energies* **2016**, *9*, 716. [CrossRef]

98. Fayaz, M.; Kim, D. A prediction methodology of energy consumption based on deep extreme learning machine and comparative analysis in residential buildings. *Electronics* **2018**, *7*, 222. [CrossRef]

99. Li, C.; Ding, Z.; Zhao, D.; Yi, J.; Zhang, G. Building energy consumption prediction: An extreme deep learning approach. *Energies* **2017**, *10*, 1525. [CrossRef]

100. Moon, J.W.; Jung, S.K.; Kim, Y.; Han, S.H. Comparative study of artificial intelligence-based building thermal control methods–Application of fuzzy, adaptive neuro-fuzzy inference system, and artificial neural network. *Appl. Therm. Eng.* **2011**, *31*, 2422–2429. [CrossRef]

101. Ahn, J.; Cho, S.; Chung, D.H. Analysis of energy and control efficiencies of fuzzy logic and artificial neural network technologies in the heating energy supply system responding to the changes of user demands. *Appl. Energy* **2017**, *190*, 222–231. [CrossRef]

102. Fong, K.; Hanby, V.; Chow, T. HVAC system optimization for energy management by evolutionary programming. *Energy Build.* **2006**, *38*, 220–231. [CrossRef]

103. Pedrasa, M.A.A.; Spooner, T.D.; MacGill, I.F. Scheduling of demand side resources using binary particle swarm optimization. *IEEE Trans. Power Syst.* **2009**, *24*, 1173–1181. [CrossRef]

104. Esmat, A.; Magdy, A.; Elkhattam, W.; Elbakly, A. A novel Energy Management System using Ant Colony Optimization for micro-grids. In Proceedings of the 2013 3rd International Conference on Electric Power and Energy Conversion Systems, Istanbul, Turkey, 2–4 October 2013; pp. 1–6. [CrossRef]

105. Zhao, Z.; Lee, W.C.; Shin, Y.; Song, K.B. An optimal power scheduling method for demand response in home energy management system. *IEEE Trans. Smart Grid* **2013**, *4*, 1391–1400. [CrossRef]

106. Nassif, N. Modeling and optimization of HVAC systems using artificial neural network and genetic algorithm. In *Building Simulation*; Springer: Berlin, Germany, 2014; Volume 7, pp. 237–245.

107. Galbazar, A.; Ali, S.; Kim, D. Optimization approach for energy saving and comfortable space using aco in building. *Int. J. Smart Home* **2016**, *10*, 47–56. [CrossRef]

108. Ali, S.; Kim, D.H. Optimized power control methodology using genetic algorithm. *Wireless Pers. Commun.* **2015**, *83*, 493–505. [CrossRef]

109. Ullah, I.; Kim, D. An improved optimization function for maximizing user comfort with minimum energy consumption in smart homes. *Energies* **2017**, *10*, 1818. [CrossRef]

110. Carlucci, S.; Cattarin, G.; Causone, F.; Pagliano, L. Multi-objective optimization of a nearly zero-energy building based on thermal and visual discomfort minimization using a non-dominated sorting genetic algorithm (NSGA-II). *Energy Build.* **2015**, *104*, 378–394. [CrossRef]

111. Garroussi, Z.; Ellaia, R.; Talbi, E.G. Appliance scheduling in a smart home using a multiobjective evolutionary algorithm. In Proceedings of the 2016 International Renewable and Sustainable Energy Conference (IRSEC), Marrakech, Morocco, 14–17 November 2016; pp. 1098–1103.

112. Delgarm, N.; Sajadi, B.; Kowsary, F.; Delgarm, S. Multi-objective optimization of the building energy performance: A simulation-based approach by means of particle swarm optimization (PSO). *Appl. Energy* **2016**, *170*, 293–303. [CrossRef]

113. Zhang, Y.; Zeng, P.; Zang, C. Optimization algorithm for home energy management system based on artificial bee colony in smart grid. In Proceedings of the 2015 IEEE International Conference on Cyber Technology in Automation, Control, and Intelligent Systems (CYBER), Shenyang, China, 8–12 June 2015; pp. 734–740.

114. Wahid, F.; Kim, D.H. An efficient approach for energy consumption optimization and management in residential building using artificial bee colony and fuzzy logic. *Math. Probl. Eng.* **2016**, *2016*. [CrossRef]

115. Rahim, S.; Javaid, N.; Ahmad, A.; Khan, S.A.; Khan, Z.A.; Alrajeh, N.; Qasim, U. Exploiting heuristic algorithms to efficiently utilize energy management controllers with renewable energy sources. *Energy Build.* **2016**, *129*, 452–470. [CrossRef]

116. Ahmad, A.; Khan, A.; Javaid, N.; Hussain, H.M.; Abdul, W.; Almogren, A.; Alamri, A.; Azim Niaz, I. An optimized home energy management system with integrated renewable energy and storage resources. *Energies* **2017**, *10*, 549. [CrossRef]

117. Hafeez, G.; Javaid, N.; Iqbal, S.; Khan, F.A. Optimal residential load scheduling under utility and rooftop photovoltaic units. *Energies* **2018**, *11*, 611. [CrossRef]

118. Shaikh, P.H.; Nor, N.B.M.; Nallagownden, P.; Elamvazuthi, I.; Ibrahim, T. Intelligent multi-objective control and management for smart energy efficient buildings. *Int. J. Electr. Power Energy Syst.* **2016**, *74*, 403–409. [CrossRef]

119. Ahmed, M.S.; Mohamed, A.; Khatib, T.; Shareef, H.; Homod, R.Z.; Abd Ali, J. Real time optimal schedule controller for home energy management system using new binary backtracking search algorithm. *Energy Build.* **2017**, *138*, 215–227. [CrossRef]

120. Butt, A.A.; Rahim, M.H.; Khan, M.; Zahra, A.; Tariq, M.; Ahmad, T.; Javaid, N. Energy efficiency using genetic and crow search algorithms in smart grid. In *International Conference on P2p, Parallel, Grid, Cloud and Internet Computing*; Springer: Berlin, Germany, 2017; pp. 63–75.

121. Aslam, S.; Iqbal, Z.; Javaid, N.; Khan, Z.A.; Aurangzeb, K.; Haider, S.I. Towards efficient energy management of smart buildings exploiting heuristic optimization with real time and critical peak pricing schemes. *Energies* **2017**, *10*, 2065. [CrossRef]

122. Javaid, N.; Ullah, I.; Akbar, M.; Iqbal, Z.; Khan, F.A.; Alrajeh, N.; Alabed, M.S. An intelligent load management system with renewable energy integration for smart homes. *IEEE Access* **2017**, *5*, 13587–13600. [CrossRef]

123. Ren, Y.; Wu, H.; Yang, H.; Yang, S.; Li, Z. A Method for Load Classification and Energy Scheduling Optimization to Improve Load Reliability. *Energies* **2018**, *11*, 1558. [CrossRef]

124. Li, S.; Yang, J.; Song, W.; Chen, A. A Real-Time Electricity Scheduling for Residential Home Energy Management. *IEEE Internet Things J.* **2019**, *6*, 2602–2611. [CrossRef]

125. Zhang, Z.; Wang, J.; Zhong, H.; Ma, H. Optimal scheduling model for smart home energy management system based on the fusion algorithm of harmony search algorithm and particle swarm optimization algorithm. *Sci. Technol. Built Environ.* **2020**, *26*, 42–51. [CrossRef]

126. Park, J.Y.; Dougherty, T.; Fritz, H.; Nagy, Z. LightLearn: An adaptive and occupant centered controller for lighting based on reinforcement learning. *Build. Environ.* **2019**, *147*, 397–414. [CrossRef]

127. Lee, S.; Choi, D.H. Energy Management of Smart Home with Home Appliances, Energy Storage System and Electric Vehicle: A Hierarchical Deep Reinforcement Learning Approach. *Sensors* **2020**, *20*, 2157. [CrossRef]

128. Ullah, I.; Rasheed, M.B.; Alquthami, T.; Tayyaba, S. A Residential Load Scheduling with the Integration of On-Site PV and Energy Storage Systems in Micro-Grid. *Sustainability* **2020**, *12*, 184. [CrossRef]

129. Tushar, M.H.K.; Zeineddine, A.W.; Assi, C. Demand-side management by regulating charging and discharging of the EV, ESS, and utilizing renewable energy. *IEEE Trans. Ind. Inf.* **2017**, *14*, 117–126. [CrossRef]

130. Mohsenian-Rad, A.; Wong, V.W.S.; Jatskevich, J.; Schober, R.; Leon-Garcia, A. Autonomous Demand-Side Management Based on Game-Theoretic Energy Consumption Scheduling for the Future Smart Grid. *IEEE Trans. Smart Grid* **2010**, *1*, 320–331. [CrossRef]

131. Nguyen, H.K.; Song, J.B.; Han, Z. Distributed Demand Side Management with Energy Storage in Smart Grid. *IEEE Trans. Parallel Distrib. Syst.* **2015**, *26*, 3346–3357. [CrossRef]

132. Zhu, Z.; Lambotharan, S.; Chin, W.H.; Fan, Z. A Game Theoretic Optimization Framework for Home Demand Management Incorporating Local Energy Resources. *IEEE Trans. Ind. Inf.* **2015**, *11*, 353–362. [CrossRef]
133. Kim, B.G.; Ren, S.; Van Der Schaar, M.; Lee, J.W. Bidirectional energy trading for residential load scheduling and electric vehicles. In Proceedings of the 2013 IEEE INFOCOM, Turin, Italy, 14–19 April 2013; pp. 595–599.
134. Yaagoubi, N.; Mouftah, H.T. User-Aware Game Theoretic Approach for Demand Management. *IEEE Trans. Smart Grid* **2015**, *6*, 716–725. [CrossRef]
135. Chen, Z.; Wu, L.; Fu, Y. Real-time price-based demand response management for residential appliances via stochastic optimization and robust optimization. *IEEE Trans. Smart Grid* **2012**, *3*, 1822–1831. [CrossRef]
136. Wang, J.; Li, P.; Fang, K.; Zhou, Y. Robust optimization for household load scheduling with uncertain parameters. *Appl. Sci.* **2018**, *8*, 575. [CrossRef]
137. Qayyum, F.A.; Naeem, M.; Khwaja, A.S.; Anpalagan, A.; Guan, L.; Venkatesh, B. Appliance Scheduling Optimization in Smart Home Networks. *IEEE Access* **2015**, *3*, 2176–2190. [CrossRef]
138. Melhem, F.Y.; Grunder, O.; Hammoudan, Z.; Moubayed, N. Optimal residential load scheduling model in smart grid environment. In Proceedings of the 2017 IEEE International Conference on Environment and Electrical Engineering and 2017 IEEE Industrial and Commercial Power Systems Europe (EEEIC/I&CPS Europe), Milan, Italy, 6–9 June 2017; pp. 1–6.
139. Nagpal, H.; Staino, A.; Basu, B. Application of Predictive Control in Scheduling of Domestic Appliances. *Appl. Sci.* **2020**, *10*, 1627. [CrossRef]
140. Erdinc, O.; Paterakis, N.G.; Mendes, T.D.; Bakirtzis, A.G.; Catalao, J.P. Smart household operation considering bi-directional EV and ESS utilization by real-time pricing-based DR. *IEEE Trans. Smart Grid* **2014**, *6*, 1281–1291. [CrossRef]
141. Tsui, K.M.; Chan, S.C. Demand response optimization for smart home scheduling under real-time pricing. *IEEE Trans. Smart Grid* **2012**, *3*, 1812–1821. [CrossRef]
142. Vivekananthan, C.; Mishra, Y.; Li, F. Real-Time Price Based Home Energy Management Scheduler. *IEEE Trans. Power Syst.* **2015**, *30*, 2149–2159. [CrossRef]
143. Affonso, C.M.; da Silva, R.V. Demand side management of a residential system using simulated annealing. *IEEE Latin America Trans.* **2015**, *13*, 1355–1360. [CrossRef]
144. Ma, J.; Chen, H.H.; Song, L.; Li, Y. Cost-efficient residential load scheduling in smart grid. In Proceedings of the 2014 IEEE International Conference on Communication Systems, Macau, China, 19–21 November 2014; pp. 590–594.
145. Bahrami, S.; Parniani, M.; Vafaeimehr, A. A modified approach for residential load scheduling using smart meters. In Proceedings of the 2012 3rd IEEE PES Innovative Smart Grid Technologies Europe (ISGT Europe), Berlin, Germany, 14–17 October 2012; pp. 1–8.
146. Squartini, S.; Boaro, M.; De Angelis, F.; Fuselli, D.; Piazza, F. Optimization algorithms for home energy resource scheduling in presence of data uncertainty. In Proceedings of the 2013 Fourth International Conference on Intelligent Control and Information Processing (ICICIP), Beijing, China, 9–11 June 2013; pp. 323–328.
147. Anvari-Moghaddam, A.; Monsef, H.; Rahimi-Kian, A. Optimal Smart Home Energy Management Considering Energy Saving and a Comfortable Lifestyle. *IEEE Trans. Smart Grid* **2015**, *6*, 324–332. [CrossRef]
148. Wang, J.; Li, Y.; Zhou, Y. Interval number optimization for household load scheduling with uncertainty. *Energy Build.* **2016**, *130*, 613–624. [CrossRef]
149. Khalid, A.; Javaid, N.; Guizani, M.; Alhussein, M.; Aurangzeb, K.; Ilahi, M. Towards dynamic coordination among home appliances using multi-objective energy optimization for demand side management in smart buildings. *IEEE Access* **2018**, *6*, 19509–19529. [CrossRef]
150. Jindal, A.; Bhambhu, B.S.; Singh, M.; Kumar, N.; Naik, K. A Heuristic-Based Appliance Scheduling Scheme for Smart Homes. *IEEE Trans. Ind. Inf.* **2019**, *16*, 3242–3255. [CrossRef]
151. Yahia, Z.; Pradhan, A. Multi-objective optimization of household appliance scheduling problem considering consumer preference and peak load reduction. *Sustain. Cities Soc.* **2020**, *55*, 102058. [CrossRef]
152. Fayaz, M.; Kim, D. Energy consumption optimization and user comfort management in residential buildings using a bat algorithm and fuzzy logic. *Energies* **2018**, *11*, 161. [CrossRef]
153. Fanitabasi, F.; Pournaras, E. Appliance-Level Flexible Scheduling for Socio-Technical Smart Grid Optimization. *IEEE Access* **2020**, *8*, 119880–119898. [CrossRef]
154. Koczy, L.T.; Medina, J.; Reformat, M.; Wong, K.W.; Yoon, J.H. Computational intelligence in modeling complex systems and solving complex problems. *Complexity* **2019**, *2019*, 7606715. [CrossRef]
155. Bhalshankar, S.S.; Thorat, C. Integration of smart grid with renewable energy for energy demand management: Puducherry case study. In Proceedings of the 2016 International Conference on Signal Processing, Communication, Power and Embedded System (SCOPES), Paralakhemundi, India, 3–5 October 2016; pp. 1–5.
156. Al-Ali, A.; El-Hag, A.; Bahadiri, M.; Harbaji, M.; El Haj, Y.A. Smart home renewable energy management system. *Energy Procedia* **2011**, *12*, 120–126. [CrossRef]
157. Venizelou, V.; Makrides, G.; Efthymiou, V.; Georghiou, G.E. Methodology for deploying cost-optimum price-based demand side management for residential prosumers. *Renew. Energy* **2020**, *153*, 228–240. [CrossRef]
158. Suryanarayanan, S.; Devadass, M.A.; Hansen, T.M. A load scheduling algorithm for the smart home using customer preferences and real time residential prices. *IFAC-PapersOnLine* **2015**, *48*, 126–131. [CrossRef]

159. Rasheed, M.; Javaid, N.; Ahmad, A.; Jamil, M.; Khan, Z.; Qasim, U.; Alrajeh, N. Energy Optimization in Smart Homes Using Customer Preference and Dynamic Pricing. *Energies* **2016**, *9*, 593. [CrossRef]
160. Xiang, S.; Chang, L.; Cao, B.; He, Y.; Zhang, C. A Novel Domestic Electric Water Heater Control Method. *IEEE Trans. Smart Grid* **2020**, *11*, 3246–3256. [CrossRef]
161. Amer, M.; Naaman, A.; M'Sirdi, N.; El-Zonkoly, A. A hardware algorithm for PAR reduction in smart home. In Proceedings of the 2014 International Conference on Applied and Theoretical Electricity (ICATE), Craiova, Romania, 23–25 October 2014; pp. 1–6.
162. Arun, S.; Selvan, M. Smart residential energy management system for demand response in buildings with energy storage devices. *Front. Energy* **2019**, *13*, 715–730. [CrossRef]
163. Chakraborty, N.; Mondal, A.; Mondal, S. Intelligent scheduling of thermostatic devices for efficient energy management in smart grid. *IEEE Trans. Ind. Inf.* **2017**, *13*, 2899–2910. [CrossRef]
164. Du, P.; Lu, N. Appliance Commitment for Household Load Scheduling. *IEEE Trans. Smart Grid* **2011**, *2*, 411–419. [CrossRef]
165. Freire, R.Z.; Oliveira, G.H.; Mendes, N. Predictive controllers for thermal comfort optimization and energy savings. *Energy Build.* **2008**, *40*, 1353–1365. [CrossRef]
166. Ostadijafari, M.; Dubey, A.; Yu, N. Linearized Price-Responsive HVAC Controller for Optimal Scheduling of Smart Building Loads. *IEEE Trans. Smart Grid* **2020**, *1*, 3131–3145. [CrossRef]
167. Perez, K.X.; Baldea, M.; Edgar, T.F. Integrated HVAC management and optimal scheduling of smart appliances for community peak load reduction. *Energy Build.* **2016**, *123*, 34–40. [CrossRef]
168. Scherer, H.F.; Pasamontes, M.; Guzmán, J.L.; Álvarez, J.; Camponogara, E.; Normey-Rico, J. Efficient building energy management using distributed model predictive control. *J. Process Control* **2014**, *24*, 740–749. [CrossRef]
169. Baniasadi, A.; Habibi, D.; Bass, O.; Masoum, M.A.S. Optimal Real-Time Residential Thermal Energy Management for Peak-Load Shifting With Experimental Verification. *IEEE Trans. Smart Grid* **2019**, *10*, 5587–5599. [CrossRef]
170. Chen, C.; Wang, J.; Heo, Y.; Kishore, S. MPC-Based Appliance Scheduling for Residential Building Energy Management Controller. *IEEE Trans. Smart Grid* **2013**, *4*, 1401–1410. [CrossRef]
171. Chew, I.; Kalavally, V.; Oo, N.W.; Parkkinen, J. Design of an energy-saving controller for an intelligent LED lighting system. *Energy Build.* **2016**, *120*, 1–9. [CrossRef]
172. Li, W.T.; Yuen, C.; Hassan, N.U.; Tushar, W.; Wen, C.K.; Wood, K.L.; Hu, K.; Liu, X. Demand response management for residential smart grid: From theory to practice. *IEEE Access* **2015**, *3*, 2431–2440. [CrossRef]
173. Boynuegri, A.R.; Yagcitekin, B.; Baysal, M.; Karakas, A.; Uzunoglu, M. Energy management algorithm for smart home with renewable energy sources. In Proceedings of the 4th International Conference on Power Engineering, Energy and Electrical Drives, Istanbul, Turkey, 13–17 May 2013; pp. 1753–1758.
174. Lopez, J.M.G.; Pouresmaeil, E.; Canizares, C.A.; Bhattacharya, K.; Mosaddegh, A.; Solanki, B.V. Smart residential load simulator for energy management in smart grids. *IEEE Trans. Ind. Electr.* **2018**, *66*, 1443–1452. [CrossRef]
175. Pipattanasomporn, M.; Kuzlu, M.; Rahman, S. An Algorithm for Intelligent Home Energy Management and Demand Response Analysis. *IEEE Trans. Smart Grid* **2012**, *3*, 2166–2173. [CrossRef]
176. Missaoui, R.; Joumaa, H.; Ploix, S.; Bacha, S. Managing energy smart homes according to energy prices: Analysis of a building energy management system. *Energy Build.* **2014**, *71*, 155–167. [CrossRef]
177. Althaher, S.; Mutale, J. Management and control of residential energy through implementation of real time pricing and demand response. In Proceedings of the 2012 IEEE Power and Energy Society General Meeting, San Diego, CA, USA, 22–26 July 2012; pp. 1–7.
178. Pilloni, V.; Floris, A.; Meloni, A.; Atzori, L. Smart home energy management including renewable sources: A qoe-driven approach. *IEEE Trans. Smart Grid* **2016**, *9*, 2006–2018. [CrossRef]
179. Zhuang, P.; Liang, H. Energy storage management in smart homes based on resident activity of daily life recognition. In Proceedings of the 2015 IEEE International Conference on Smart Grid Communications (SmartGridComm), Miami, FL, USA, 2–5 November 2015; pp. 641–646.

 *sustainability*

Article

PV/Wind-Integrated Low-Inertia System Frequency Control: PSO-Optimized Fractional-Order PI-Based SMES Approach

Md Shafiul Alam [1,*], Fahad Saleh Al-Ismail [1,2,3,4] and Mohammad Ali Abido [1,2,4]

[1]   K.A.CARE Energy Research & Innovation Center, King Fahd University of Petroleum & Minerals (KFUPM), Dhahran 31261, Saudi Arabia; fsalismail@kfupm.edu.sa (F.S.A.-I.); mabido@kfupm.edu.sa (M.A.A.)

[2]   Department of Electrical Engineering, King Fahd University of Petroleum & Minerals (KFUPM), Dhahran 31261, Saudi Arabia

[3]   Center for Environment & Marine Studies, Research Institute, King Fahd University of Petroleum & Minerals (KFUPM), Dhahran 31261, Saudi Arabia

[4]   Center for Renewable Energy & Power Systems, King Fahd University of Petroleum & Minerals (KFUPM), Dhahran 31261, Saudi Arabia

*   Correspondence: mdshafiul.alam@kfupm.edu.sa; Tel.: +966-559105462

Citation: Alam, M.S.; Al-Ismail, F.S.; Abido, M.A. PV/Wind-Integrated Low-Inertia System Frequency Control: PSO-Optimized Fractional-Order PI-Based SMES Approach. *Sustainability* **2021**, *13*, 7622. https://doi.org/10.3390/su13147622

Academic Editors: Thanikanti Sudhakar Babu and Marc A. Rosen

Received: 17 May 2021
Accepted: 25 June 2021
Published: 7 July 2021

Publisher's Note: MDPI stays neutral with regard to jurisdictional claims in published maps and institutional affiliations.

Abstract: A paradigm shift in power engineering transforms conventional fossil fuel-based power systems gradually into more sustainable and environmentally friendly systems due to more renewable energy source (RES) integration. However, the control structure of high-level RES integrated system becomes complex, and the total system inertia is reduced due to the removal of conventional synchronous generators. Thus, such a system poses serious frequency instabilities due to the high rate of change of frequency (RoCoF). To handle this frequency instability issue, this work proposes an optimized fractional-order proportional integral (FOPI) controller-based superconducting magnetic energy storage (SMES) approach. The proposed FOPI-based SMES technique to support virtual inertia is superior to and more robust than the conventional technique. The FOPI parameters are optimized using the particle swarm optimization (PSO) technique. The SMES is modeled and integrated into the optimally designed FOPI to support the virtual inertia of the system. Fluctuating RESs are considered to show the effectiveness of the proposed approach. Extensive time-domain simulations were carried out in MATLAB Simulink with different load and generation mismatch levels. Systems with different inertia levels were simulated to guarantee the frequency stability of the system with the proposed FOPI-based SMES control technique. Several performance indices, such as overshoot, undershoot, and settling time, were considered in the analysis.

Keywords: virtual inertia control; renewable energy resources; solar and wind energy; superconducting magnetic energy storage (SMES); fractional-order proportional integral (FOPI); frequency response

1. Introduction

Due to the continuous depletion of fossil fuels, increased government incentives, technological advancements, and price drops, the utilization of renewable energy sources (RESs) as distributed generators (DGs) has increased dramatically in recent years. In power systems, several technical issues, such as low reserve generation, fault ride through capability, inertia, and high fault current, have arisen because of high-level RES integration [1]. Thus, the frequency stability issue of high-level RES-integrated systems is greatly affected. Moreover, the two main sources of renewable energy, solar and wind, are highly unpredictable. The intermittent and unpredictable RESs can be modeled with sophisticated methods to lower the risk of instability in power systems [2]. A high share of RESs complicates grid-balancing and market operations. Several dedicated devices can be installed in a RES-integrated system to provide ancillary services such as power variations, congestion reduction, grid balancing, and primary reserve [3,4]. The technical issues of RES integration with a power system could also be handled with different cutting edge technologies such as modern control and optimization techniques, energy storage devices including batteries and supercapacitors, and fault current limiting devices [5].

The overall inertia of a power system is decreased greatly as a result of the integration of low-inertia wind and inertia-less PV systems [6]. Power electronic converter decoupling between the wind generator and the power system is responsible for the low inertia. As a result, such a low-inertia wind system cannot properly maintain the frequency stability of the power system. Moreover, solar PV with no inertia is highly responsible for the frequency deviation of the system. Therefore, high-level PV and wind penetration reduces the total inertia and augments the rate of change of frequency (RoCoF), which are responsible for the unexpected load-shedding controller activation even at small generation–load mismatch [7]. In addition, reserve power reduction due to high-level PV/wind integration causes frequency deviation [8]. In summary, inertia emulation controllers need to be designed to improve the frequency stability of RES-integrated power systems.

In order to minimize the frequency excursion of a low-inertia system, several methods have been presented in the literature, such as the auxiliary load frequency (LFC) control technique, the inertia emulation technique, the deloading technique, the droop technique, and the energy storage-based technique [9–12]. In [13], an auxiliary LFC technique was presented to control the frequency of the Egyptian grid considering high-level PV and wind integration employing the proportional-integral-derivative (PID) controller. However, the LFC technique does not consider the detailed model of the Egyptian grid; instead, it excludes tie line power flow, which needs further investigation. In general, conventional PI and PID controllers, the parameters of which were fine-tuned experimentally or tuned by Ziegler–Nichols methods, were employed in system frequency control [14,15]. However, the conventional tuning methods of PI/PID controllers may not provide satisfactory performance. In [16], a virtual inertia support technique was presented for a low-inertia microgrid with a particle swarm optimization (PSO)-based PI controller.

The superconducting magnetic energy storage (SMES) is considered a promising device for the low-inertia issue of the microgrid system in [17]. The conventional derivative approach for the virtual inertia control loop was implemented. The detailed design of feedback and proportion gains, however, were not discussed in this work. Another energy storage, the battery, was presented in [18] for frequency support of the doubly fed induction generator (DFIG)-based wind system. The battery was connected to the DC link of DFIG and controlled with the droop technique in order to reduce frequency deviation by scheduling active power exchange during system disturbances. In [19], a self-adaptive virtual inertia fuzzy controller was adopted for a high-level renewable integrated system. The proportional virtual gain was adapted by the fuzzy system, which uses the deviation of real power and frequency as it inputs. In this scheme, however, the generalized energy storage system (ESS) was considered a simple first-order system. Since the specific ESS was neither discussed nor modeled, the presented frequency support scheme needs further improvement or investigation. The sharing of active power from different energy storage devices were scheduled based on their abilities in [20] for frequency control of renewable sources. In this capability-coordinated frequency control (CCFC) approach, the total error signal was forwarded to the primary control loop of each unit based on its capabilities. The LFC for mass-less inertia PV systems was presented in [21] with PI controllers. The parameters were optimized with the hybrid optimization technique in the case of different step load changes. In order to stabilize the low-inertia PV system, another virtual inertia synthetization using a synchronverter was reported in [22] with the learning technique. The optimized virtual inertia frequency control and protection schemes were developed in [23,24] for a low-frequency interconnected power system. The combination of SMES and thyristor-controlled phase shifters (TCPS) [25] was applied in a low-inertia utility grid with the adaptive neuro-fuzzy system (ANFIS) controller. The detailed design of SMES negative feedback and proportional gains, however, was not considered. The main advantage of SMES is the quick charging/discharging ability to react to sudden changes in system dynamics. Thus, the fast-response capability of SMES could be the most effective countermeasure against frequency deviations in a power system. The voltage and frequency stability issues of a power system are addressed in some of the literature with SMES [26–28]. Furthermore, the transient stability issues are also handled with the application of SMES [29–31]. Based on a comprehensive literature survey on SMES device applications in power systems, it is concluded that further study on virtual inertia control topologies using SMES is imperative.

In recent years, several theoretical and applied studies have been conducted on fractional-order controllers [32,33]. Better system performance is observed with fractional-order controllers over conventional PI controllers because the fractional-order controller involves additional real parameters [34]. However, in general, there is no hard and fast rule for tuning the parameters of fractional-order controllers. The tuning of fractional-order proportional integral (FOPI) controller parameters with the artificial bee colony (ABC) [35] technique has been presented, which is complex in objective function evaluation and low convergence speed. The parameter-tuning task of FOPI is formulated as an optimization problem and solved with the seeker optimization algorithm (SOA) in [36]. The harmony search (HS) algorithm is reported in [37] for FOPI parameter optimization to control the power-switched reluctance motor. However, there are no conclusive studies on the application of the virtual inertia technique using an SMES topology-based FOPI controller.

Based on several studies [17–19,25,38], it is identified that the detailed design of the PSO-optimized SMES is missing the FOPI controller to support virtual inertia for RESs. Thus, in this paper, we propose a PSO-optimized FOPI-SMES controller design approach for a two-area power system. The proposed approach can support the virtual inertia of the high-level renewable energy integrated system. The addition of this virtual inertia makes the system stable over a wide range of load–generation mismatches. Since the FOPI controller is superior to the conventional PI, the proposed technique performs better when reducing system frequency deviation. However, the design of FOPI is challenging compared to the conventional PI. Thus, this work introduces a detailed model of FOPI, SMES, and a two-area power system to find the design parameters. The dynamic model of the system presented along with SMES and FOPI is utilized to develop the frequency deviation-based cost function for the PSO algorithm. To validate the proposed optimized FOPI controller-based SMES, several case studies were considered and simulated for a wide range of load profile variations. The robustness of the proposed virtual inertia control scheme was tested under reduced system inertia. The proposed controller was compared with the conventional controller, where the improvements in several indices, such as total frequency deviation, overshoot, undershoot, and settling time, were observed. Furthermore, the performance of the non-optimized FOPI was compared with the PSO-optimized FOPI.

The manuscript is organized as follows. The dynamic model of the system including RESs is given in Section 2. The SMES modeling and PSO-based FOPI-SMES design techniques are discussed in Section 3. The simulation results are discussed in Section 4. Finally, the conclusions of this study are given in Section 5.

2. High-Level PV/Wind-Integrated System Modeling

The fractional-order PI controller for superconducting magnetic energy storage (SMES) is designed to virtually support inertia for a high-level solar PV- and wind-integrated two-area power system. An interconnected power system with low inertia due to a high-level integration of PV and wind energy sources, as shown in Figure 1, is considered in this study. The areas are connected by a tie-line, and both of them consist of thermal generating units, an industrial load, a residential load, solar PV, wind, and SMES. The measured frequency and tie-line signals are accumulated in the control and monitoring center. Since the system faces low inertia, it is expected to support the inertia via the control center, which sends control signals to the controllable energy storage devices of both areas if the communication network is available. However, in absence of a communication link, local controllers such as decentralized control, primary control, and droop control can be employed. The net power (P_{net}) in each area in Figure 1 can be calculated using the power of (1) the thermal unit (P_{TH}), (2) the solar array (P_{SA}), (3) the wind farm (P_{WF}), (4) SMES (P_{SMES}), (5) combined industrial and other loads (P_L), and (6) the tie-line (P_{tie12}). The expression for P_{net} is given below.

$$P_{net} = P_{TH} + P_{SA} + P_{WF} - P_L \pm P_{SMES} \pm P_{tie} \tag{1}$$

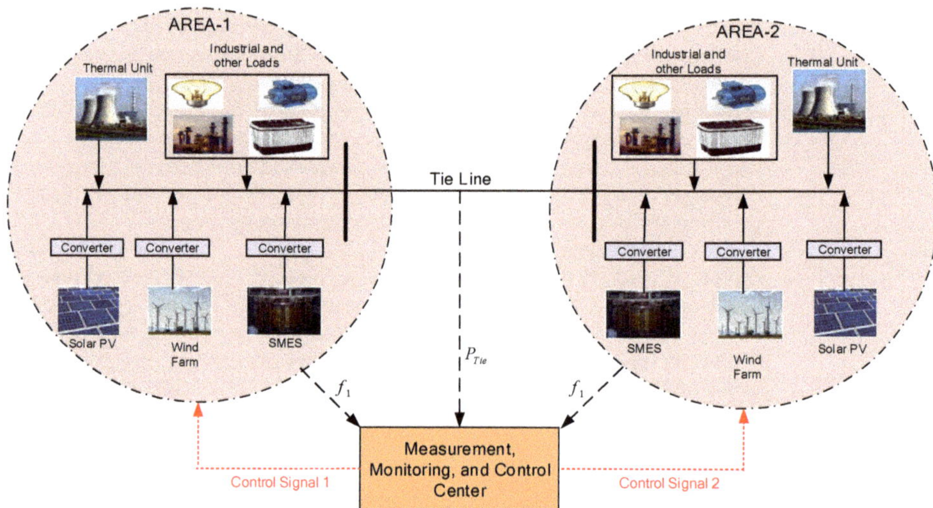

Figure 1. Two-area low inertia interconnected power system.

In general, higher-order models for thermal generating units, wind systems, solar PV, and converters, with nonlinearity are considered to precisely demonstrate the dynamic behaviors of the interconnected system. For large power systems with power electronic converters, however, simplified dynamic models are employed to study the frequency stability. The interested readers can find more details on such dynamic modeling in [39–41]. The simplified dynamic model of the two-area system can be developed as shown in Figure 2 for frequency stability analysis.

From the dynamic model, as shown in Figure 2, the frequency deviation for the kth area can be written as follows.

$$\Delta f_k = \frac{1}{2H_k s + D_k}(\Delta P_{TH,k} + \Delta P_{SA,k} + \Delta P_{WF,k} - \Delta P_{L,k} + \Delta P_{tie,k}) \tag{2}$$

where,

$$\Delta P_{TH,k} = \frac{1}{1 + sT_{t,k}}(\Delta P_{g,k}) \tag{3}$$

$$\Delta P_{g,k} = \frac{1}{1 + sT_{g,k}}\left(\Delta P_{AEC,k} - \frac{1}{R_k}\Delta f_k\right) \tag{4}$$

$$\Delta P_{WT,k} = \frac{1}{1 + sT_{wind,k}}(\Delta P_{wind,k}) \tag{5}$$

$$\Delta P_{SA,k} = \frac{1}{1 + sT_{pv,k}}(\Delta P_{pv,k}) \tag{6}$$

where H_k is the inertia constant in area k, D_k is the damping constant in area k, $\Delta P_{TH,k}$ is the incremental power of the thermal unit in area k, $\Delta P_{SA,k}$ is the incremental power of solar farm in area k, $P_{WF,k}$ is the incremental power of wind farm in area k, $T_{t,k}$ is the turbine time constant in area k, $T_{g,k}$ is the governor time constant in area k, $T_{wind,k}$ is the wind turbine time constant in area k, and $T_{pv,k}$ is the solar system time constant in area k.

Two physical constraints, governor dead band (GDB) and generation rate constraint (GRC), affect the dynamic performance of the power system. The thermal units consist of rotating mass, which inherentlyhas mechanical inertia; thus, it puts a constraint/limit on the output power change, which is known as GRC. The controller designed without GRC may not perform well in practical applications. To handle this issue, GRC is considered for the virtual inertia controller design in this work, as shown in Figure 2. Furthermore, the governor cannot change its valve position within a specific range of speed variation. Due to this dead-band, the tie-line power oscillation with a natural frequency of 0.5 Hz is observed. The dead-band for governor is also taken into consideration in this study to reflect the practical implementation case. The solar PV, wind, and different loads are modeled as disturbances in the dynamic model. The interested readers are directed to the literature [39] for more details on dynamic modeling of PV/wind integrated system.

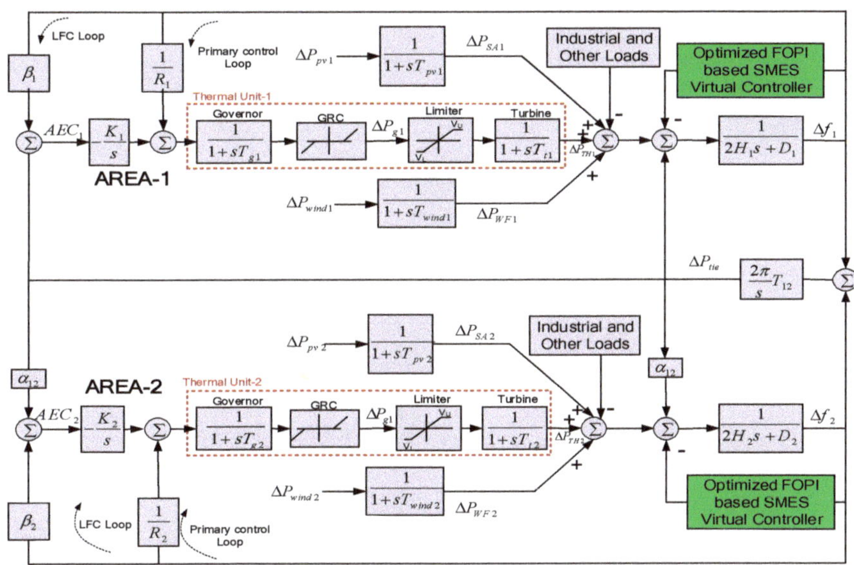

Figure 2. The dynamic model of low inertia system with the proposed controller.

3. SMES Model with FOPI Controller

SMES is a promising device for dynamic stability improvement of power systems. The SMES has several components: thte power conversion system (PCS), consisting of the inverter/rectifier, and the superconducting coil which is kept under extremely low temperature [25]. The PCS also consists of three-phase transformers to allow for energy exchange between the AC grid and the superconducting coil. The harmonic contents of the signals are filtered by two cascaded six pulse bridges, as shown in Figure 3. The capability of SMES to exchange huge power within a very short duration has drawn the attention of researchers in the power system application.

In normal conditions, the SMES coil charges quickly to its pre-defined peak value. As the coil temperature is maintained below the critical value, it conducts the current with nearly zero loss. During contingencies, as the power demand is initiated by the power system, the SMES discharges power through the PCS to the grid almost instantly. While the governors of the generators support the power demand after contingencies, the SMES again charges at its preset value. The inductor DC voltage is given by the equation below [25,42].

$$E_d = 2V_{d0}cos\alpha - 2I_D R_D \tag{7}$$

where V_{d0} is the maximum voltage of the bridge circuit, α is the triac firing angle, I_D is the superconducting coil current, and R_D is the damping resistor. Thus, the DC voltage appearing across the superconducting coil can be controlled with the variation of the triac firing angle α. If α is above 90°, the energy stored in the superconducting coil is released to the grid. In contrast, the superconducting coil charges if α is below the 90°. In this way, the superconducting coil charges and discharges through the bidirectional converter system to absorb or provide energy.

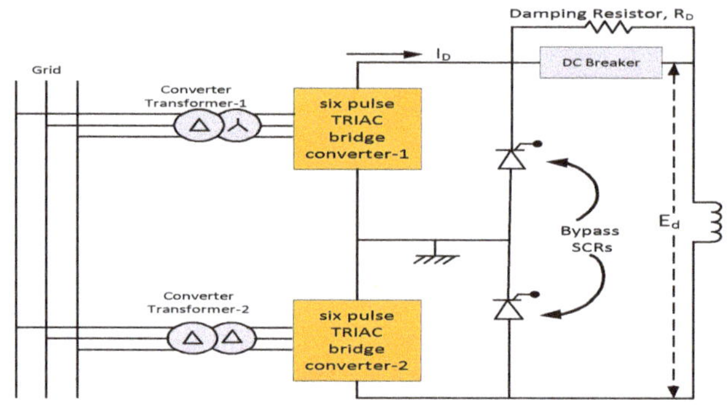

Figure 3. SMES basic configuration.

The detailed dynamic model of SMES for frequency stability studies along with the FOPI controller are shown in Figure 4. During excessive system loading, the load surpasses the generation, the E_D becomes negative, while the current I_D maintains the same direction. The incremental change in E_D is written as

$$\Delta E_D = \frac{K_{SMES}\Delta E - K_{ID}\Delta I_D}{1 + sT_{DC}} \tag{8}$$

where K_{SMES} is the SMES gain, ΔE_D is the output of the FOPI controller, K_{ID} is the negative feedback gain, ΔI_D is the incremental change in superconducting coil current, and T_{DC} is the converter delay time. The incremental change in inductor current I_D is written as

$$\Delta I_D = \frac{\Delta E_D}{sL} \tag{9}$$

The active power of SMES can be derived as follows based on Equations (8) and (9).

$$\Delta P_{SMES} = \frac{K_{SMES}(1 + sT_{DC})sL}{(1 + sT_{DC})[(1 + sT_{DC})sL + K_{ID}]}(I_D + I_{D0}) \tag{10}$$

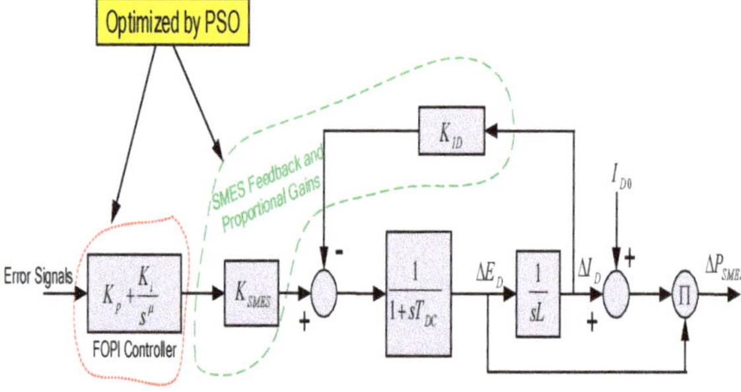

Figure 4. The dynamic SMES model along with the FOPI controller.

3.1. Controller Design

This study focuses on the optimal FOPI-SMES design based on the PSO algorithm to augment the frequency stability of the two-area power system. The fractional-order calculus involves generalized differentiation and integration of non-integer order [33,34]. The fractional-order controller is applied in several engineering fields such as automatic control and power systems due to its superiority over conventional integer order controllers.

The time domain FOPI controller can be represented as

$$u(t) = K_p.e(t) + \int_t^\lambda K_i.e(t) \tag{11}$$

where $e(t)$ is the error signal, K_p is the proportional gain, K_i is the integral gain, and λ is a fractional order and real number that lies between 0 and 2. The Laplace transformation gives the following transfer function for the FOPI controller.

$$C(s) = K_p + \frac{K_i}{s^\lambda}, \lambda \in (0,2) \tag{12}$$

The conventional integer order PI and the FOPI can be understood using Figure 5 in the λ axis. The integer order controller is represented by two points on the λ axis. However, the FOPI controller can be represented by the infinite number of points between 0 and 2. Thus, it gives more degree of freedom and flexibility over the conventional integer order controller.

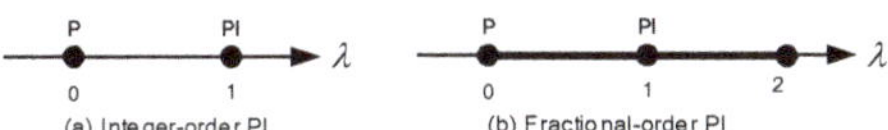

Figure 5. PI controller (fractional order and integer order).

As presented in Figure 4, the SMES virtual inertia based on FOPI is developed in this study to support the frequency of the low-inertia interconnected system. The feedback and proportional gains of SMES along with the FOPI's proportional gain, integral gain, and fractional parameter are optimized with PSO. The following subsections describe the objective function formulation and solution system with the PSO.

3.2. Description of Cost Function

The appropriate cost function is vital in the application of nature-inspired and heuristic optimization techniques in power systems. In general, the cost function is defined to minimize or maximize some variables. In this work, several FOPI gains, fractional orders, SMES feedback gains, and proportional gains are designed based on tie-line power fluctuation and area frequency deviation. For better comprehension of the optimization process, the following cost function is considered.

$$\text{Minimize: } ISE = \int_0^T (|\Delta f_1|^2 + |\Delta f_2|^2 + |\Delta P_{tie}|^2)dt \tag{13}$$

$$\text{Decision Variables: } K_{p1}, K_{i1}, \lambda_1, K_{p2}, K_{i2}, \lambda_2, K_1, K_2, K_{ID1}, K_{SMES1}, K_{ID2}, K_{SMES2} \tag{14}$$

$$\text{Constraints: } K_{p12min} \leq K_{p12} \geq K_{p12max}, K_{i12min} \leq K_{i12} \geq K_{i12max}, K_{1min} \leq K_1 \geq K_{1max},$$
$$K_{1min} \leq K_1 \geq K_{1max}, K_{ID12min} \leq K_{ID12} \geq K_{ID2max}, K_{SMES12min} \leq K_{SMES12} \geq K_{SMES12max} \tag{15}$$

where subscripts 1 and 2 are to denote area 1 and area 2 for the interconnected power system. T is the simulation time, Δf is the frequency deviation, ΔP_{tie} is tie-line power deviation, K_p is the FOPI proportional gain, K_i is the FOPI integral gain, K_{ID} is the SMES negative feedback gain, and K_{SMES} is the SMES proportional gain. Mainly, the upper and lower limits of Equation (15) are selected based on knowledge/experience of FOPI and SMES applications in power system. The optimization algorithm is coded in a MATLAB script (.m files) environment and linked with the MATLAB Simulink (.slx files) environment.

3.3. Solution Approach with PSO

This study proposes a FOPI-based SMES virtual inertia approach in which the minimization problem described by Equation (13) is solved by the PSO. PSO, a heuristic optimization technique, was inspired by the sociological behavior of birds flocking [43]. In the PSO algorithm, several random particles that move in a search space to find the best minimum or maximum value of the cost function based on the minimization or maximization problem, respectively, are initially generated. The PSO shows outstanding performance compared to the other algorithms, as follows [44–46]:

- Since the PSO uses a numerical valued cost function, it is suitable for any nonderivative cost function optimization.
- The PSO facilitates more flexible and robust control frameworks as it uses probability rules.
- It does not fall into premature convergence.
- It has great flexibility for use in online optimization.
- It requires less time compared to other algorithms.
- It provides accurate results with very simple operations.

In recent years, the PSO has been implemented successfully to solve several power system problems such as that presented in [47,48]. The position and velocity vectors in a multi-dimensional solution space for PSO algorithm are mainly described by two equations as follows [49]:

$$v_i^k = c\left\{v_i^{k-1} + c_1 r_1(p_i^{k-1} - x_i^{k-1}) + c_2 r_2(p_g^{k-1} - x_i^{k-1})\right\} \tag{16}$$

$$x_i^k = x_i^{k-1} + v_i^k \tag{17}$$

where v_i^k and x_i^k are the velocities of ith particle for the kth iteration in a multi-dimension search space and the position of ith particle for the kth iteration in a multi-dimension search space, respectively; p_i^{k-1} and p_g^{k-1} are the individual best and global best, respectively, for the ith particle of the $(k-1)$th iteration; r_1 and r_2 are the uniformly distributed random numbers in [0 1]; and c_1 and c_2 are the learning factors used to obtain the best solution. In addition, the c is the constriction factor that is calculated from the values of c_1 and c_2, as follows:

$$c = \frac{1}{\left|2 - (c_1 + c_2) - \sqrt{(c_1 + c_2)^2 - 4(c_1 + c_2)}\right|} \tag{18}$$

The maximum velocity and minimum velocity of each particle can be calculated as follows:

$$v_i^{max,min} = \pm(x_i^{max} - x_i^{min})/N \tag{19}$$

where v_i^{max} and v_i^{min} are the maximum and minimum velocities of the ith particle, respectively; x_i^{max} and x_i^{min} are the maximum and minimum limits of the ith particle, respectively; and N is a number that takes a value between 5–10. The PSO solution steps for solving the optimization problem formulated in Section 3.2 is described below.

Step 1: Initialization of the limits of several variables and particle velocity, as described by Equations (15) and (19), respectively.
Step 2: Selection of the PSO initial parameters including $c1$, $c2$, maximum iteration, population size, etc.
Step 3: Generation of the initial population within the limits.
Step 4: Running the time domain simulation and determining the value of the objective function described by Equation (13).

Step 5: Storing the local best, the best of the current population, and the global best, the best of the total population.
Step 6: Updating the velocity of all populations using Equation (16).
Step 7: Generating a new population based on the updated particle position calculated by Equation (17).
Step 8: Stopping the optimization if the termination criteria are met. Otherwise, returning to step 4.

The overall flowchart for the PSO algorithm to design FOPI and SMES parameters is shown in Figure 6.

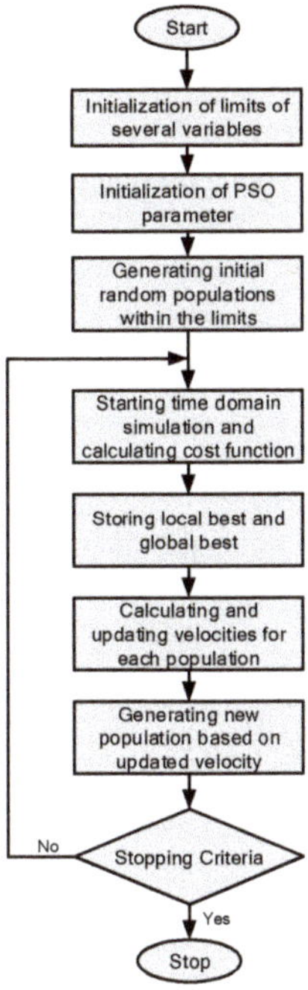

Figure 6. The PSO flowchart for optimizing control parameters.

4. Results and Discussion

The effectiveness and robustness of the proposed optimized FOPI controller in improving the frequency stability are presented in this section. The dynamic model of the system presented in Figure 2 is considered for analytical analysis. The system parameters listed in Table 1 are used to conduct computer simulations and to facilitate analyses. The total generation capacity of the two-area power is 55 MW. The rating of the energy storage device is 6 MW. The proposed energy storage with only 10.9% of the total plant capacity is capable of maintaining frequency stability in case of several load–generation mismatches. The simulations were conducted in MATLAB Simulink considering several scenarios such as light loading, medium loading, heavy loading, and reduced inertia. The system dynamic model was built in Simulink and linked with the PSO optimization code to optimally design the SMES and FOPI parameters. PSO algorithm convergence for the proposed cost function is depicted in Figure 7. As shown in Figure 7, the optimization algorithm converges at the iteration number 20 for several runs, and the corresponding optimized parameters are listed in Table 2.

The system was tested under several step load variations in both areas of the system. The frequency deviations in both areas were plotted for three cases such as (i) without any inertia controller, (ii) with a conventional SMES controller, and (iii) with the PSO optimized FOPI-based SMES controller.

Table 1. System parameters for simulation.

Parameters	Value	
	Area-1	Area-2
Inertia (p.u. MW s)	0.079	0.11
Damping constant (p.u. MW/Hz)	0.016	0.017
Time constant of solar system (s)	1.2	1.2
Time constant of wind system (s)	1.4	1.4
Frequency bias factor (p.u. MW/Hz)	0.3585	0.3928
Valve gate maximum limit (p.u. MW)	0.5	0.5
Valve gate minimum limit (p.u. MW)	-0.5	-0.5
Synchronizing coefficient (p.u. MW/Hz)	0.09	0.09
Area capacity ratio	-0.055	-0.055
Generation rate constraint (GRC)	0.3	0.3
Thermal generator (MW)	12	15
Wind generator (MW)	8	8
PV generator (MW)	6	6
Energy storage power rating (MW)	3	3
Energy storage inductor (H)	2.65	2.65
Energy storage time constant (s)	0.05	0.05
Energy storage reference current (kA)	4.5	4.5

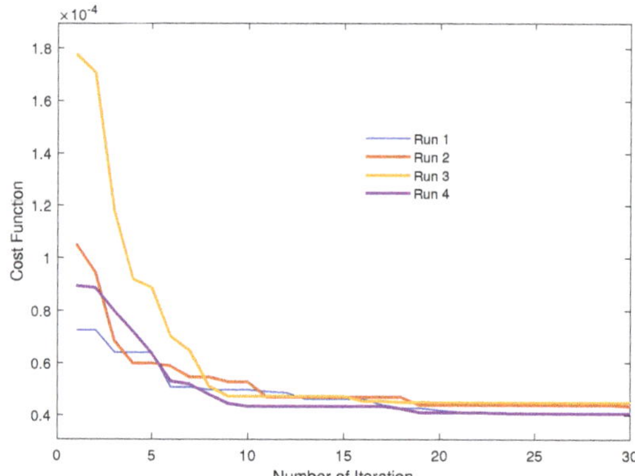

Figure 7. The convergence of the cost function.

Table 2. PSO-optimized FOPI and SMES parameters.

Optimzed Parameters			
Name	Value	Name	Value
K_{p1}	6.8650	K_1	0.1931
K_{i1}	117.60	K_2	0.5000
λ_1	0.9603	K_{ID1}	2.9782
K_{p2}	21.660	K_{SMES1}	0.1300
K_{i2}	78.820	K_{ID2}	2.4879
λ_2	0.6855	K_{SMES2}	0.0410

4.1. Frequency Response Study for Step Load Change in Area 1

In this case, the studied system is simulated for default inertia (100%), as shown in Table 1. The frequency deviations for both areas are depicted in Figure 8 for low, medium, and high step load changes in area 1. The positive effect of the proposed controller is visualized through the reduction in frequency deviations. As visualized in Figure 8a, a step load change of 0.1 p.u. in area 1 causes a significant frequency deviation in area 1 without a virtual inertia controller. The frequency deviation is around 0.42 Hz without any auxiliary controller. The conventional controller-based SMES improves the deviation to about 0.035 Hz. However, the proposed PSO optimization-based FOPI controller for SMES greatly improves the frequency deviation in area 1, which is around 0.005 Hz. It is noteworthy that the settling time is slightly increased for conventional SMES controllers while the frequency deviation is improved. However, the proposed optimized FOPI-based SMES significantly improves all indices, such as settling time, maximum undershoot, and maximum overshoot. Likewise, the frequency deviation in area 2 is very high, around 0.02 Hz, without any inertia controller, as depicted in Figure 8b. The conventional SMES controller improves frequency deviation to some extent. However, the proposed optimized FOPI-based SMES controller reduces the frequency deviation to almost zero. It is observed that, for a large step load change (0.35 p.u.) in area 1, the system cannot maintain stable operation. As visualized in Figure 8e,f, the frequency deviations in both areas continue to increase, leading to instability in the system. The application of the conventional SMES controller can maintain stable operation with some frequency deviation. On the other hand, our proposed techniques stabilize the system with almost zero frequency deviations in both areas. Thus, the system response for several load disturbances in area 1 using the proposed controller is faster, has a very small steady-state error, and is better in terms of overshoot and undershoot compared to other control strategies. The frequency deviations for several scenarios in area 1 and area 2 are given in Table 3 to clearly show the positive impact of the proposed FOPI-based SMES controller on system performance.

Table 3. Reduction in frequency deviations in area 1 and area 2 with the proposed controller.

		Frequency Deviation					
		Δf_1 **(Hz)**			Δf_2 **(Hz)**		
Area	Step Load Change in p.u.	without Inertia Controller	Conventional SMES Controller	FOPI Based SMES	without Inertia Controller	Conventional SMES Controller	FOPI Based SMES
	0.1	0.420	0.035	0.005	0.020	0.004	4.0×10^{-5}
Area-1	0.2	0.850	0.080	0.010	0.040	0.008	3.2×10^{-5}
	0.35	unstable	0.120	0.040	unstable	0.030	2.1×10^{-5}
	0.1	0.395	0.025	1.9×10^{-5}	0.320	0.030	0.010
Area-2	0.2	0.840	0.060	1.2×10^{-5}	0.630	0.090	0.012
	0.35	2.650	0.150	1.0×10^{-5}	2.125	0.106	0.025

4.2. Frequency Response Study for Step Load Change in Area 2

The load disturbances, ranging from the low to high levels, are also applied in area 2 with the system default inertia. It is noticed that the system frequency oscillates over a wide range without any inertia controller. In some cases, the oscillations are beyond the acceptable limits; thus, it requires the system frequency protection relay to operate. As depicted in Figure 9, the frequency deviation in area 1 is 0.395 Hz without any virtual inertia controller for a step load change of 0.1 p.u. The conventional SMES controller reduces the frequency deviation to 0.025 Hz, whereas the proposed optimized FOPI controller is capable of maintaining almost zero frequency deviation. Similarly, the frequency deviation in area 2 is 0.32 Hz without any auxiliary controller. The conventional SMES controller is capable of reducing the frequency deviation by 90.6%. However, the proposed optimized FOPI-based SMES controller reduces the frequency deviation by 96.87%. For the medium and high step load changes in area 2, at 0.2 p.u. and 0.35 p.u., respectively, the frequencies of both areas fall below the under-frequency relay operating setpoint of 59.5 Hz [50] without any virtual inertia controller.

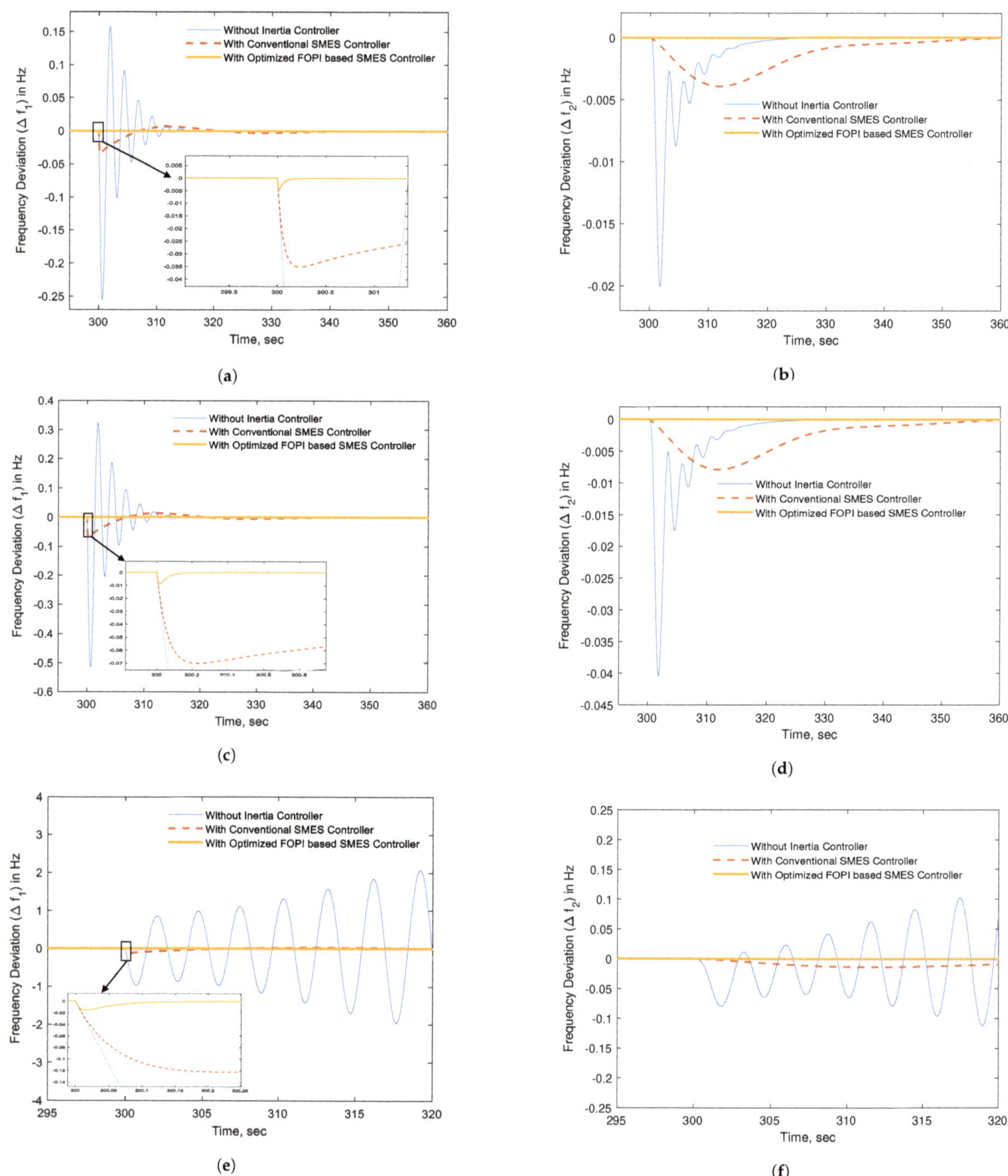

Figure 8. Performance improvement with the proposed controller for load disturbances in area 1. (**a**) Area 1 frequency response for a 0.1 p.u. step load change. (**b**) Area 2 frequency response for a 0.1 p.u. step load change. (**c**) Area 1 frequency response for a 0.2 p.u. step load change. (**d**) Area 2 frequency response for a 0.2 p.u. step load change. (**e**) Area 1 frequency response for a 0.35 p.u. step load change. (**f**) Area 2 frequency response for a 0.35 p.u. step load change.

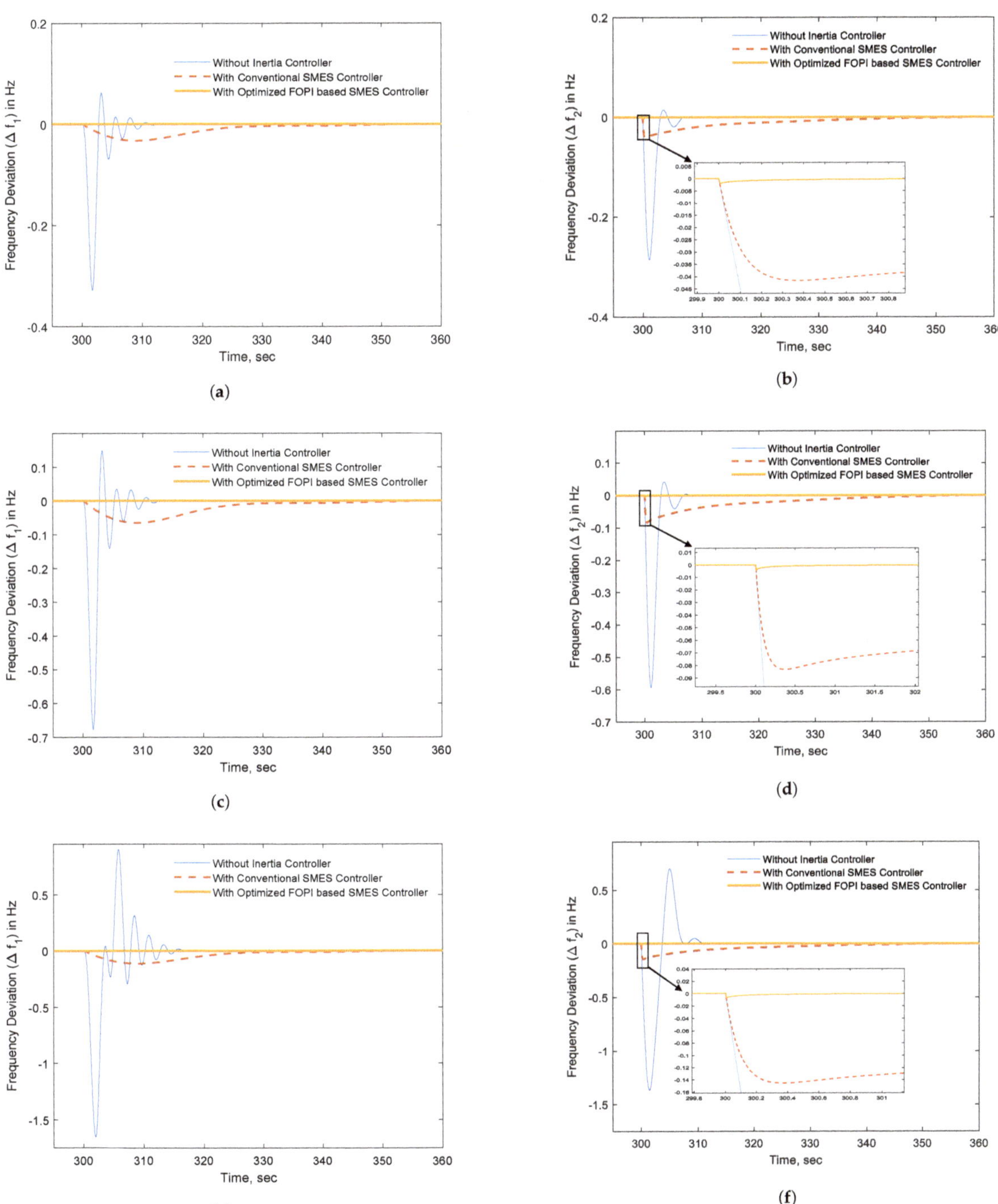

Figure 9. Performance improvement with the proposed controller for load disturbances in area 2. (**a**) Area 1 frequency response for a 0.1 p.u. step load change. (**b**) Area 2 frequency response for a 0.1 p.u. step load change. (**c**) Area 1 frequency response for a 0.2 p.u. step load change. (**d**) Area 2 frequency response for a 0.2 p.u. step load change. (**e**) Area 1 frequency response for a 0.35 p.u. step load change. (**f**) Area 2 frequency response for a 0.35 p.u. step load change.

However, the frequency deviation is well below the under-frequency relay operating point with the conventional SMES controller, as depicted in Figure 9c–f. In these figures, it is visualized that the proposed controller is capable of maintaining the frequency deviations in both areas at almost zero. Thus, the system stability and reliability are guaranteed with the proposed FOPI-based SMES controller. The overall frequency deviations for several cases are listed in Table 3.

4.3. Controller Performance with Solar PV and Wind Power Fluctuations

The effectiveness of the proposed controller was also tested with fluctuating solar and wind power in both areas. The intermittent solar and wind power disturbances considered in this study are depicted in Figure 10a,b, respectively. The solar and wind powers have mean values of 0.05 p.u. and 0.15 p.u., respectively. The solar power is integrated in area 1 at 50 s during the 150 s simulation time, which continues to inject fluctuating power during the entire simulation period. On the other hand, the intermittent wind generating unit is connected at 75 s, which is kept connected throughout the entire simulation period. As shown in Figure 10c,d, the connection of varying solar and wind powers has a detrimental effect on system frequency response without any auxiliary controller.

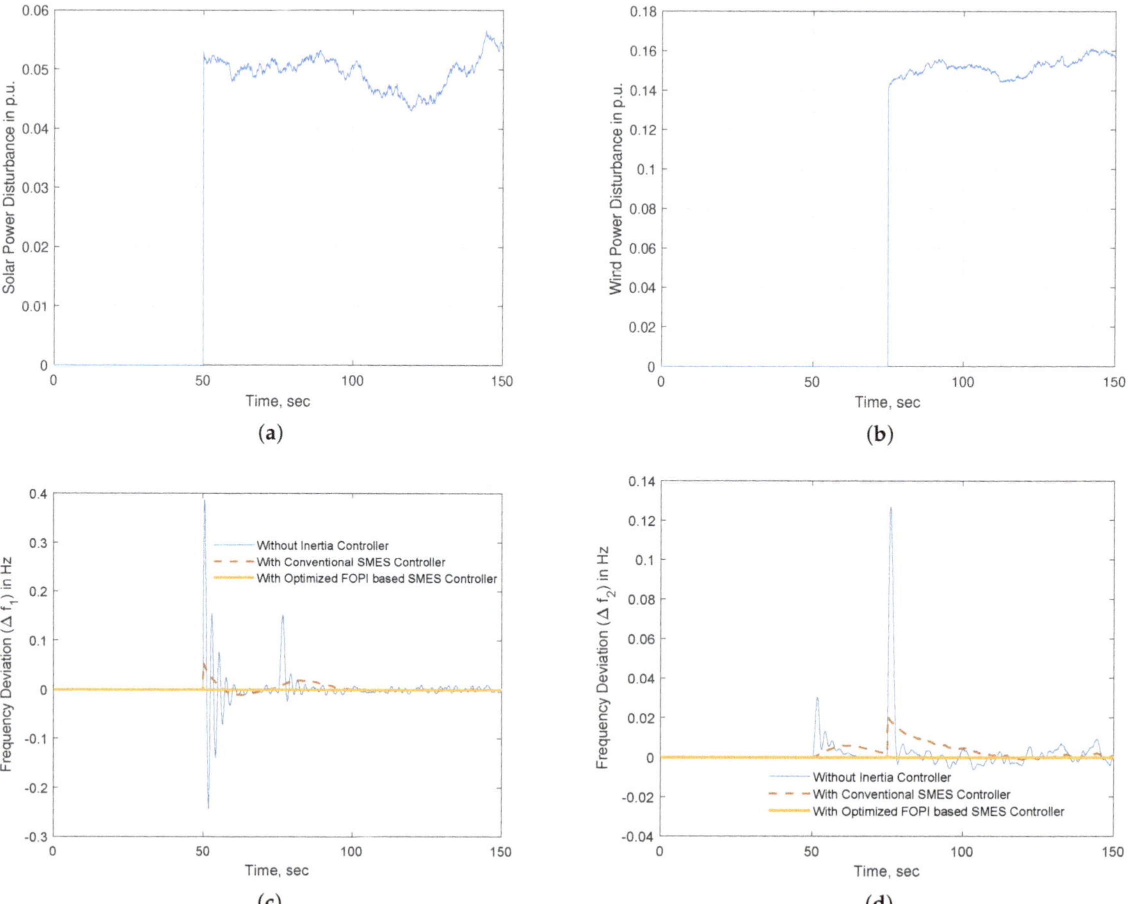

Figure 10. The frequency response for wind generation addition at 50 s and solar generation addition at 75 s. (**a**) Solar power disturbance. (**b**) Wind power disturbance. (**c**) Area 1 frequency response for intermittent solar and wind power. (**d**) Area 2 frequency response for intermittent solar and wind power.

The frequency of the system continues to vary during the entire simulation period and does not settle to a steady-state value. The conventional SMES controller slightly improves the system frequency response. On the other hand, the proposed controller performance is superior, in terms of settling time, overshoot, and undershoot, to the conventional SMES controller. The improvement of several performance indices is listed in Table 4 to demonstrate the superiority of the proposed controller.

Table 4. Improvement of the performance indices for intermittent solar and wind power integration.

Performance Indices	without Inertia Controller	Conventional SMES	Optimized FOPI Based SMES
Maximum Overshoot, Hz	0.39	0.05	0.0100
Maximum Undershoot, Hz	0.25	0.01	0.0001
Settling Time, Sec	inf	50	2

4.4. Frequency Response Analysis for Multiple Load Changes

The effectiveness and robustness of the proposed control technique for virtual control of low inertia systems were also tested with multiple load change scenarios. Several step load changes were considered, as shown in Figure 11a, to investigate the system capability to bring back the frequency deviation to zero before the next changes. Better performance of the proposed FOPI-based SMES is visible from the system frequency response, as seen in Figure 11b, following the first step load change of 0.1 p.u. at 25 s. The proposed controller is faster at eliminating the frequency deviation before the beginning of the second step load change of 0.15 p.u. at 50 s compared to conventional techniques. The frequency deviations in area 1 are very high at all points of step changes without a virtual inertia controller. Although the conventional SMES controller improves the frequency response slightly, a notable improvement is achieved with the proposed technique. In this case, also, the proposed method provides a much better performance in terms of overshoot, undershoot, and settling time. The frequency response for area 2 as visualized in Figure 11c shows better performance with the proposed control technique.

4.5. The Robust Performance of the Proposed Controller with the Reduced System Inertia

In this scenario, the robustness of the proposed controller is verified with the system inertia variations. The inertia in both areas is reduced by 50%, and a step load change of 0.15 p.u. is applied in area −1 at 50 s. The frequency response for this load change is depicted in Figure 12a,b. As depicted in Section 4.1, the system is capable of maintaining stable operation with a step load change of 0.15 p.u. in the case of default inertia (100%). However, Figure 12a,b show that the frequency deviations in both areas gradually increase, leading to instability. The system without SMES requires the under-frequency relay to start operation within 1 second of the load variation since the frequency deviation goes below 0.5 Hz, as depicted in Figure 12a. Although the area 2 frequency takes a longer time to operate under frequency relay, it is also unstable, as depicted by the increasing frequency oscillation in Figure 12b. The conventional SMES controller reduces the frequency deviations and stabilizes the system. However, the proposed control method augments the system stability greatly by reducing frequency deviations to almost zero even with 50% system inertia. The model presented in Figure 2 was also tested for very low inertia with a step load change of 0.1 p.u. in area 1. As shown in Figure 13, the controller is capable of stabilizing the model of Figure 2 for these low inertia. Furthermore, the robustness of the proposed optimized FOPI controller is compared with the non-optimized FOPI controller. The frequency deviation for the system with 15% inertia is plotted in Figure 14 with the optimized FOPI and non-optimized FOPI controller. Thus, the proposed controller is more robust compared to the conventional technique. The main limitation of the proposed technique is that the SMES is a costly solution. Further studies may be conducted on FOPI-based hybrid energy storage devices such as SMES, battery, and supercapacitor for the frequency control of low inertia PV/wind-integrated systems.

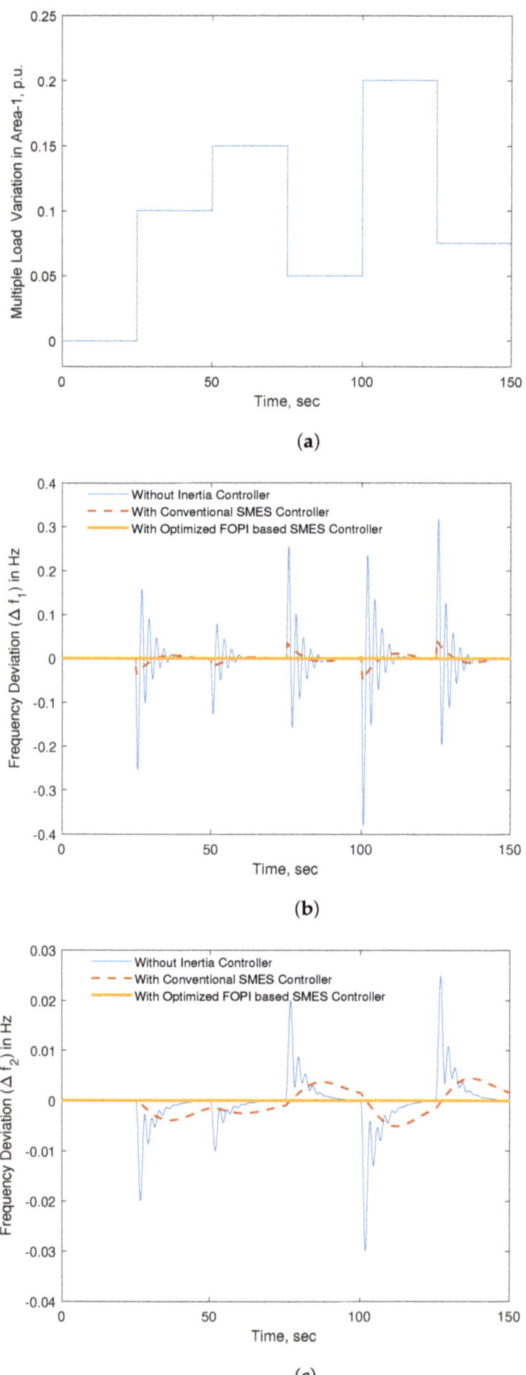

Figure 11. (**a**) Multiple load variations in area 1. (**b**) Frequency response in area 1. (**c**) Frequency response in area 2.

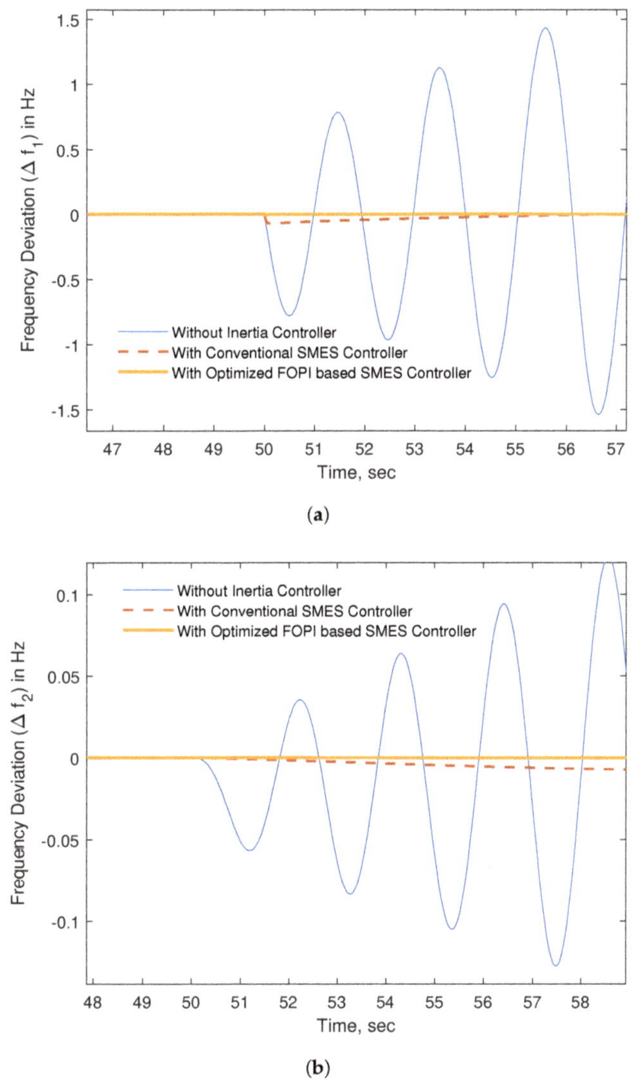

Figure 12. System response for 50% inertia (**a**) Area 1 frequency response. (**b**) Area -2 frequency response.

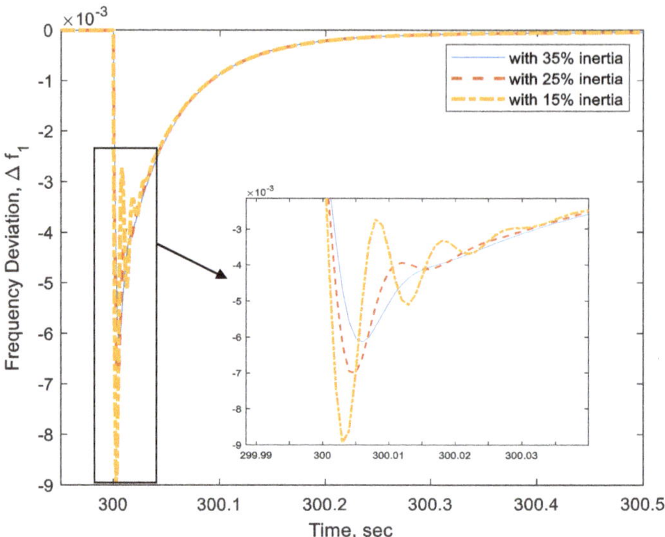

Figure 13. The system response for very low inertia.

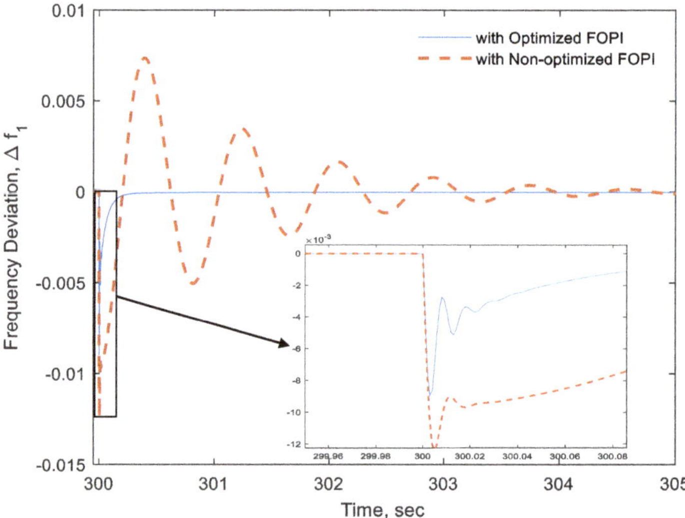

Figure 14. The frequency response comparison for the optimized and non-optimized FOPI controllers.

5. Conclusions

In this work, an optimized FOPI-based SMES virtual inertia controller is designed for a highly renewable energy integrated system. The dynamic model of the system is developed with FOPI to facilitate analysis and design of optimal parameters using PSO. The system response was analyzed with the designed virtual inertia controller considering highly fluctuating solar PV and wind energy. The system and the associated controllers were simulated in MATLAB Simulink. Small, medium, and large load disturbances were applied in the system to prove the effectiveness of the proposed energy storage-based virtual inertia control strategy. The system with default inertia and reduced inertia were tested under single and multiple load disturbances to guarantee the robustness of the proposed controller. The simulated results show promising performance in reducing system frequency deviations and in improving the frequency stability of the system. The proposed controller is superior to the conventional controller in reducing settling time, overshoot, and undershoot, as evident from the analysis. Moreover, the simulation outcomes prove the potential benefits of FOPI controller-based energy storage in high-level renewable energy integration and endorse the green efforts to improve sustainability. Finally, a detailed large-scale DFIG offshore wind farm model with FOPI-based hybrid energy storage virtual inertia controller can be studied as future work.

Author Contributions: Conceptualization, M.S.A.; methodology, M.S.A.; formal analysis, M.S.A., F.S.A.-I. and M.A.A.; writing—original draft preparation, M.S.A.; writing—review and editing, M.S.A. and F.S.A.-I.; supervision, M.A.A.; and funding acquisition, M.S.A. and F.S.A.-I. All authors have read and agreed to the published version of the manuscript.

Funding: This research received no external funding.

Institutional Review Board Statement: Not applicable.

Informed Consent Statement: Not applicable.

Data Availability Statement: Not applicable.

Acknowledgments: The authors acknowledge the support provided by King Fahd University of Petroleum & Minerals (KFUPM) through the directly funded project No. DF201022. The authors also acknowledge the funding support provided by K.A.CARE Energy Research & Innovation Center (ERIC), KFUPM, Dhahran, Saudi Arabia.

Conflicts of Interest: The authors declare no conflicts of interest.

Abbreviations

The following abbreviations are used in this manuscript:

RESs	Renewable energy sources
RoCoF	Rate of change of frequency
FOPI	Fractional-order proportional integral
SMES	Superconducting magnetic energy storage
PV	Photovoltaic
RE	Renewable Energy
PI	Proportional integral
PID	Proportional integral derivative
DFIG	Doubly fed induction generator
CCFC	Capacity constrained frequency control
LFC	Load frequency control
AEC	Area control error
PSO	Particle swarm optimization
GDB	Governor dead band
ANFIS	Adaptive neuro fuzzy system
TCPS	Thyristor controlled phase shifter
ABC	Ant bee colony
SOA	Seeker optimization algorithm
HS	Harmony search
WF	Wind farm
GRC	Generation rate constraint
PCS	Power conversion system
SA	Solar array
ISE	Integral squared error

References

1. Yongning, C.; Yan, L.; Zhen, W.; Zhen, L.; Hongzhi, L.; Zhankui, Z. The Characteristics and Challenge on Large-Scale Renewable Energy Grid Integration. In Proceedings of the 2019 IEEE Innovative Smart Grid Technologies-Asia (ISGT Asia), Chengdu, China, 21–24 May 2019; pp. 1393–1398.
2. Kumar, K.P.; Saravanan, B. Recent techniques to model uncertainties in power generation from renewable energy sources and loads in microgrids—A review. *Renew. Sustain. Energy Rev.* **2017**, *71*, 348–358. [CrossRef]

3. Samani, A.E.; D'Amicis, A.; De Kooning, J.D.; Bozalakov, D.; Silva, P.; Vandevelde, L. Grid balancing with a large-scale electrolyser providing primary reserve. *IET Renew. Power Gener.* **2020**, *14*, 3070–3078. [CrossRef]
4. Dadkhah, A.; Bozalakov, D.; De Kooning, J.D.; Vandevelde, L. On the optimal planning of a hydrogen refuelling station participating in the electricity and balancing markets. *Int. J. Hydrog. Energy* **2021**, *46*, 1488–1500. [CrossRef]
5. Alam, M.S.; Al-Ismail, F.S.; Salem, A.; Abido, M.A. High-Level Penetration of Renewable Energy Sources Into Grid Utility: Challenges and Solutions. *IEEE Access* **2020**, *8*, 190277–190299. [CrossRef]
6. Morren, J.; De Haan, S.W.; Kling, W.L.; Ferreira, J. Wind turbines emulating inertia and supporting primary frequency control. *IEEE Trans. Power Syst.* **2006**, *21*, 433–434. [CrossRef]
7. Fini, M.H.; Golshan, M.E.H. Frequency control using loads and generators capacity in power systems with a high penetration of renewables. *Electr. Power Syst. Res.* **2019**, *166*, 43–51. [CrossRef]
8. Ulbig, A.; Borsche, T.S.; Andersson, G. Impact of low rotational inertia on power system stability and operation. *IFAC Proc. Vol.* **2014**, *47*, 7290–7297. [CrossRef]
9. Saxena, P.; Singh, N.; Pandey, A.K. Enhancing the dynamic performance of microgrid using derivative controlled solar and energy storage based virtual inertia system. *J. Energy Storage* **2020**, *31*, 101613. [CrossRef]
10. Pradhan, C.; Bhende, C. Enhancement in primary frequency contribution using dynamic deloading of wind turbines. *IFAC-PapersOnLine* **2015**, *48*, 13–18. [CrossRef]
11. Ma, H.; Chowdhury, B. Working towards frequency regulation with wind plants: Combined control approaches. *IET Renew. Power Gener.* **2010**, *4*, 308–316. [CrossRef]
12. Khooban, M.H.; Niknam, T. A new intelligent online fuzzy tuning approach for multi-area load frequency control: Self Adaptive Modified Bat Algorithm. *Int. J. Electr. Power Energy Syst.* **2015**, *71*, 254–261. [CrossRef]
13. Magdy, G.; Mohamed, E.A.; Shabib, G.; Elbaset, A.A.; Mitani, Y. SMES based a new PID controller for frequency stability of a real hybrid power system considering high wind power penetration. *IET Renew. Power Gener.* **2018**, *12*, 1304–1313. [CrossRef]
14. Cominos, P.; Munro, N. PID controllers: Recent tuning methods and design to specification. *IEE Proc. Control Theory Appl.* **2002**, *149*, 46–53. [CrossRef]
15. Yang, L.; Hu, Z.; Xie, S.; Kong, S.; Lin, W. Adjustable virtual inertia control of supercapacitors in PV-based AC microgrid cluster. *Electr. Power Syst. Res.* **2019**, *173*, 71–85. [CrossRef]
16. Magdy, G.; Shabib, G.; Elbaset, A.A.; Mitani, Y. A novel coordination scheme of virtual inertia control and digital protection for microgrid dynamic security considering high renewable energy penetration. *IET Renew. Power Gener.* **2018**, *13*, 462–474. [CrossRef]
17. Kerdphol, T.; Watanabe, M.; Mitani, Y.; Phunpeng, V. Applying Virtual Inertia Control Topology to SMES System for Frequency Stability Improvement of Low-Inertia Microgrids Driven by High Renewables. *Energies* **2019**, *12*, 3902. [CrossRef]
18. Gomez, L.; Grilo, A.P.; Salles, M.; Sguarezi Filho, A. Combined Control of DFIG-Based Wind Turbine and Battery Energy Storage System for Frequency Response in Microgrids. *Energies* **2020**, *13*, 894. [CrossRef]
19. Kerdphol, T.; Watanabe, M.; Hongesombut, K.; Mitani, Y. Self-Adaptive Virtual Inertia Control-Based Fuzzy Logic to Improve Frequency Stability of Microgrid With High Renewable Penetration. *IEEE Access* **2019**, *7*, 76071–76083. [CrossRef]
20. Kim, J.; Muljadi, E.; Gevorgian, V.; Mohanpurkar, M.; Luo, Y.; Hovsapian, R.; Koritarov, V. Capability-coordinated frequency control scheme of a virtual power plant with renewable energy sources. *IET Gener. Transm. Distrib.* **2019**, *13*, 3642–3648. [CrossRef]
21. Abo-Elyousr, F.K.; Abdelaziz, A.Y. A Novel Modified Robust Load Frequency Control for Mass-Less Inertia Photovoltaics Penetrations via Hybrid PSO-WOA Approach. *Electr. Power Compon. Syst.* **2019**, *47*, 1744–1758. [CrossRef]
22. Yap, K.Y.; Sarimuthu, C.R.; Lim, J.M.Y. Grid Integration of Solar Photovoltaic System Using Machine Learning-Based Virtual Inertia Synthetization in Synchronverter. *IEEE Access* **2020**, *8*, 49961–49976. [CrossRef]
23. Rahman, F.S.; Kerdphol, T.; Watanabe, M.; Mitani, Y. Optimization of virtual inertia considering system frequency protection scheme. *Electr. Power Syst. Res.* **2019**, *170*, 294–302. [CrossRef]
24. Zhang, X.; Zhu, Z.; Fu, Y.; Li, L. Optimized virtual inertia of wind turbine for rotor angle stability in interconnected power systems. *Electr. Power Syst. Res.* **2020**, *180*, 106157. [CrossRef]
25. Pappachen, A.; Fathima, A.P. Load frequency control in deregulated power system integrated with SMES–TCPS combination using ANFIS controller. *Int. J. Electr. Power Energy Syst.* **2016**, *82*, 519–534. [CrossRef]
26. Ali, M.H.; Wu, B.; Dougal, R.A. An overview of SMES applications in power and energy systems. *IEEE Trans. Sustain. Energy* **2010**, *1*, 38–47. [CrossRef]
27. Shayeghi, H.; Jalili, A.; Shayanfar, H. A robust mixed H2/Hinf based LFC of a deregulated power system including SMES. *Energy Convers. Manag.* **2008**, *49*, 2656–2668. [CrossRef]
28. Ngamroo, I.; Karaipoom, T. Cooperative control of SFCL and SMES for enhancing fault ride through capability and smoothing power fluctuation of DFIG wind farm. *IEEE Trans. Appl. Supercond.* **2014**, *24*, 1–4. [CrossRef]
29. Ali, M.H.; Park, M.; Yu, I.K.; Murata, T.; Tamura, J.; Wu, B. Enhancement of transient stability by fuzzy logic-controlled SMES considering communication delay. *Int. J. Electr. Power Energy Syst.* **2009**, *31*, 402–408. [CrossRef]
30. Chen, L.; Chen, H.; Yang, J.; Yu, Y.; Zhen, K.; Liu, Y.; Ren, L. Coordinated control of superconducting fault current limiter and superconducting magnetic energy storage for transient performance enhancement of grid-connected photovoltaic generation system. *Energies* **2017**, *10*, 56. [CrossRef]

31. Abu-Siada, A.; Islam, S. Application of SMES unit in improving the performance of an AC/DC power system. *IEEE Trans. Sustain. Energy* **2010**, *2*, 109–121. [CrossRef]
32. Podlubny, I. Fractional-order systems and $PI^{\lambda}D^{\mu}$ controllers. *IEEE Trans. Autom. Control* **1999**, *44*, 208–214. [CrossRef]
33. Li, H.; Luo, Y.; Chen, Y. A fractional order proportional and derivative (FOPD) motion controller: Tuning rule and experiments. *IEEE Trans. Control Syst. Technol.* **2009**, *18*, 516–520. [CrossRef]
34. Komathi, C.; Umamaheswari, M. Design of Gray Wolf Optimizer Algorithm-Based Fractional Order PI Controller for Power Factor Correction in SMPS Applications. *IEEE Trans. Power Electron.* **2019**, *35*, 2100–2118. [CrossRef]
35. Kesarkar, A.A.; Selvaganesan, N. Tuning of optimal fractional-order PID controller using an artificial bee colony algorithm. *Syst. Sci. Control Eng.* **2015**, *3*, 99–105. [CrossRef]
36. Kumar, M.R.; Deepak, V.; Ghosh, S. Fractional-order controller design in frequency domain using an improved nonlinear adaptive seeker optimization algorithm. *Turk. J. Electr. Eng. Comput. Sci.* **2017**, *25*, 4299–4310. [CrossRef]
37. Mosaad, M.I. Direct power control of SRG-based WECSs using optimised fractional-order PI controller. *IET Electr. Power Appl.* **2020**, *14*, 409–417. [CrossRef]
38. Alam, M.; Alotaibi, M.A.; Alam, M.A.; Hossain, M.; Shafiullah, M.; Al-Ismail, F.S.; Rashid, M.; Ur, M.; Abido, M.A. High-Level Renewable Energy Integrated System Frequency Control with SMES-Based Optimized Fractional Order Controller. *Electronics* **2021**, *10*, 511. [CrossRef]
39. Lee, D.J.; Wang, L. Small-signal stability analysis of an autonomous hybrid renewable energy power generation/energy storage system part I: Time-domain simulations. *IEEE Trans. Energy Convers.* **2008**, *23*, 311–320. [CrossRef]
40. Bevrani, H.; Habibi, F.; Babahajyani, P.; Watanabe, M.; Mitani, Y. Intelligent frequency control in an AC microgrid: Online PSO-based fuzzy tuning approach. *IEEE Trans. Smart Grid* **2012**, *3*, 1935–1944. [CrossRef]
41. Bevrani, H. *Robust Power System Frequency Control*; Springer: Boston, MA, USA, 2009; Volume 85.
42. Tripathy, S.; Balasubramanian, R.; Nair, P. Effect of superconducting magnetic energy storage on automatic generation control considering governor deadband and boiler dynamics. *IEEE Trans. Power Syst.* **1992**, *7*, 1266–1273. [CrossRef]
43. Eberhart, R.; Kennedy, J. A new optimizer using particle swarm theory. In Proceedings of the MHS'95, Sixth International Symposium on Micro Machine and Human Science, Nagoya, Japan, 4–6 October 1995; pp. 39–43.
44. Tasgetiren, M.F.; Sevkli, M.; Liang, Y.C.; Gencyilmaz, G. Particle swarm optimization algorithm for permutation flowshop sequencing problem. In Proceedings of the International Workshop on Ant Colony Optimization and Swarm Intelligence, Brussels, Belgium, 5–8 September 2004; pp. 382–389.
45. Engelbretch, A. *Fundamentals of Computational Swarm Intelligence*; John Wiley & Sons Ltd.: Oxford, UK, 2005; pp. 5–129.
46. Abdel-Magid, Y.L.; Abido, M.A. AGC tuning of interconnected reheat thermal systems with particle swarm optimization. In Proceedings of the 10th IEEE International Conference on Electronics, Circuits and Systems, Sharjah, United Arab Emirates, 14–17 December 2003; Volume 1, pp. 376–379.
47. Hossain, M.A.; Pota, H.R.; Squartini, S.; Zaman, F.; Guerrero, J.M. Energy scheduling of community microgrid with battery cost using particle swarm optimisation. *Appl. Energy* **2019**, *254*, 113723. [CrossRef]
48. Aguilar, M.E.B.; Coury, D.V.; Reginatto, R.; Monaro, R.M. Multi-objective PSO applied to PI control of DFIG wind turbine under electrical fault conditions. *Electr. Power Syst. Res.* **2020**, *180*, 106081. [CrossRef]
49. Kanwar, N.; Gupta, N.; Niazi, K.; Swarnkar, A.; Bansal, R. Simultaneous allocation of distributed energy resource using improved particle swarm optimization. *Appl. Energy* **2017**, *185*, 1684–1693. [CrossRef]
50. Obaid, Z.A.; Cipcigan, L.M.; Abrahim, L.; Muhssin, M.T. Frequency control of future power systems: Reviewing and evaluating challenges and new control methods. *J. Mod. Power Syst. Clean Energy* **2019**, *7*, 9–25. [CrossRef]

Article

A Machine Learning-Based Communication-Free PV Controller for Voltage Regulation

Shabib Shahid [1,*], Saifullah Shafiq [2], Bilal Khan [1], Ali T. Al-Awami [1,3] and Muhammad Omair Butt [2]

[1] Electrical Engineering Department, King Fahd University of Petroleum & Minerals, Dhahran 31261, Saudi Arabia; g201707470@kfupm.edu.sa (B.K.); aliawami@kfupm.edu.sa (A.T.A.-A.)
[2] Electrical Engineering Department, Prince Mohammad Bin Fahd University, Khobar 31952, Saudi Arabia; sshafiq@pmu.edu.sa (S.S.); mbutt@pmu.edu.sa (M.O.B.)
[3] Interdisciplinary Research Center for Smart Mobility and Logistics, King Fahd University of Petroleum & Minerals, P.O. Box 5067, Dhahran 31261, Saudi Arabia
* Correspondence: g201806380@kfupm.edu.sa

Abstract: Due to the recent advancements in the manufacturing process of solar photovoltaics (PVs) and electronic converters, solar PVs has emerged as a viable investment option for energy trading. However, distribution system with large-scale integration of rooftop PVs, would be subjected to voltage upper limit violations, unless properly controlled. Most of the traditional solutions introduced to address this problem do not ensure fairness amongst the on-line energy sources. In addition, other schemes assume the presence of communication linkages between these energy sources. This paper proposes a control scheme to mitigate the over-voltages in the distribution system without any communication between the distributed energy sources. The proposed approach is based on artificial neural networks that can utilize two locally obtainable inputs, namely, the nodal voltage and node voltage sensitivity and control the PV power. The controller is trained using extensive data generated for various loading conditions to include daily load variations. The control scheme was implemented and tested on a 12.47 kV feeder with 85 households connected on the 220 V distribution system. The results demonstrate the fair control of all the rooftop solar PVs mounted on various houses to ensure the system voltage are maintained within the allowed limits as defined by the ANSI C84.1-2016 standard. Furthermore, to verify the robustness of the proposed PV controller, it is tested during cloudy weather condition and the impact of integration of electric vehicles on the proposed controller is also analyzed. The results prove the efficacy of the proposed controller.

Keywords: photovoltaic; autonomous control; electric vehicles

Citation: Shahid, S.; Shafiq, S.; Khan, B.; Al-Awami, A.T.; Butt, M.O. A Machine Learning-Based Communication-Free PV Controller for Voltage Regulation. *Sustainability* **2021**, *13*, 12280. https://doi.org/10.3390/su132112208

Academic Editor: Thanikanti Sudhakar Babu

Received: 20 August 2021
Accepted: 30 September 2021
Published: 5 November 2021

Publisher's Note: MDPI stays neutral with regard to jurisdictional claims in published maps and institutional affiliations.

1. Introduction

Most of the governments around globe have set ambitious targets for reducing carbon emissions. In order to achieve these targets, a significant amount of small and medium scale renewable energy resources need to be integrated into the power grids. Hence, the traditional power systems are observing an ongoing transition that focus on environmental concerns such as smart grid initiatives, etc. Considering the renewables as an integral part of the power grids, the concept of unidirectional flow of power is not applicable anymore. Furthermore, the solar power being available in abundance and easy to harvest energy from, is allowing prominent integration of photovoltaic (PV) generators in power systems and enabling bi-directional power flow [1].

Among various renewable energy resources, the PV panels and wind turbines (WT) are considered as most suitable options for distributed power generation. Both the PVs and WTs tend to provide an economic and environment-friendly solution, besides being readily available [2]. However, the performance of these resources is inconsistent and is highly dependent on climatic factors such as location, time of the day, weather, etc. Thus, it is highly likely that the power generated through these renewable resources will not follow power variations in load. Consequently, introducing an uncertainty in the power quality and reliability of overall power system [3–6]. A net negative demand may result in the power system, due to inclusion of large number of renewables which can lead to voltage, thermal and other technical problems [7]. Hence, a controlling mechanism is required for all the system elements along with respective participants, to ensure effective integration of renewables in the power distribution systems [8].

Different machine-learning based controllers have been developed in the recent years, which are different in scope and objectives [9–12]. Artificial neural networks (ANN) are used by researchers to demonstrate faster and more accurate maximum power point tracking for solar PVs [9]. In [10], a multi-layer feed-forward ANN is used to improve the power quality in a power system with wide-spread EV chargers. Neural network-based vector controller is designed in [11,12] for the integration of residential solar PV with a utility grid. This intelligent control mechanism has several advantages over conventional controllers. These controllers are capable of mapping non-linear relationships and provide accurate solutions for multivariable problems. Hence, ANN is used to construct the controller used in this research work.

Active control strategies can be classified into model-based [13,14] or model-free [15] from the perspective of network operation modelling. In addition, control schemes can be defined as stochastic [16] and robust [17] when the relating uncertainties from renewable generation and communication networks are taken into account. An important system aspect is observability when classifying the control strategies. In this regard, approaches can be grouped into centralized, decentralized and fully autonomous [18].

A centralized control scheme requires an extensive observation platform to remotely monitor the distribution system parameters [19], lacking by current weaker or older distribution systems. Additionally, to process this huge amount of data, higher processing computing resources are needed in the central control unit. This centrally processed data is used to calculate power of the PV units. Therefore, this strategy turns out to be suboptimal for distribution systems which lack established communication links [20]. While the decentralized control procedures rely on reduced communication linkages, they also apply less computational resources as reduced data is transacted [20]. In spite of that, the need for communication networks, make these strategies less desirable.

Contrary to the presented strategies, autonomous schemes rely on locally available information. In addition, these local controllers avoid nearby communication dependence and are fast and less expensive to deploy. Thus, these techniques are most suitable solutions pre-widespread communication dependence. Autonomous techniques also have better thermal management since the amount of power curtailment reduce system components congestions [21]. However, there is a highly non-linear relationship between PV's output and the system voltages.

In centralized and decentralized strategies, the system voltages are maintained by regulating PV's active and reactive power outputs. However, in the literature, few autonomous control schemes are discussed. For voltage regulation reference [21] employs a voltage sensitivity concept to control PVs active and reactive power outputs. Although the concept of voltage sensitivity-based regulation is promising but the sensitivity calculation method is not robust enough to adapt to the system changes. Since configuration changes occur frequently in a typical distribution system. Moreover, the PVs fail to contribute equally for voltage regulation since the farthest PVs participate unfairly. Embedded inverter features, such as volt/var and volt/watt curves, can also be put in service to regulate the voltages in the absence of communication infrastructure [22]. In [23], optimal volt-var curves are found offline for rooftop PV inverters which are connected in the system taking count of load and PV's active power scenarios. However, this approach has a major shortcoming as it cannot integrate system changes. Additionally, the present optimal volt/var curves may become worst due to resulting changes in the system configuration.

Generally, the rooftop PVs are installed to maximize the monetary profits and therefore, most of the PVs are installed with controllers that drive them towards unity power factor [24]. However, smart inverters are required to operate within a range of selected power factors to support voltage regulation, as per the IEEE 1547 Standard for distributed energy resources (DERs) interconnection [25]. Thus, all PVs should contribute equally when system support is needed, especially during over-voltage events. In this context, similar contribution from the PVs available at disparage locations in the distribution system is called *'fairness'*. An autonomous PV controller is designed that ensures fairness among the PVs, but it is limited to a small-scale distribution system [26].

In this article, a machine learning-based autonomous PV controller framework is presented. The training data are generated for changing loading conditions to include daily, monthly and yearly load variations. Nodal voltage and its sensitivity to changes in the load are input signals to the controller and the determined output is the applicable demand. The proposed controller thresholds the active power output of the PVs to maintain the voltages within the allowable range as given by ANSI C84.1-2016 standard [27]. A noteworthy feature of the proposed controller is ensuring fairness among PVs installed at various locations in the secondary distribution system. A PV-rich test distribution system is employed to evaluate efficacy of the proposed controller. Moreover, response of the controller is further tested under cloudy weather and it fairly controls all the PVs irrespective of their locations. To further test the robustness of the PV controller, many EVs are connected in the network. Results show that the implemented controller regulates the system voltages effectively. Moreover, all the PVs fairly contribute when the system requires support. The following are specific contributions of this article:

(1)   A communication-free PV controller is proposed that determines the power cap for each PV based on the local measurements, such as nodal voltage and nodal sensitivity.
(2)   The proposed controller effectively regulates the system voltages as defined by ANSI C84.1-2016 standard without the need for any communication infrastructure.
(3)   The most attractive feature of the proposed controller is to fairly control the PV power outputs irrespective of their nodal placements in distribution system.

2. Proposed Methodology

2.1. PV Voltage Control

The proposed controller adjusts the power injection of the solar PVs to keep the system voltage within an acceptable range. These PV power plants are connected at various locations in the distribution system, some are closer to and others farther away from the distribution transformer. Each of these solar PVs must be restricted to generate power not more than the upper generation limit termed as power generation cap (PGcap).

In order to circumvent the communication requirements between the distributed solar power producers, it is important to estimate the PGcap using local measurements only. Although voltage at the point of connection (POC) is an important basis for estimating PGcap, but it is not sufficient. Because voltages of the different nodes may behave differently to the power injections. Some of the nodes may violate the upper voltage limits, especially when PV generates more than the connected demand. In this case, downstream or farther nodes in the system will have higher voltages as compared to the nodes at the upstream. In fact, the downstream nodes which are farther from the feeding point are more sensitive to load/generation changes as compared to the upstream nodes. Therefore, the nodal sensitivity can be used along with the nodal voltages, to regulate the power generated by PVs. Moreover, the local voltage sensitivity can also be estimated remotely [18,20]. In addition, the method for its computation is described by Algorithm 1.

Algorithm 1 uses the local nodal voltage (V_{PV}), electric load (P_{load}) and the power generated by the solar PV (P_{gen}). These inputs are used to compute the local voltage sensitivity (δ_{PV}) for each instant. However, in some instances when the change in $Pgen_{net}$ at a particular node is less than the threshold β, the sensitivity value is not updated because these events may result in the incorrect sensitivity calculations. In fact, the change in load/generation at a particular node would have more impact on the voltage of that node as compared to the other nodes. So, if the sensitivity is calculated for the node during the instances when the change in load/generation is low at that specific node, the change in the voltage at that node may have the more influence of the significant changes in load/generation at the nearby nodes. Hence, these instances may result in wrong sensitivity calculations and are ignored. A tuning parameter β is used to filter out these values. Note that β may vary from one distribution network to another.

Algorithm 1 Proposed voltage sensitivity calculation

Input: $V_{PV,t}$, $V_{PV,t-1}$, $P_{gen,t}$, $P_{gen,t-1}$, $P_{load,t}$, $P_{load,t-1}$, $\delta_{PV,t-1}$
Output: $\delta_{PV,t}$
Variables: $Pgen_{net,t}$, $Pgen_{net,t-1}$
1: $Pgen_{net,t} = P_{gen,t} - P_{load,t}$
2: $Pgen_{net,t-1} = P_{gen,t-1} - P_{load,t-1}$
3: **if** $\left(\frac{Pgen_{net,t} - Pgen_{net,t-1}}{Pgen_{net,t-1}} \right) \geq \beta$ **then**
4: $\qquad \delta_{PV,t} = \frac{V_{PV,t} - V_{PV,t-1}}{Pgen_{net,t} - Pgen_{net,t-1}}$
5: **else**
6: $\qquad \delta_{PV,t} = \delta_{PV,t-1}$
7: **End if**

The value of the voltage sensitivity coefficient at any node and the respective voltage at the POC, gives a good description of the nodes position in the distribution system. Therefore, a machine learning-based approach is designed to use these inputs to determine the PGcap fairly for all the solar power plant. Once the objective and input to the machine learning-based strategy is identified, the strategy is further grouped in offline training and an online application module as given in Figure 1.

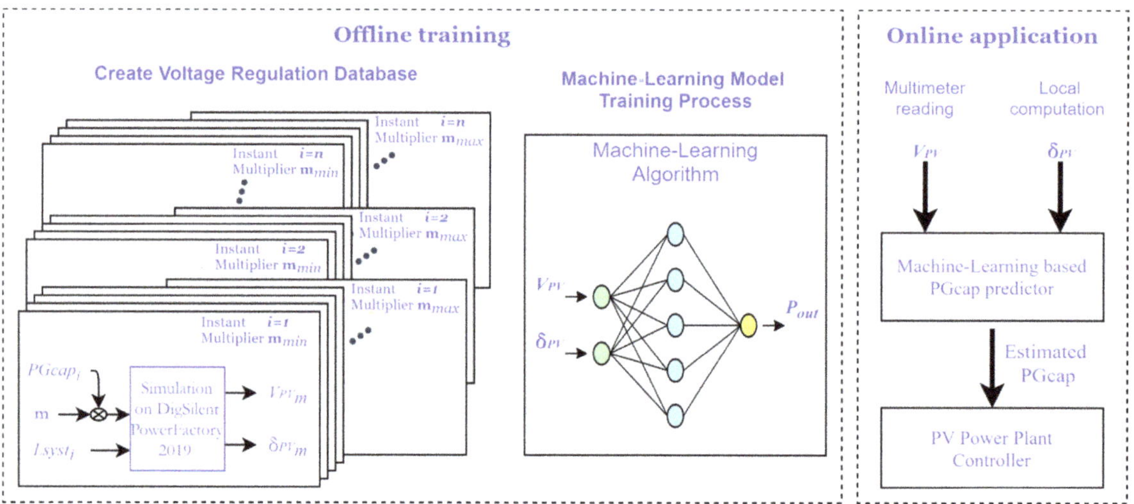

Figure 1. Flow chart brief overview.

2.2. Create Voltage Regulation Database

For any machine learning algorithm to achieve good performance in online application, the training set needs to cover operation conditions in the field. In this application, this means realistic ranges of system loads (*Lsyst*) and solar power generations needs to be simulated.

To generate a required data set, load is set at all the system nodes and then the PGcap is found by an iterative process as shown in Figure 2. At the start, all the PVs are selected to generate at their maximum rated capacity. Then, the power flow analysis is performed using a simulator and subsequently power outputs of PVs are reduced to the level that all the system over-voltages are eliminated. This procedure is repeated for various conditions of electrical load in the system to simulate load variations ranging from hours to seasons. Based on the power flow results of each loading condition, the PGcap that needs to be applied at each node is determined so that the acceptable voltage profiles are obtained.

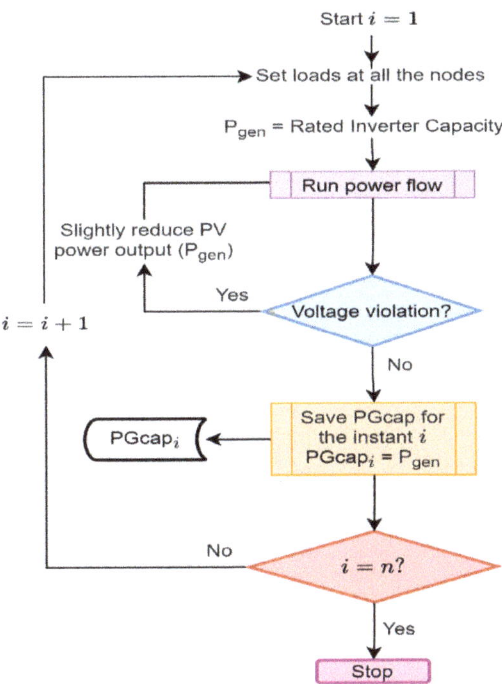

Figure 2. Flow chart for determining PGcap for different system loads.

Furthermore, the additional training data points are required to train the controller's response to high voltage and low voltage scenarios. The high voltages may appear in the system when the power generated by the PVs is greater than the PGcap at any particular instant. Likewise, low voltages may occur when the PV power output is less than the PGcap. These events are very common for the distribution systems with variable renewable energy sources and load excursions. To imitate these events, power generated by the PVs is varied and the power-flow is simulated for all the high/low voltage scenarios. These sub-optimal power generations are represented by $(m \times PGcap)$, where m is a multiplier and it is varied between m_{min} and m_{max}, as shown in Figure 1. Training data points for low-voltage and high-voltage scenarios are obtained when m is less than 100% and greater than 100%, respectively. The response of the controller in these sub-optimal power generation conditions is explained in Section 2.3.

Voltages of the nodes in the distribution system change differently with the changes in power generations. For instance, consider the LV distribution system shown in Figure 3. Houses available at different levels are connected to the nearby LV transformer. Note that the transformer bus is regulated at the voltage of 1.01 p.u., as mentioned in Figure 3. In addition, note that the house at level 1 is less sensitive to the changes in load/generation as compared to the house at level 4. The voltages and sensitivities of these houses for different power generations $(m \times PGcap)$, are shown in Figure 4. Note that the Figure 4 is plotted at a specific loading condition. It can be seen that the slope of the curve for level 4 house is steeper than the curve of the level 1 house. That means, the house at level 4 is more sensitive when compared to the level 1 house. Another important fact is that when the generation (P_{gen}) becomes equal to the load (P_{load}), the net power generation ($Pgen_{net}$) becomes zero and the voltages of the nodes become the same as the voltage at the transformer bus (i.e., 1.01 p.u.). For a specific loading condition of this example system, $Pgen_{net}$ becomes zero when $m = 50\%$ as shown in Figure 4. It is important to note that this event can happen at different loading conditions for smaller/bigger distribution systems. It can also be observed from the Figure 4 that for the positive $Pgen_{net}$ (i.e., to the right of the marked circle) the nodes with higher sensitivities have higher voltages than the other less sensitive nodes.

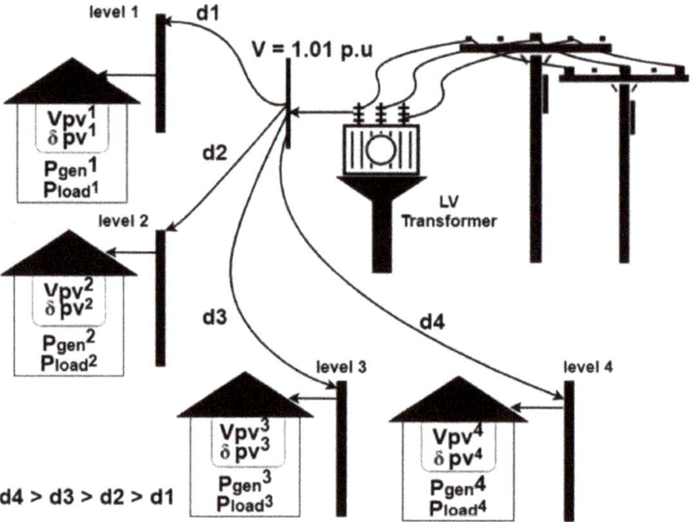

Figure 3. Example distribution system.

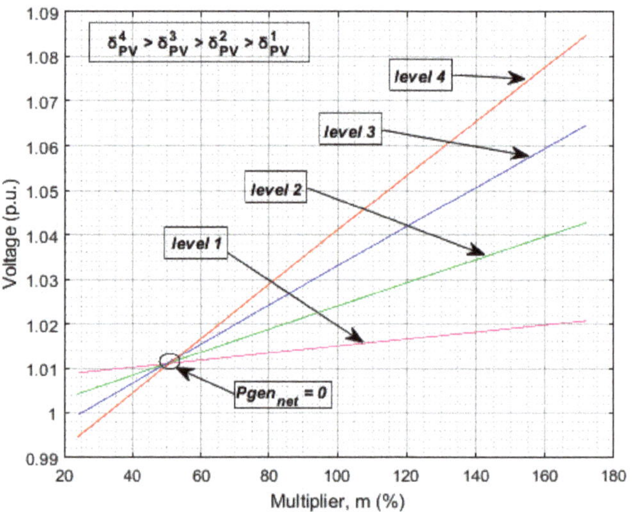

Figure 4. Voltage and sensitivities in extreme conditions.

2.3. Training Controller Response

A neural network is a non-linear statistical model that endeavors to recognize underlying relationships in a set of data. It can represent the complex behaviors of natural or engineered processes. They can adapt to changing input; so, the network generates the best possible result without needing to redesign the output criteria. Hence, a fully connected multi-layer perceptron (MLP) neural network is used to design the control formula for the PV power output. This network is trained with the generated voltage regulation database to calculate the parameters associated with MLP neural network. The concept of transfer learning is employed for optimal training of the controller.

In this ML approach, the inputs are the nodal voltages (V) and their corresponding sensitivities (δ) and the output is the desired PV power generation (P_{out}^d). Both the inputs V and δ, were obtained during the voltage regulation database generation. By design, the database included the input points which represent the over/under power generation ($m \times PGcap$) instances. Now, to secure the stability of the system in these extreme scenarios, the desired outputs of the controller (P_{out}^d) must improve the voltages in steps. For instance, if the voltages are high then the controller is expected to reduce the power generation. Similarly, for lower voltages the power generation needs to be increased. The relationship between P_{out}^d and $m \times PGcap$ is calculated using (1).

$$P_{out}^d = PGcap[1 + \alpha_r(1 - m)] \tag{1}$$

where

$$0 \leq \alpha_r \leq 1$$

$$m_{\min} < m < m_{\max}$$

P_{out}^d is the desired power output, $PGcap$ is the power generation cap, m is the multiplier and α_r is the gradient of the controller response. Note that when the multiplier is 1, the P_{out}^d becomes equal to $PGcap$. However, when the multiplier is not equal to 1, then P_{out}^d is adjusted accordingly to improve the system voltages.

As mentioned earlier, the output of PVs may vary since they are highly variable. Moreover, the output of neural network-based controller may vary a bit from the desired power output depending on the input data. Hence, the data points must be generated in such a way that the desired controller response is achieved. Note that α_r is the parameter that defines the behaviour of the controller response. It is important to understand that if the value of α_r is close to 0, then the controller output would not change significantly during the extreme voltage conditions. In other words, the controller response would be highly conservative. On the other hand, if the value of α_r is close to 1, then the controller output would quickly respond to the changes in the load/generation but may oscillate between the extreme voltage conditions. Therefore, the best controller response would be achieved when the value of α_r is between 0 and 1. In this work, the best controller response is achieved when the α_r is set to 0.37. However, it may vary depending on the controller requirements.

Let us demonstrate the controller behaviour as described in (1). Figure 5 shows the relationship between $P_{out}^d/PGcap$ and m (as defined in Equation (1)), with $\alpha_r = 0.37$. For example, if the multiplier m is 160% (i.e., point A in Figure 5), then the voltage reaches to 1.078 p.u. at the downstream bus (i.e., level 4 house) as shown in Figure 4. Since the voltage violation is severe, the controller would reduce the power output to 78%. Correspondingly, the controller will move to the point B (see Figure 5) and the voltage would become 1.028 p.u. (see Figure 4). At the subsequent instants, the controller will move to 108% (i.e., point C in Figure 5). Eventually, the controller response will reach to PGcap (point D) and regulate the system voltages within the allowed limits. It is important to note that the controller will mostly operate in the normal range, that means, the data points generated for the low/high voltages will be only followed during these extreme events.

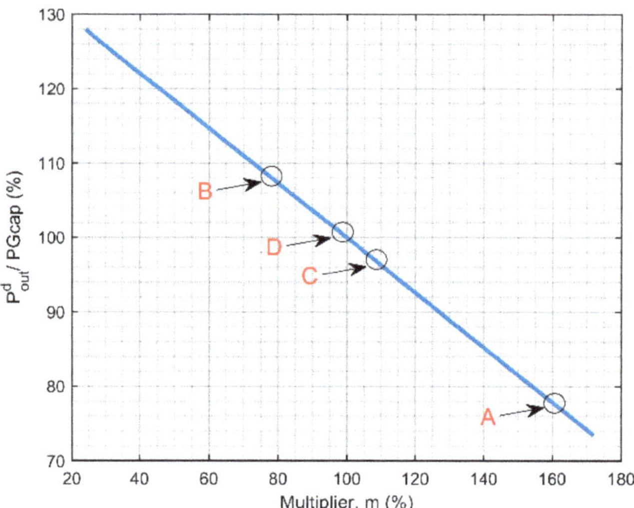

Figure 5. Controller response in extreme conditions.

2.4. Neural Network Structure

In this control strategy only two inputs are used, namely, voltage and the nodal sensitivity. Therefore, a small ANN structure is sufficient to be used for effective PV power control. Figure 1 shows the general overview of the proposed control structure. The purpose of the controller is to predict the PV power output depending on the inputs.

Note that the activation function of the output layer is linear, as provided in (2). However, the sigmoid activation function, given in (3), is used in the input layer. The associated weights and biases are represented by w and b, respectively. Moreover, N represents the total number of neurons in the hidden layer.

$$P_{out} = \sum_{i=1}^{N} w_i^{output} \times v_i + b_i^{output} \tag{2}$$

$$v_i = tanh\left(w_{i,voltage}^{input} \times V_{PV} + w_{i,sensitivity}^{input} \times \delta_{PV} + b_i^{input}\right) \tag{3}$$

The network is obtained through training by using the Levenberg-Marquardt algorithm. The training of neural network means adjusting the weights of layers and biases to get the target values. During the training process, weights and biases are adjusted and the target values are tracked continuously until the squared error between the actual and the desired outputs is minimized. The performance function of ANN is the mean squared error (MSE), as described in (4).

$$MSE(v) = \sum_{j=1}^{\substack{Total \\ samples}} \left(P_{out,j} - P_{out,j}^d\right)^2 \tag{4}$$

Neural networks have the ability to adapt to the distribution function and this makes them more likely to find the non-linear relationships between the input measurements and the output. Nevertheless, this ability to adapt may result in the neural network that largely overfits the training data, producing the effect called *'overfitting'* [28]. To avoid this, a split of the data must be carried out between the training model and the test model (e.g., 80%–20%), with the aim of obtaining low MSE on the test data. In this work, 80% of the data is used to train the controller while the controller is tested on the rest of the 20% data.

3. Test System

The distribution network used to test the effectiveness of the proposed PV controller is shown in Figure 6 [20,29]. This is a balanced three-phase medium voltage (MV) system with 17 primary nodes. Where each node is connected to a low voltage (LV) radial distribution system. The accumulated load and rated PV capacity at each of these LV systems is $300(60 \times 5)$ kW and $560(140 \times 4)$ kW, respectively. The system parameters of the primary and secondary nodes are provided in Table 1.

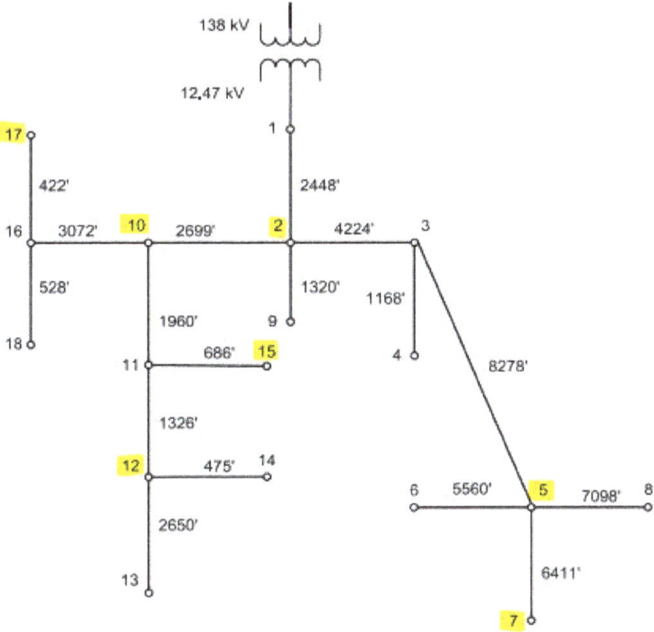

Figure 6. Test primary distribution system.

Table 1. Primary and secondary distribution system parameters.

Parameter	Value
Primary conductor	ACSR 2
Max. current for primary conductors	180 A
Distribution service transformer	150 kVA
Secondary conductor	350 Al, 4/0
No. of customers per node	5
PV capacity per secondary bus	140 kW
System frequency	60 Hz

Primary nodes have a three-phase 12.47/0.22 kV secondary distribution transformer. Each transformer feeds 20 houses at each phase through five laterals, as shown in Figure 7. The PVs are installed on four buses in the secondary systems, which are labelled as SB 02, SB 03, SB 04 and SB 05 (see Figure 7). Note that the secondary systems connected to the highlighted primary nodes in Figure 6 have solar PVs installed.

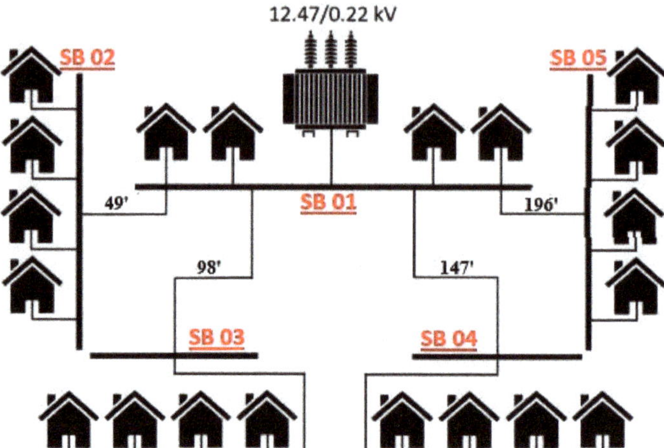

Figure 7. Secondary distribution network topology.

The communication-free PV controller is installed with each rooftop PV. The voltage regulation database (explained in Section 2.2) is used to train the controller to their respective output values determined by Equation (1) as mentioned in Section 2.3. Since there are only two inputs and one output, a small ANN with 5 hidden neurons is chosen. This network is trained using "nntool" in MATLAB. The network is obtained through training by using the Levenberg–Marquardt algorithm. In this control strategy, the controller is trained in such a way that it can be installed at any location in the distribution system. The relationship between inputs and the output is shown in Figure 8. It can be observed that there are four different groups of data points. These groups correspond to the different levels of houses in the secondary distribution system. The red plane shown in Figure 8 shows the response of the neural network controller.

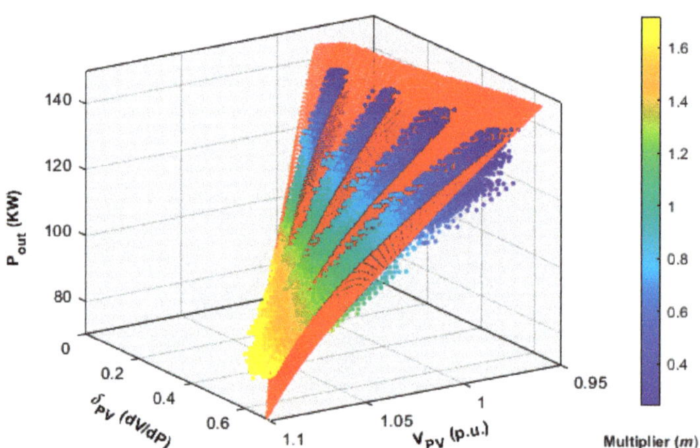

Figure 8. Relationship between inputs and the output.

4. Results and Discussion

The purpose of the proposed controller is to modulate the power injected by the rooftop solar PVs connected in the distribution system. These controllers are expected to regulate the system voltages within an acceptable range, defined by the ANSI C84.1-2016 standard. Additionally, fairness among the distributed PVs is also an important requisite for customer satisfaction. The machine learning-based PV controller presented in this article is designed to work in the absence of any communication linkages between the controllers. The trained network only uses the local voltage measurements and the sensitivity estimations as inputs (see Algorithm 1).

The test distribution system under consideration has loads on all the $(17 \times 5) = 85$ secondary buses. A typical load profile and the average load profile of these loads are depicted in Figure 9 for illustrative purposes. It can be noticed that the load is highly variable, this is due to the on/off actions of various household appliances. These high load variations in each LV node of the system will make the voltage regulation problem more challenging and, hence, test the controller's effectiveness in the abnormal or extreme conditions.

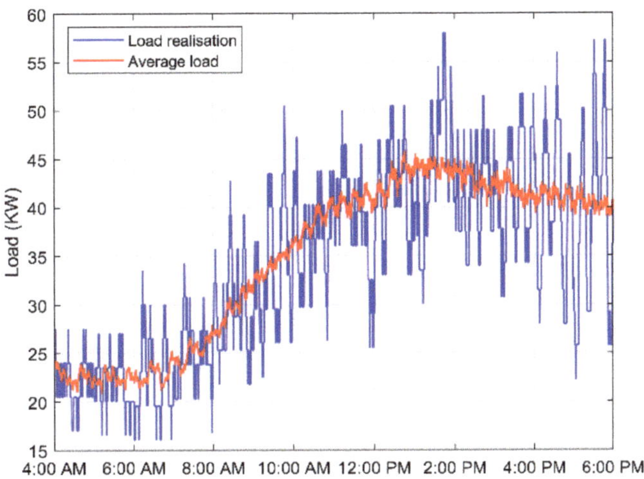

Figure 9. A typical load profile.

The voltage regulation performances of different communication-free PV controllers are compared and they are arranged in increasing order of their benefits. These control schemes include, opportunistic maximum power point tracking controller, droop-based and voltage-based on-off controllers and lastly the proposed PV controller. Note that, considering space limitations, results are presented for some specific nodes. To present node-specific results, two extreme nodes, i.e., Node-10 (upstream) and Node-7 (downstream), are selected since they can provide sufficient performance details.

4.1. Conventional System (i.e., without PVs)

In order to showcase the impact of increasing the solar power penetration in the present electric power infrastructure, for comparison the test system without PVs is shown. For this case, bus voltages of the secondary system connected to Node-7 and Node-10 are shown in Figure 10.

It is clearly visible that there is not much of a difference between the voltage profiles of the upstream and downstream MV nodes. The voltages of the downstream LV system (Figure 10a) are only slightly lower than that of the upstream LV system (Figure 10b). However, there is a significant voltage difference between the upstream and downstream LV busses. Voltage of the upstream bus (i.e., SB 02) is close to 1 p.u. while the voltage at the most downstream bus (i.e., SB 05) is much lower yet within the allowable voltage range as defined by the ANSI C84.1-2016 standard. A bus connected closer to the MV/LV transformer (i.e., strong/upstream bus) is less sensitive to change in load than that connected farther from the transformer (i.e., weak/downstream node). That is why, SB 05 has the highest voltage variations. This elucidates the importance of voltage regulation for the power sources connected to the LV busses.

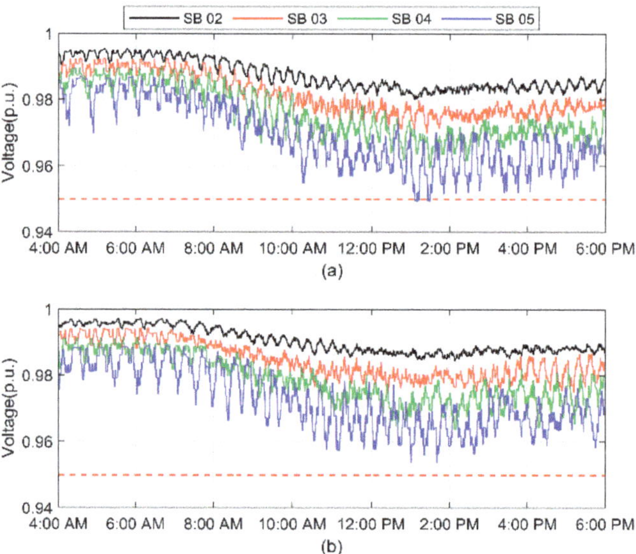

Figure 10. No PV system voltage profiles at (**a**) Node-7, (**b**) Node-10.

4.2. Base Case (i.e., PVs without Any Controller)

This case represents moderate levels of solar power integration with the system loads shown in the previous sub-section. The solar power is injected into the power system without any control. PV power profiles for a particular sunny day in the summer season is considered which is shown in Figure 11. In a sunny day during the peak sunlight hours, there is more energy being produced by the PVs than consumed by the loads. Hence, in this section the impact of this high penetration of widespread PVs in the distribution system is studied.

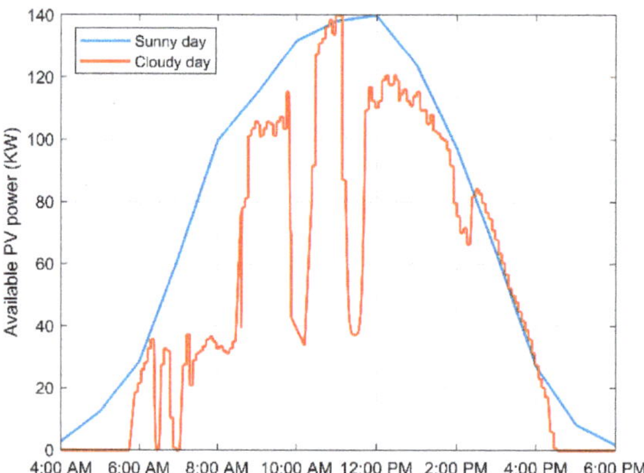

Figure 11. Real power available from the PV systems without controllers.

The base case voltage profiles are shown in Figure 12 with uncontrolled PV integration. The buses that are farther away from the transformer (i.e., SB 04 and SB 05) have voltage upper limit violations; however, buses that are closer to the transformer (i.e., SB 02 and SB 03) have the voltages within the allowable limits. It is noteworthy to observe that the voltage at the downstream bus is much higher as compared to the upstream bus. However, it was much lower as compared to the upstream bus when no PVs were installed (see

Figures 10 and 12). It is due to the fact that the downstream buses are often more sensitive to changes in the load and/or power generation when compared to the upstream buses.

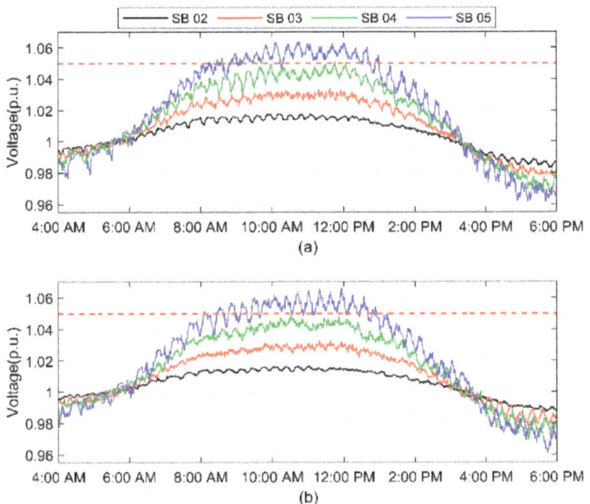

Figure 12. Base case voltage profile at (**a**) Node-7, (**b**) Node-10.

4.3. On-Off Controller

In this case, an on/off switch controller is tested. A reference voltage, $v_r = 1.045$ p.u. is selected. Based on the POC bus voltage, the controller decides to ramp down the power generation if the voltage is higher than v_r and ramp up the power generation if the voltage is lower than v_r. Ramp limits are selected to be 1 kW/10 s to avoid abrupt variations.

In Figures 13 and 14, the voltage profiles and power generation profiles of the secondary buses are shown. It is evident that the power is only curtailed for downstream buses (i.e., SB 04 and SB 05), whereas the upstream buses are allowed to generate all the available power. Moreover, due to the frequent switching the SB 05 bus has fluctuating voltage problem, which may lead to other power system problems [30].

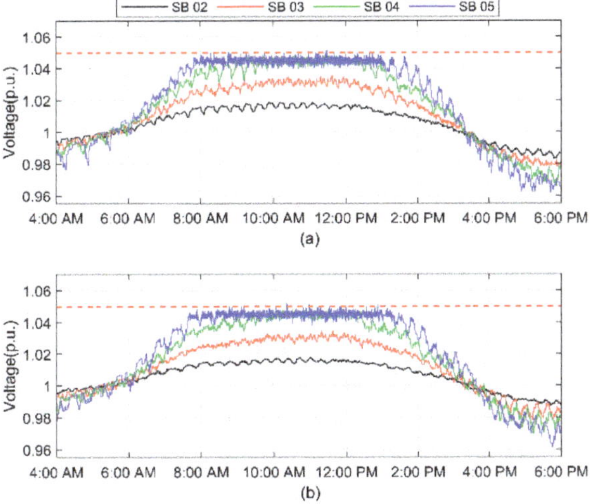

Figure 13. On-off case voltage profile at (**a**) Node-7, (**b**) Node-10.

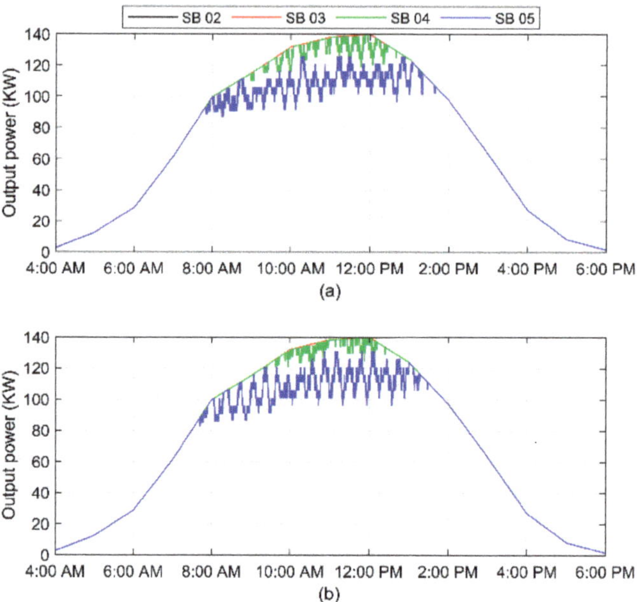

Figure 14. On-off case output power generated at (**a**) Node-7, (**b**) Node-10.

4.4. Droop-Based Controller

In this section, a prevalent control scheme based on droop characteristics is shown for comparison. This scheme relies on the static droop P-V characteristics curve of each node in the distribution system. This relationship between the voltage and the output power is defined by the piecewise linear function as provided in (5). In [31], settings are done based on the location and the system loads. However, due to system reconfigurations this scheme would require communication, which is absent in present-time distribution systems.

A simpler form of the droop-based controller is tested with same droop constants, k_d. A critical system voltage, $v_{critical}$, is set to be 1.05, which is the maximum allowable voltage defined by the ANSI C84.1-2016 standard. Considering a safety margin, the controller is activated at 1.02 p.u. The droop constant, k_d, is set at 21.

$$P_{out}(t) = PG_{available} \times \begin{cases} k_d(v_{critical} - v_i(t)), & v_i(t) \geq 1.03 \text{ p.u.} \\ 1, & v_i(t) < 1.03 \text{ p.u.} \end{cases} \tag{5}$$

where:

$$k_d = \frac{140 \text{ kW}}{(1.05 - 1.02)(220\text{V})} = 21.2 \text{ kW/v} \tag{6}$$

The system voltages are shown in Figure 15. It can be seen that all the voltages are within the allowed limits. However, the buses located downstream inject less power than the buses closer to the transformer as illustrated in Figure 16. The buses closer to the transformer can export full PV power available while the downstream buses are undesirably restricted due to their higher voltages. Thus, the PV generated revenues for these customers are lower as compared to other customers.

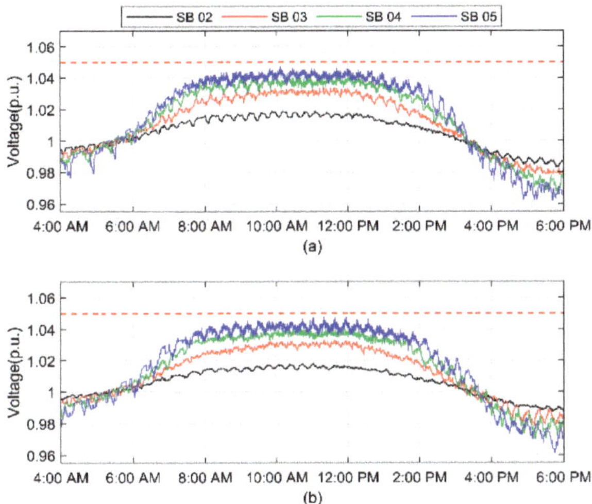

Figure 15. Droop-based controller voltage profile at (**a**) Node-7, (**b**) Node-10.

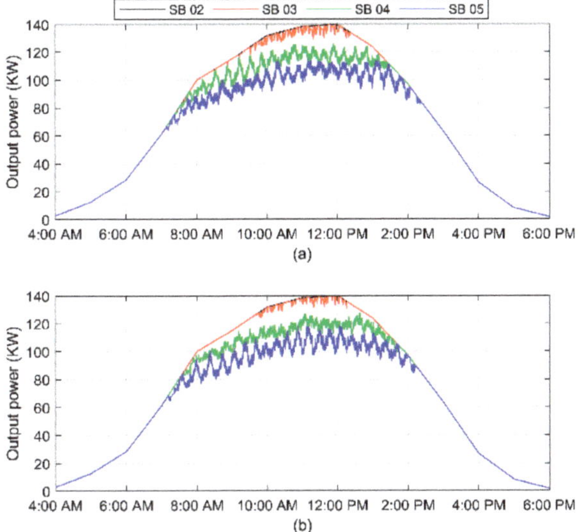

Figure 16. Droop-based controller power generated at (**a**) Node-7, (**b**) Node-10.4.5. Proposed voltage-and-sensitivity-based controller.

The uncontrolled integration of PVs results in voltage-rise problems, as it is shown in Section 4.2. While other communication-free controllers (Sections 4.3 and 4.4) are not able to act fairly among all the PVs connected in the system. Additionally, these controllers are not able to provide a stable voltage profile at the PVs point of connection. It is because they only use voltage readings to control the power injected and there are no integral or derivative control modules.

In order to mitigate the voltage-rise problem, a machine learning-based autonomous PV controller is presented. The controller utilizes both, the nodal voltage and its sensitivity to throttle the power output of the solar PV. Each solar PV in the distribution system is controlled independently by these controllers. The resulting voltage profiles are provided in Figure 17 when the proposed controllers are implemented. It can be clearly seen that the controller is able to effectively regulate the system voltages with good system voltage stability. Throughout the day, the voltages are maintained within the permissible range, except between 8:00 A.M. and 9:00 A.M., where some negligible over-voltages are recorded. This is because of the time required by the controller to respond to the significant variations in load/generation (as shown in Figure 5 and explained in Section 2.3).

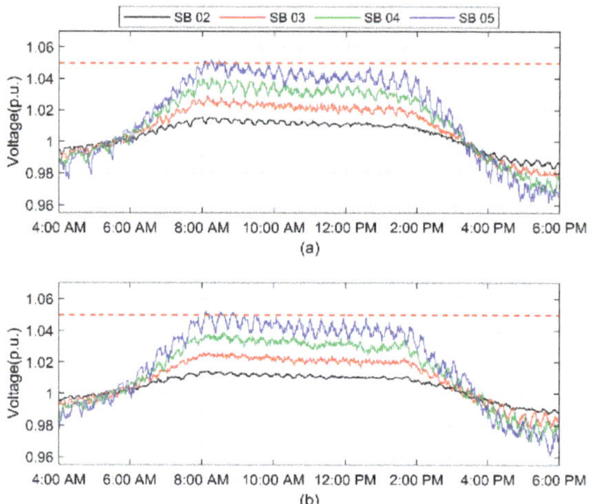

Figure 17. Proposed controller voltage profile at (**a**) Node-7, (**b**) Node-10.

The most interesting feature of the proposed controller is fair power curtailment, despite the fact that there is no communication among the PVs connected at various locations in the distribution network. The power generated in the presence of the proposed controller is shown in Figure 18. The proposed controller curtails similar amount of power from all the PVs when the system support is required. The daily energy produced by the proposed and the droop-based controllers are provided in Table 2. It can be observed that the secondary downstream bus (i.e., SB 05) produces around 240 kWh less energy than the upstream secondary bus (i.e., SB 02) when the droop-based controllers are utilized. However, for the proposed controller similar amount of energy is being produced at all the secondary buses in the system. These results indicate the efficacy of the proposed controller.

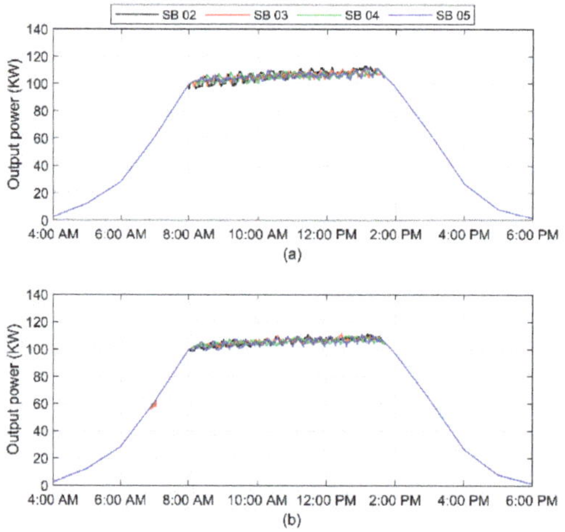

Figure 18. Proposed controller power generated at (**a**) Node-7, (**b**) Node-10.

Table 2. Comparison of daily energy production (kWh).

Control Mode	P *	SB 02	SB 03	SB 04	SB 05
Droop-based	D	1090	1077	940	847
	U	1086	1079	948	850
Proposed	D	937	936	932	933
	U	935	938	935	931

* P represents the primary node, whereas D and U represent downstream node (i.e., Node-7) and upstream node (i.e., Node-10), respectively.

5. Control Strategy Performance

In this section, performance of the proposed voltage and sensitivity-based control strategy is further examined. Table 3 summarizes the fairness for all control modes and effectiveness of these methods to avoid voltage limit violations. Total energy produced by each primary node in a day are also included. As anticipated, the uncontrolled integration of solar power results in the maximum energy production; however, the voltage violations are extreme and persist for longer duration (i.e., for about 286 min). On the other hand, the voltages are appreciably improved for on-off and droop-based controllers, but the fairness issue was gravely compromised. It is evident that the downstream buses (SB 04 and SB 05) experience most of the energy curtailment; therefore, these control techniques are not suitable for fair power integration. However, the proposed controller fairly curtails the energy from each PV system. Moreover, it drastically improves the system voltages as compared to the uncontrolled case. Note that the proposed controller curtails a little higher amount of the generated power in comparison to the other controllers. However, this slightly higher curtailment allows all the distributed PVs to participate fairly in the voltage improvement process.

Table 3. Comparison of controller performances.

| Control Mode | P * | Energy Produced in a Day (kWh) | % Energy Curtailed | | | | | Fairness Ad-dressed? | Maximum Voltage (p.u.) | Duration of Voltage Violations (Minutes) | Voltage Accept-able? |
			Total (%)	SB 02	SB 03	SB 04	SB 05				
Uncontrolled	D	4198	-	-	-	-	-	Yes	1.067	286.5	No
	U	4198	-	-	-	-	-		1.069	286.5	
On-Off	D	4082.5	2.75	0	0	0.31	2.44	No	1.052	0.333	Yes
	U	4094.7	2.46	0	0	0.16	2.30		1.052	0.333	
Droop-based	D	3954.1	5.81	0	0.14	2.03	3.64	No	1.046	0	Yes
	U	3963.7	5.58	0	0.08	1.90	3.60		1.045	0	
Proposed	D	3737.9	10.96	2.70	2.72	2.78	2.76	Yes	1.051	19.167	Yes
	U	3739.2	10.93	2.72	2.69	2.73	2.79		1.052	19.167	

* P represents the primary node, whereas D and U represent downstream node (i.e., Node-7) and upstream node (i.e., Node-10), respectively. The bus will be reported as a high voltage violation if the voltage exceeds 1.05 p.u. % Energy curtailed is calculated using energy produced in uncontrolled case.

In addition, the performance of the proposed control strategy is assessed with solar resource variability. This can happen for many reasons, including the cloud covering the solar PVs. The PV's output generation is highly varied on a cloudy day compared to a sunny day, as shown in Figure 11. The voltage and power generation profiles of the secondary buses are shown in Figures 19 and 20, respectively, when the proposed controller is tested during cloudy day. The results illustrate that the controller response to fluctuating available solar power is quick. Despite that the change in output power is not restricted (no ramp limit), power is always fairly curtailed to minimize the over voltage violations.

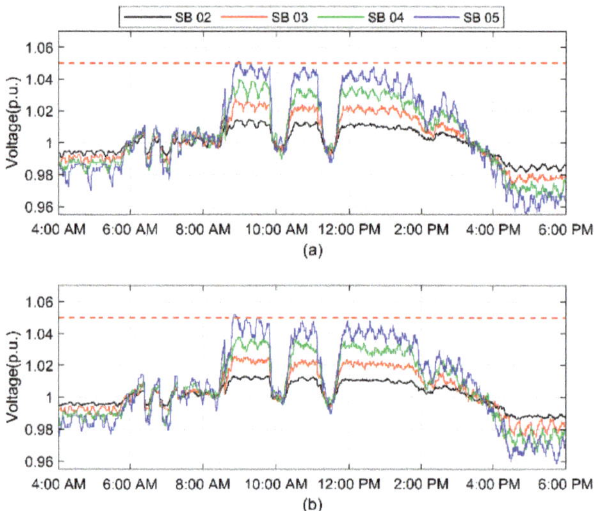

Figure 19. Voltage profiles in a cloudy day at (**a**) Node-7, (**b**) Node-10.

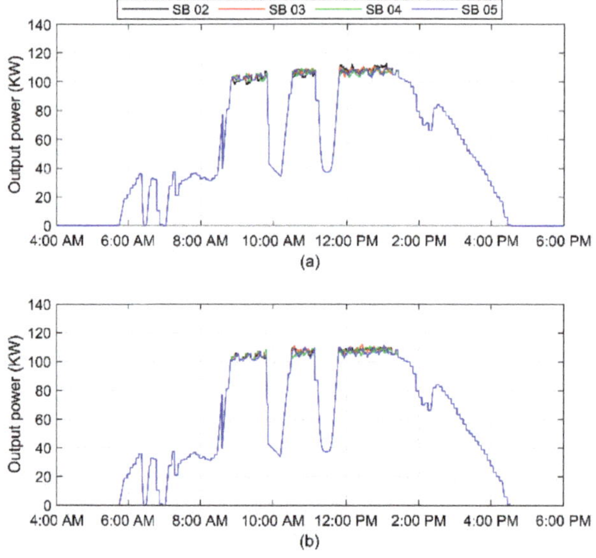

Figure 20. Power generation profiles in a cloudy day at (**a**) Node-7, (**b**) Node-10.

To further assess the robustness of the proposed autonomous PV controller, it is assumed that one of every two houses has an electric vehicle (EV). Nissan Leafs having 40 kWh battery pack model [32] are used to test the proposed controller when they are connected to the secondary buses. The EVs are assumed to charge at the rate of 7 kW when plugged in. That means, each EV takes 4 h to charge the battery from 30% to 100%. EVs are expected to plug-in between 6:30 a.m. and 10:30 a.m. since the impact of EVs charging on the PV control needs to be studied. The state of charge (SOC) for the EVs connected in primary Node-10 are shown in Figure 21.

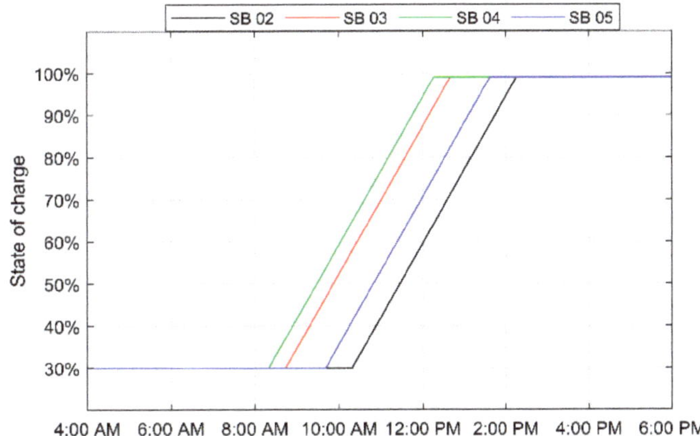

Figure 21. State of charge of EVs connected at secondaries of Node-10.

When two EVs get connected to a bus at any given time, the load at that bus increases by 14 kW. Consequently, the voltage at that bus changes which affects the PV power generation. Figure 22 shows the PV power output when the proposed controller is being used during these events. Note that the EVs only connected at SB 03 and SB 04 are charging during the period from 9:00 a.m. until 9:30 a.m. During this time period, the power generated at these buses is only 3 to 5 kW more than the power generated at other buses. Similarly, from 12:30 p.m. until 1:25 p.m. the EVs only connected to SB 01 and SB 05 are charging, while the others are already fully charged. During this time, the PV power output for the buses with connected EVs is only a little more than the other buses. The voltage profiles of the EVs charging points are shown in Figure 23. Note that the voltages of the buses decrease when the EVs start charging, this happens because of the instantaneous increase in the load.

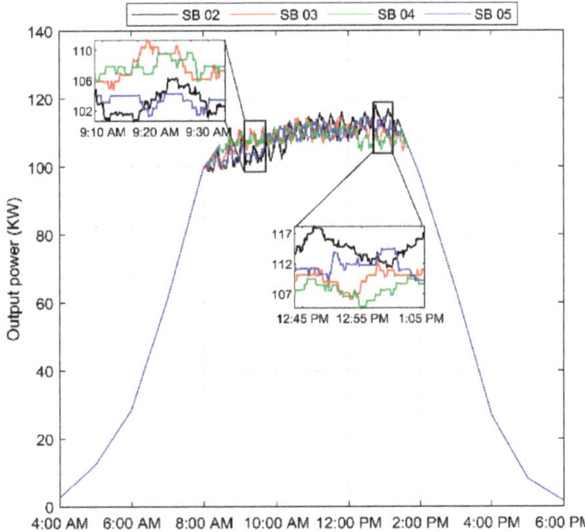

Figure 22. Power output comparison, with EVs charging at different times.

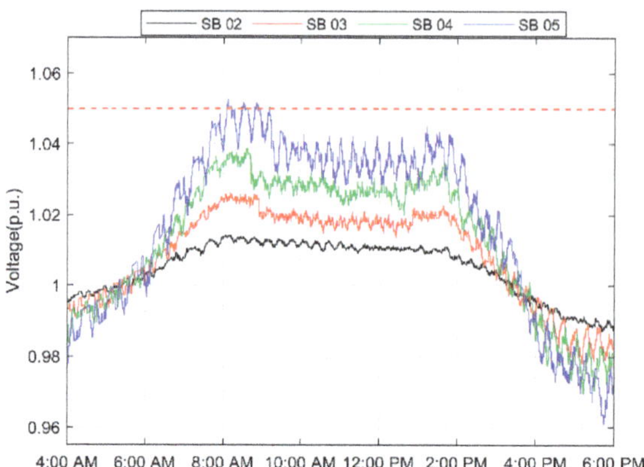

Figure 23. Voltage profiles at EV connection points.

Table 4 shows the overall PV energy produced in a day at the secondary buses of Node-7 and Node-10 when the EVs are being charged. It can be seen that the daily energy production has been increased when there is extra load connected in the system (see Tables 3 and 4). In fact, the system voltages drop when more load is connected which allows the PVs to generate relatively more power. These results indicate that the proposed controller successfully balances the trade-off between high penetration of renewable energy and fair power curtailment, so that the voltages remain within the allowable range as defined by ANSI C84.1-2016 standard.

Table 4. Energy produced in a day with EV charging from 30% to 100% (kWh).

Scenarios	P *	SB 02	SB 03	SB 04	SB 05	Total
EVs not charging	D	937	936	932	933	3738
	U	935	938	935	931	3739
EVs charging during day	D	953	962	955	957	3827
	U	953	962	955	957	3827

* P represents the primary node, whereas D and U represent downstream node (i.e., Node-7) and upstream node (i.e., Node-10), respectively.

6. Conclusions

This paper proposes a local voltage regulation control scheme for distributed solar PVs connected in the secondary network of the distribution system. The proposed technique is aimed at fairly controlling the power injections of all the PVs without the need of any communication infrastructure. The proposed machine-learning based controller uses two local measurements as inputs, namely, the nodal voltage and its sensitivity, to determine the PV power output. The performance of the proposed controller is validated in a PV-rich MV/LV test distribution system. Even though the controllers in the system work independently, they fairly controlled the power injections of all the PVs to keep the network voltages in an acceptable range. The distinctiveness of the proposed technique is highlighted by comparing a few other communication-free controllers as presented in the previously published literature. In addition, the robustness of the controller is verified by considering a cloudy day. In addition, the electric vehicles (EVs) are integrated into the secondary distribution system to further confirm the efficacy of the proposed controller in the future distribution systems.

The contributions of the proposed approach are evident since the machine learning-based controller is relatively inexpensive yet easy to deploy solution, which only requires the local voltage measurements at the point of connection. Additionally, this method is computationally fast because of the minimal hardware requirement and there are no communication delays. The fast computational time ensures the suitability of the approach for solving real-time voltage problems.

Author Contributions: Conceptualization, S.S. (Shabib Shahid), S.S. (Saifullah Shafiq) and A.T.A.-A.; methodology, S.S. (Shabib Shahid), S.S. (Saifullah Shafiq) and B.K.; software, A.T.A.-A.; validation, S.S. (Saifullah Shafiq), B.K. and A.T.A.-A.; formal analysis, S.S. (Shabib Shahid) and S.S. (Saifullah Shafiq); investigation, A.T.A.-A.; resources, A.T.A.-A.; data curation, S.S. (Shabib Shahid) and B.K.; writing—original draft preparation, S.S. (Shabib Shahid), S.S. (Saifullah Shafiq), B.K. and M.O.B.; writing—review and editing, S.S. (Shabib Shahid) and S.S. (Saifullah Shafiq); visualization, S.S. (Shabib Shahid), B.K. and M.O.B.; supervision, A.T.A.-A.; project administration, S.S. (Saifullah Shafiq) and M.O.B.; funding acquisition, S.S. (Saifullah Shafiq) and M.O.B. All authors have read and agreed to the published version of the manuscript.

Funding: This research received no external funding.

Institutional Review Board Statement: Not applicable.

Informed Consent Statement: Not applicable.

Acknowledgments: The authors would like to acknowledge the support provided by World Futures Studies Federation (WFSF), Prince Mohammad Bin Fahd University (PMU), Prince Mohammad Bin Fahd Center for Futuristic Studies (PMFCFS), Khobar, Saudi Arabia.

Conflicts of Interest: The authors declare no conflict of interest.

References

1. Farag, H.E.; Abdelaziz, M.; El-Saadany, E.F. Voltage and Reactive Power Impacts on Successful Operation of Islanded Microgrids. *IEEE Trans. Power Syst.* **2012**, *28*, 1716–1727. [CrossRef]
2. Khalid, M.; Akram, U.; Shafiq, S. Optimal Planning of Multiple Distributed Generating Units and Storage in Active Distribution Networks. *IEEE Access* **2018**, *6*, 55234–55244. [CrossRef]
3. Khatamianfar, A.; Khalid, M.; Savkin, A.V.; Agelidis, V.G. Improving Wind Farm Dispatch in the Australian Electricity Market with Battery Energy Storage Using Model Predictive Control. *IEEE Trans. Sustain. Energy* **2013**, *4*, 745–755. [CrossRef]
4. Shafiq, S.; Al-Awami, A.T. Reliability and economic assessment of renewable micro-grid with V2G electric vehicles coordination. In Proceedings of the 2015 IEEE Jordan Conference on Applied Electrical Engineering and Computing Technologies (AEECT), Amman, Jordan, 3–5 November 2015; pp. 1–6. [CrossRef]
5. Khalid, M.; Ahmadi, A.; Savkin, A.V.; Agelidis, V.G. Minimizing the energy cost for microgrids integrated with renewable energy resources and conventional generation using controlled battery energy storage. *Renew. Energy* **2016**, *97*, 646–655. [CrossRef]
6. Akram, U.; Khalid, M. A Coordinated Frequency Regulation Framework Based on Hybrid Battery-Ultracapacitor Energy Storage Technologies. *IEEE Access* **2018**, *6*, 7310–7320. [CrossRef]
7. Walling, R.A.; Saint, R.; Dugan, R.C.; Burke, J.; Kojovic, L.A. Summary of Distributed Resources Impact on Power Delivery Systems. *IEEE Trans. Power Deliv.* **2008**, *23*, 1636–1644. [CrossRef]
8. Ochoa, L.F.; Dent, C.J.; Harrison, G.P. Distribution network capacity assessment: Variable DG and active networks. *IEEE Trans. Power Syst.* **2010**, *25*, 87–95. [CrossRef]
9. Gowid, S.; Massoud, A. A robust experimental-based artificial neural network approach for photovoltaic maximum power point identification considering electrical, thermal and meteorological impact. *Alex. Eng. J.* **2020**, *59*, 3699–3707. [CrossRef]
10. Shafiq, S.; Khan, B.; Raussi, P.; Al-Awami, A.T. A novel communication-free charge controller for electric vehicles using machine learning. *IET Smart Grid* **2021**, *4*, 334–345. [CrossRef]
11. Sun, Y.; Li, S.; Lin, B.; Fu, X.; Ramezani, M.; Jaithwa, I. Artificial Neural Network for Control and Grid Integration of Residential Solar Photovoltaic Systems. *IEEE Trans. Sustain. Energy* **2017**, *8*, 1484–1495. [CrossRef]
12. Arora, A.; Singh, A. Design and implementation of Legendre-based neural network controller in grid-connected PV systems. *IET Renew. Power Gener.* **2019**, *13*, 2783–2792. [CrossRef]
13. Alnaser, S.; Ochoa, L. Advanced Network Management Systems: A Risk-Based AC OPF Approach. *IEEE Trans. Power Syst.* **2015**, *30*, 409–418. [CrossRef]
14. Sansawatt, T.; Ochoa, L.; Harrison, G. Smart Decentralized Control of DG for Voltage and Thermal Constraint Management. *IEEE Trans. Power Syst.* **2012**, *27*, 1637–1645. [CrossRef]
15. Olivier, F.; Aristidou, P.; Ernst, D.; Van Cutsem, T. Active management of low-voltage networks for mitigating over voltages due to photovoltaic units. *IEEE Trans. Smart Grid* **2016**, *7*, 926–936. [CrossRef]
16. Donkers, M.C.F.; Heemels, W.P.M.H.; Bernardini, D.; Bemporad, A.; Shneer, V. Stability analysis of stochastic networked control systems. *Automatica* **2012**, *48*, 917–925. [CrossRef]
17. Yue, D.; Han, Q.; Lam, J. Network-based robust HŒ control of systems with uncertainty. *Automatica* **2005**, *41*, 999–1007. [CrossRef]

18. Shafiq, S.; Al-Awami, A.T. A Novel Communication-Free Charge Controller for Electric Vehicles. In Proceedings of the 2018 IEEE Industry Applications Society Annual Meeting (IAS), Portland, OR, USA, 23–27 September 2018; pp. 1–6.
19. Vovos, P.N.; Kiprakis, A.; Wallace, A.R.; Harrison, G. Centralized and Distributed Voltage Control: Impact on Distributed Generation Penetration. *IEEE Trans. Power Syst.* **2007**, *22*, 476–483. [CrossRef]
20. Shafiq, S.; Al-Awami, A.T. An Autonomous Charge Controller for Electric Vehicles Using Online Sensitivity Estimation. *IEEE Trans. Ind. Appl.* **2020**, *56*, 22–33. [CrossRef]
21. Zhang, Z.; Ochoa, L.F.; Valverde, G. A Novel Voltage Sensitivity Approach for the Decentralized Control of DG Plants. *IEEE Trans. Power Syst.* **2018**, *33*, 1566–1576. [CrossRef]
22. Seal, B. *Common Functions for Smart Inverters*, 4th ed.; Tech. Rep. 3002008217; EPRI: Palo Alto, CA, USA, December 2016.
23. O'Connell, A.; Keane, A. Volt–var curves for photovoltaic inverters in distribution systems. *IET Gener. Transm. Distrib.* **2017**, *11*, 730–739. [CrossRef]
24. Antoniadou-Plytaria, K.E.; Kouveliotis-Lysikatos, I.N.; Georgilakis, P.S.; Hatziargyriou, N.D. Distributed and Decentralized Voltage Control of Smart Distribution Networks: Models, Methods, and Future Research. *IEEE Trans. Smart Grid* **2017**, *8*, 2999–3008. [CrossRef]
25. *IEEE Standard for Interconnecting Distributed Resources with Electric Power Systems—Amendment 1*; IEEE Standard 1547a; IEEE: Piscataway, NJ, USA, May 2014.
26. Shafiq, S.; Khan, B.; Al-Awami, A.T. An Autonomous DG Controller Using Artificial Intelligence Approach for Voltage Control. In Proceedings of the 2020 2nd International Conference on Smart Power & Internet Energy Systems (SPIES), Bangkok, Thailand, 15–18 September 2020; pp. 446–451.
27. *American National Standard for Electrical Power Systems and Equipment—Voltage Ratings (60 Hertz)*; ANSI Std. C84.1-2020; National Electrical Manufacturers Association: Arlington, VA, USA, 2020.
28. Jin, L.; Kuang, X.; Huang, H.; Qin, Z.; Wang, Y. Study on the Overfitting of the Artificial Neural Network Forecasting Model. *J. Meteorol. Res.* **2005**, *19*, 216–225.
29. Al-Awami, A.T.; Sortomme, E.; Akhtar, G.M.A.; Faddel, S. A Voltage-Based Controller for an Electric-Vehicle Charger. *IEEE Trans. Veh. Technol.* **2015**, *65*, 4185–4196. [CrossRef]
30. Wiczynski, G. Analysis of Voltage Fluctuations in Power Networks. *IEEE Trans. Instrum. Meas.* **2008**, *57*, 2655–2664. [CrossRef]
31. Tonkoski, R.; Lopes, L.A.C.; El-Fouly, T. Coordinated Active Power Curtailment of Grid Connected PV Inverters for Overvoltage Prevention. *IEEE Trans. Sustain. Energy* **2010**, *2*, 139–147. [CrossRef]
32. Nissan. Leaf Specs. Available online: https://www.nissan.co.uk/vehicles/new-vehicles/leaf/range-charging.html (accessed on 15 January 2021).

 *sustainability*

Article

Adaptive Nonsingular Fast Terminal Sliding Mode Control for Maximum Power Point Tracking of a WECS-PMSG

Muhammad Maaruf [1], Md Shafiullah [2,*], Ali T. Al-Awami [3,4] and Fahad S. Al-Ismail [2,4,5,6]

1   Systems Engineering Department, King Fahd University of Petroleum & Minerals, P.O. Box 5067, Dhahran 31261, Saudi Arabia; g201705070@kfupm.edu.sa
2   Interdisciplinary Research Center for Renewable Energy and Power Systems, Research Institute, King Fahd University of Petroleum & Minerals, P.O. Box 8088, Dhahran 31261, Saudi Arabia; fsalismail@kfupm.edu.sa
3   Interdisciplinary Research Center for Smart Mobility and Logistics, Research Institute, King Fahd University of Petroleum & Minerals, P.O. Box 5038, Dhahran 31261, Saudi Arabia; aliawami@kfupm.edu.sa
4   Electrical Engineering Department, King Fahd University of Petroleum & Minerals, P.O. Box 5038, Dhahran 31261, Saudi Arabia
5   Centre for Environment and Marine Studies, King Fahd University of Petroleum & Minerals, P.O. Box 5026, Dhahran 31261, Saudi Arabia
6   K.A. CARE Energy Research and Innovation Center, King Fahd University of Petroleum & Minerals, P.O. Box 94293, Dhahran 31261, Saudi Arabia
*   Correspondence: shafiullah@kfupm.edu.sa

Abstract: This paper investigates maximum power extraction from a wind-energy-conversion system (WECS) with a permanent magnet synchronous generator (PMSG) operating in standalone mode. This was achieved by designing a robust adaptive nonsingular fast terminal sliding mode control (ANFTSMC) for the WECS-PMSG. The proposed scheme guaranteed optimal power generation and suppressed the system uncertainties with a rapid convergence rate. Moreover, it is independent of the upper bounds of the system uncertainties as an online adjustment algorithm was utilized to estimate and compensate them. Finally, four case studies were carried out, which manifested the remarkable performance of ANFTSMC in comparison to previous methods reported in the literature.

Keywords: adaptive control; maximum power point tracking; nonsingular fast terminal sliding mode control; permanent magnet synchronous generator; wind-energy-conversion system

Citation: Maaruf, M.; Shafullah, M.; Al-Awami, A.T.; Al-Ismail, F.S. Adaptive Nonsingular Fast Terminal Sliding Mode Control for Maximum Power Point Tracking of a WECS-PMSG. *Sustainability* **2021**, *13*, 13427. https://doi.org/10.3390/su132313427

Academic Editor: Domenico Mazzeo

Received: 7 August 2021
Accepted: 17 November 2021
Published: 3 December 2021

Publisher's Note: MDPI stays neutral with regard to jurisdictional claims in published maps and institutional affiliations.

1. Introduction

Renewable energy resources (RERs) have certainly been viewed as a potential alternative energy source, as traditional fossil fuels are limited and the main contributors to greenhouse gas (GHG) emissions. They not only provide cleaner energy, but have also become cost-competitive in recent years. Amongst various sources of RE, wind energy is one of the most desirable sources, which offers plenty of advantages including abundance and broad distribution [1–4]. The capacity of global wind power installed exceeded 651 GW in 2019, with a 10% increase compared to 2018 [5]. Generally, the variable-speed operation of wind turbine systems is based largely on double-fed induction generators (DFIGs) [6], squirrel cage induction generators (SCIGs) [7], and permanent magnet synchronous generators (PMSGs) [8]. During the past few years, the application of PMSGs has significantly expanded due to their high-performance efficiency, low noise, high reliability, and gear-less design. Besides, the efficiency of the PMSG has been increased by around 10% due to its wide operating speed range and the absence of a direct-current (DC)-excitation system [9–11].

An efficient optimal power extraction with a low cost of implementation, also known as the maximum power point tracking (MPPT) control technique, is needed for operating performance improvement of the WECS [12,13]. Vector control incorporated with proportional-integral (PI) loops has been the most commonly used control method due to its simplicity and ease of implementation [14]. Its control architecture is primarily based on a linearized model at a particular operating point; therefore, the controllability of such a method may drastically degrade or even contribute to system instability as the system operating conditions can change frequently due to weather conditions and wind speeds. To tackle this problem, a self-tuning PI controller was suggested in [15]. Metaheuristic algorithms and machine-learning tools are very popular in power systems' application for optimizing the controller parameters [16–24]. For instance, metaheuristic algorithms such as the bacterial foraging algorithm [17] and grey wolf optimization [19] were used to tune the gains of the PI controller for PMSG applications. However, they are based on either generations or iterations that delay the optimization process; therefore, they cannot be used for online tuning controller parameters. In response, real-time tuning of the PI controller parameters was proposed in [23] where a wavelet neural network was employed for gain adjustment. The machine-learning-based approaches require adequate data, training, and testing to achieve satisfactory performance, and the lack of sufficient data may sometimes hinder their application.

To deal with the challenges mentioned above, nonlinear control strategies have been widely explored and investigated [25]. For instance, feedback linearization controllers capable of globally linearizing the system nonlinearities were reported in [26,27] to attain MPPT for PMSGs. Besides, the backstepping controls are also popular nonlinear control methods that are based on step-by-step approaches. A backstepping control was presented in [28] for maximum power extraction from the wind. However, both backstepping and feedback linearization approaches required exact system parameters, and their performance deteriorates in the presence of dynamic uncertainties [29]. In response, the sliding mode control (SMC) methods offer promising solutions to handle the uncertainties [25]. The SMC methods have gained much attention in the control of WECSs because of their robustness, low sensitivity to parameter changes, simplicity, and fast response [29]. A wide range of SMC methods including the passivity-based SMC [30], fractional-order SMC (FOSMC) [31], fuzzy-logic-based SMC (FOSMC) [32], second-order SMC [33], super-twisting SMC (ST-SMC) [34], terminal SMC (TSMC) [35], second-order TSMC [36], and super-twisting fractional-order terminal SMC (ST-FOTSMC) [37] have been used for maximum power extraction from the WECS-PMSG. The mentioned control schemes are mainly based on the fact that the upper bounds of the uncertainties and the disturbances are known. However, in practical applications, it may be difficult to determine the upper bounds because of the complexity of the PMSG system. Therefore, several controllers combined with adaptation schemes were proposed to solve the unknown upper bounds of the disturbances. In [38], an adaptive SMC was utilized to capture the maximum wind energy from the PMSG with perturbation. In [39], an adaptive STSMC was designed for ocean current turbine-driven PMSG. In [40], a robust adaptive TSMC was developed to deal with uncertainties in the PMSG system while capturing the maximum power. The adaptive backstepping control scheme was proposed for a PMSG with unknown perturbation to achieve MPPT in [41]. It is worth noting that most of the mentioned SMC strategies were formulated with known upper bounds of the disturbances. In addition, the TSMC methods in [34–36] cannot guarantee the avoidance of singularities. In [42], a piecewise function was used to avoid singularities while extracting maximum energy from the WECS. However, the piecewise function introduces other challenges, e.g., a sharp jump while controlling the inputs beyond a certain boundary.

Considering the above-mentioned aspects, adaptive nonsingular fast terminal sliding mode control (ANFTSMC) has gained popularity in recent years and has been used to control quadrotors [43] and robotic manipulators [44]. The fundamental advantage of deploying ANFTSMC is avoiding singularities, strong robustness against the system disturbances and uncertainties, and fast convergence when the states of the system are far from the equilibrium point. Therefore, the authors propose ANFTSMC for maximum power extraction from the WECS-PMSG. To the best of our knowledge, this is the first time

ANFTSMC has been proposed for WECS-PMSGs. The main contributions of this article are as follows:

1. Utilization of a Lyapunov-based adaptation approach for the estimation of the unknown upper bounds of the system uncertainties;
2. Elimination of any unwanted singularities in WECS-PMSGs while extracting the maximum energy;
3. Accomplishment of faster convergence using the proposed strategy over other strategies when the system states are far away from the origin;
4. Validation of the efficacy of ANFTSMC based on the obtained comparative results.

This article is structured as follows: Section 2 provides the mathematical modeling of the WECS-PMSG. The proposed control scheme is presented in Section 3. Section 4 presents the simulation results and discussions. Section 5 provides the concluding remarks.

2. Modeling of the WECS-PMSG

The structure of the wind energy conversion system PMSG is illustrated in Figure 1. It consists of three subsystems, namely the aerodynamic, PMSG, and shaft subsystems. The wind energy harnessed by the turbine blades is converted into mechanical energy used for the generation of electrical energy by the PMSG. The generator-side converter controls the generated power, while the grid-side converter transmits the active power to the grid at the constant DC-link voltage. This work aims to control the generator-side converter.

2.1. Aerodynamic Model

The aerodynamic equations for rotor power and torque are given by [30]:

$$P_A = \frac{1}{2}\rho\pi R^2 C_p(\lambda,\beta)V_{wind}^3 \tag{1}$$

$$T_A = \frac{P_A}{\omega_r} = \frac{1}{2\omega_r}\rho\pi R^3 C_T(\lambda,\beta)V_{wind}^2 \tag{2}$$

where ρ is the air density, R is the radius of the wind turbine, V_{wind} is the wind speed, $C_p(\lambda,\beta)$ and $C_T(\lambda,\beta)$ represent the power and torque coefficients, respectively, β is the pitch angle, and λ is the tip-speed ratio. The tip-speed ratio is a function of the rotor speed, which can be represented as:

$$\lambda = \frac{R\omega_r}{V_{wind}} \tag{3}$$

The power coefficient is a function of both the pitch angle (β) and tip-speed ratio (λ), as defined by the following expression:

$$C_p(\lambda,\beta) = 0.5176(\frac{116}{\lambda_j} - 0.4\beta - 5)e^{\frac{-21}{\lambda_j}} + 0.0068\lambda \tag{4}$$

$$\frac{1}{\lambda_j} = \frac{1}{\lambda + 0.08\beta} - \frac{0.035}{\beta^3 + 1} \tag{5}$$

$$C_{pmax} = C_p(\lambda_{opt},\beta) \tag{6}$$

The wind turbine can generate the maximum power provided that the power coefficient $C_p(\lambda,\beta)$ is maximum for any wind speed within the wide operation region of the turbine. The power coefficient can be maximized by maintaining the optimal value of the tip-speed ratio and fixed pitch angle. The relationship between $C_p(\lambda,\beta)$ and λ at different fixed values of β is illustrated in Figure 2. Therefore, the optimal reference speed applied to the WECS-PMSG is given by:

$$\omega_{r\_opt} = \frac{\lambda_{opt}}{R} V_{wind} \tag{7}$$

The maximum power extracted by the WECS-PMSG under the optimal rotor speed thus can be represented as:

$$P_{A_{opt}} = \frac{1}{2}\rho \pi R^2 C_{pmax}(\lambda_{opt}, \beta)\left(\frac{R\,\omega_{r\_opt}}{\lambda_{opt}}\right)^3 \tag{8}$$

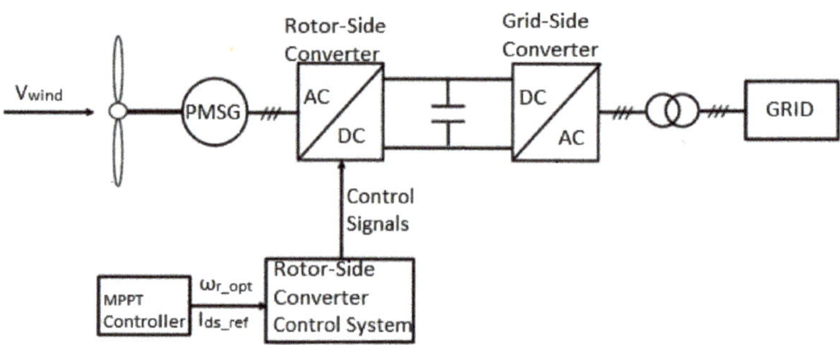

Figure 1. The structure of the PMSG wind turbine system.

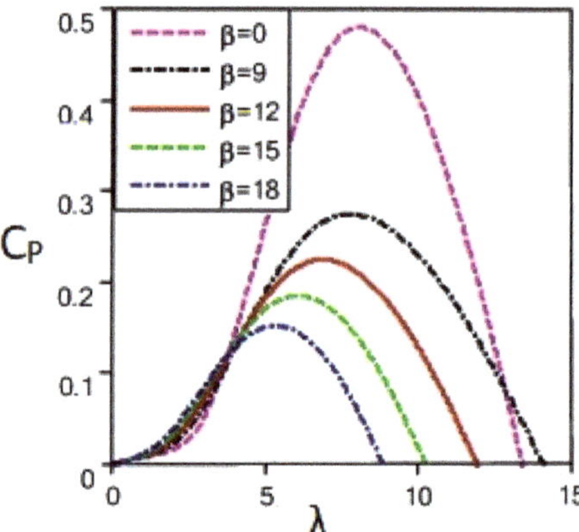

Figure 2. Power coefficient and tip-speed ratio ($C_p - \lambda$) relationship at different pitch angles.

2.2. PMSG Model

The dynamic model of the PMSG and the torque in the d-q coordinate system are formulated as [41]:

$$U_{ds} = R_s I_{ds} + L_d + \frac{dI_{ds}}{dt} - \omega_e L_q I_{qs} \tag{9}$$

$$U_{qs} = R_s I_{qs} + L_q + \frac{dI_{qs}}{dt} + \omega_e L_d I_{ds} + \omega_e \Lambda_f \tag{10}$$

$$T_e = \frac{3}{2}p[(L_d - L_q)I_{ds}I_{qs} + \Lambda_f I_{qs}] \tag{11}$$

where I_{ds} and I_{qs} are the d and q axes' stator currents, U_{ds} and U_{qs} are the stator voltages, L_d and L_q stand for inductance, R_s denotes the stator resistance, Λ_f represents the rotor flux,

$\omega_e = p\omega_r$ is the electrical speed, and T_e indicates the electromagnetic torque. If $L_d = L_q$, Equation (11) will evolve as:

$$T_e = \frac{3}{2}p\Lambda_f I_{qs} \tag{12}$$

2.3. Shaft System Model

The dynamic model of the wind turbine shaft system is expressed as [30]:

$$\frac{d\omega_r}{dt} = J^{-1}T_a - J^{-1}T_e - J^{-1}b\omega_r \tag{13}$$

where b and J indicate the friction coefficient and the total mechanical inertia, respectively.

2.4. Overall Model

The overall model of the WECS-PMSG can be written as [30]:

$$\dot{x}_1 = a_1 T_A + a_2 x_1 + a_3 x_2 + \Delta_1 \tag{14}$$
$$\dot{x}_2 = a_4 x_2 + a_5 x_3 x_1 + a_6 x_1 + g_1 U_{qs} + \Delta_2 \tag{15}$$
$$\dot{x}_3 = a_7 x_3 + a_8 x_1 x_2 + g_2 U_{ds} + \Delta_3 \tag{16}$$

where $x_1 = \omega_r$, $x_2 = I_{qs}$, $x_3 = I_{ds}$, $a_1 = J^{-1}$, $a_2 = -\frac{3}{2}p\Lambda_f$, $a_3 = -J^{-1}b$, $a_4 = -\frac{R_s}{L_q}$, $a_5 = -p\frac{L_d}{L_q}$, $a_6 = -\Lambda_f p$, $a_7 = -\frac{R_s}{L_d}$, and $a_8 = \frac{L_q}{L_d}p$.

3. Control of WECS-PMSG

The article aims to design a robust control algorithm that keeps operating the WECS-PMSG within the point of maximum power extraction. The control variables are $y_1 = x_1$ and $y_2 = x_2$. By differentiating y_1 twice and y_2 once, the following equations are obtained:

$$\ddot{y}_1 = f_1 + g_1 u_1 + \Delta_{y_1} \tag{17}$$
$$\dot{y}_2 = f_2 + g_2 u_2 + \Delta_{y_2} \tag{18}$$

where $f_1 = a_1 \dot{T}_A + a_2 \dot{x}_1 + a_3[a_4 x_2 + a_5 x_3 x_1 + a_6 x_1]$, $g_1 = a_3 b_1$, $f_2 = a_7 x_3 + a_8 x_1 x_2$, $g_2 = b_2$, $\Delta_{y_1} = \dot{\Delta}_1 + a_3\Delta_2$, and $\Delta_{y_2} = \Delta_3$

Assumption 1. *The lumped disturbances are bounded, e.g.,*

$$\Delta_{y_1} \leq M_{11}$$
$$\Delta_{y_2} \leq M_{21}$$

where M_{11} and M_{21} are the upper bounds of the disturbances and Δ_{y_1} and Δ_{y_2} are the net disturbances in the input–output dynamics of Equations (17) *and* (18).

3.1. Design of NFTSMC

In this section, NFTSMC is designed for y_1 and y_2 by assuming that the upper bounds of the lumped disturbances are known exactly.

3.1.1. Design of the NFTSMC for the Rotor Speed

If the tracking error between y_1 and ω_{r_opt} is defined as:

$$\begin{cases} e_1 &= y_1 - \omega_{r\_opt} \\ \dot{e}_1 &= \dot{y}_1 - \dot{\omega}_{r\_opt} \\ \ddot{e}_1 &= \ddot{y}_1 - \ddot{\omega}_{r\_opt} = f_1 + g_1 u_1 + \Delta_{y_1} - \ddot{\omega}_{r\_opt} \end{cases} \tag{19}$$

The NFTSMC surface is defined as [43]:

$$S_1 = e_1 + C_{11}|e_1|^{\mu} sign(e_1) + C_{12}|\dot{e}_1|^{\alpha} sign(\dot{e}_1) \tag{20}$$

where C_{11} and C_{12} are positive constants, $1 < \alpha < 2$ and $\mu > \beta$. The following equation is derived from the time derivative of S_1:

$$\begin{aligned}
\dot{S}_1 &= \dot{e}_1 + C_{11}\mu|e_1|^{\mu-1}\dot{e}_1 + C_{12}\alpha|\dot{e}_1|^{\alpha-1}\ddot{e}_1 \\
&= \dot{e}_1 + C_{11}\mu|e_1|^{\mu-1}\dot{e}_1 + C_{12}\alpha|\dot{e}_1|^{\alpha-1}\left[f_1 + g_1 u_1 + \Delta_{y_1} - \ddot{\omega}_{r_opt}\right]
\end{aligned} \tag{21}$$

By recognizing that $\dot{S}_1 = S_1 = 0$, the equivalent control input u_{1eq} is derived as:

$$u_{1eq} = g_1^{-1}\left[-f_1 - \frac{1}{C_{12}\alpha}|\dot{e}_1|^{2-\alpha}(1 + \mu C_{11}|e_1|^{\mu-1})sign(\dot{e}_1)\right] \tag{22}$$

If the system dynamics are known precisely, the equivalent control law u_1 can make the states remain on (20). In order to meet the sliding condition in the presence of the lumped disturbance, the reaching law is given by the following equation:

$$u_{1r} = g_1^{-1}[-M_{11}sign(S_1) - M_{12}S_1] \tag{23}$$

where M_{11} and M_{12} are constants. Thus, the overall control law is established by the following equation as:

$$\begin{aligned}
u_1 &= u_{1eq} + u_{1r} \\
&= g_1^{-1}\left[-f_1 - \frac{1}{C_{12}\alpha}|\dot{e}_1|^{2-\alpha}(1 + \mu C_{11}|e_1|^{\mu-1})sign(\dot{e}_1) - M_{11}sign(S_1) - M_{12}S_1\right]
\end{aligned} \tag{24}$$

Theorem 1. *Considering the output dynamics of Equation* (17)*, if it is controlled with Equation* (24)*, the state variables will converge to the surface of Equation* (20)*.*

Proof of Theorem 1. Consider the following Lyapunov function candidate:

$$V_1 = \frac{1}{2}S_1^2 \tag{25}$$

Taking the time derivative of V_1 and using Equations (21) and (24), the following equation is evolved:

$$\begin{aligned}
\dot{V}_1 = S_1\dot{S}_1 &= C_{12}\alpha|\dot{e}_1|^{\alpha-1}\left[S_1\Delta_{y_1} - M_{11}|S_1| - M_{12}S_1^2\right] \\
&\leq C_{12}\alpha|\dot{e}_1|^{\alpha-1}\left[(\Delta_{y_1} - M_{11})|S_1| - M_{12}S_1^2\right]
\end{aligned} \tag{26}$$

The following equation is obtained by considering Assumption 1.

$$\dot{V}_1 \leq -C_{12}\alpha|\dot{e}_1|^{\alpha-1}M_{12}S_1^2 \leq 0 \tag{27}$$

From the definition of Lyapunov stability theory, the output y_1 asymptotically converges to the surface $S_1 = 0$. $\square$

3.1.2. Design of the NFTSMC for the D-Component of the Stator Current

Consider the following tracking error between y_2 and I_{ds_ref}:

$$\begin{cases} e_2 = y_2 - I_{ds\_ref} \\ \dot{e}_2 = \dot{y}_2 - \dot{I}_{ds\_ref} = f_2 + g_2 u_2 + \Delta_{y_2} - \dot{I}_{ds\_ref} \end{cases} \tag{28}$$

Since the relative degree of y_2 is one, the following NFTSMC surface is introduced.

$$S_2 = e_2 + C_2 |\dot{e}_2|^\alpha sign(\dot{e}_2) \tag{29}$$

where C_2 is a positive constant. The time derivative of S_2 yields:

$$\begin{aligned} \dot{S}_2 &= \dot{e}_2 + C_2 \alpha |\dot{e}_2|^{\alpha-1} \ddot{e}_2 \\ &= \dot{e}_2 + C_2 \alpha |\dot{e}_2|^{\alpha-1} [\dot{f}_2 + g_2 \dot{u}_2 + \dot{\Delta}_{y_2} - \ddot{I}_{ds\_ref}] \end{aligned} \tag{30}$$

where $\dot{f}_2 = \frac{\partial f_2}{\partial x_1}\dot{x}_1 + \frac{\partial f_2}{\partial x_2}\dot{x}_2 + \frac{\partial f_2}{\partial x_3}\dot{x}_3$. The equivalent control law is derived as:

$$\dot{u}_{2eq} = g_2^{-1}\left[-\dot{f}_2 - \frac{1}{C_2\alpha}|\dot{e}_2|^{2-\alpha} + \ddot{I}_{ds\_ref} \right] \tag{31}$$

The reaching law is designed as $\dot{u}_{2r} = -g_2^{-1}[M_{21}sign(S_2) + M_{22}S_2]$. The control law for Equation (18) is given by:

$$\begin{aligned} u_2 = u_{2eq} + u_{2r} &= \int_0^t (\dot{u}_{2eq}(\tau) + \dot{u}_{2r}(\tau))d\tau \\ &= g_2^{-1} \int_0^t \left[-\dot{f}_2 - \frac{|\dot{e}_2|^{2-\alpha}}{C_2\alpha} + \ddot{I}_{ds\_ref} - M_{21}sign(S_2) - M_{22}S_2 \right] d\tau \end{aligned} \tag{32}$$

Theorem 2. *Considering the output dynamics of Equation (18), if it is controlled with Equation (32), the state variables will converge to the surface as shown in Equation (29).*

Proof of Theorem 2. Consider the Lyapunov candidate as:

$$V_2 = \frac{1}{2}S_2^2 \tag{33}$$

After differentiating V_2 with respect to time and using Equations (30) and (32), the following equation can be obtained:

$$\begin{aligned} \dot{V}_2 = S_2 \dot{S}_2 &= C_2 |\dot{e}_2|^{\alpha-1}\left[S_2 \Delta_{y_2} - M_{21}|S_2| - M_{22}S_2^2 \right] \\ &\leq C_2 \alpha |\dot{e}_2|^{\alpha-1}\left[(\Delta_{y_2} - M_{21})|S_2| - M_{22}S_2^2 \right] \end{aligned} \tag{34}$$

Based on Assumption 1, Equation (34) becomes:

$$\dot{V}_2 \leq -C_2\alpha|\dot{e}_2|^{\alpha-1}M_{22}S_2^2 \leq 0 \tag{35}$$

$\square$

3.2. Design of ANFTSMC

In practical applications, it is difficult to precisely obtain the upper bounds of the system lumped disturbances. As such, we developed an adaptation scheme to estimate the upper bounds and suppress the lumped disturbances, which can improve the robustness of the control system.

3.2.1. Design of the ANFTSMC for the Rotor Speed

The control law of Equation (24) is modified as:

$$u_1 = g_1^{-1}\left[-f_1 - \frac{1}{C_{12}\alpha}|\dot{e}_1|^{2-\alpha}(1 + \mu C_{11}|e_1|^{\mu-1})sign(\dot{e}_1) - \hat{M}_{11}sign(S_1) - \hat{M}_{12}S_1\right] \quad (36)$$

where $\hat{M}_{11}$ and $\hat{M}_{12}$ are the estimates of M_{11} and M_{12}, respectively. The following adaptive rules update the gains:

$$\begin{cases} \dot{\hat{M}}_{11} &= \gamma_{11}|\dot{e}_1|^{\alpha-1}|S_1| \\ \dot{\hat{M}}_{12} &= \gamma_{12}|\dot{e}_1|^{\alpha-1}S_1^2 \end{cases} \quad (37)$$

where γ_{11} and γ_{12} are positive constants.

Remark 1. *The adaptation gains γ_{11} and γ_{12} are adjusted by trial and error and then kept constant when the desired responses are achieved.*

The main results of the adaptive scheme can be expressed in the following theorem:

Theorem 3. *Considering that the upper bounds of the lumped disturbances of Equation (17) are unknown, if the NFTSMC surface is chosen as Equation (20), the adaptive controller is designed as Equation (36); then, the trajectory tracking error asymptotically converge to zero.*

Proof of Theorem 3. The Lyapunov function of Equation (25) is modified as follows:

$$V_1 = \frac{1}{2}S_1^2 + \frac{1}{2\gamma_{11}}\tilde{M}_{11}^2 + \frac{1}{2\gamma_{12}}\tilde{M}_{12}^2 \quad (38)$$

where $\tilde{M}_{11} = M_{11} - \hat{M}_{11}$, $\tilde{M}_{12} = M_{12} - \hat{M}_{12}$. Calculating the time derivative of Equation (38) yields:

$$\dot{V}_1 = S_1\dot{S}_1 - \tilde{M}_{11}\dot{\hat{M}}_{11} - \tilde{M}_{12}\dot{\hat{M}}_{12} \quad (39)$$

Based on Equations (21) and (36), the following equation can be obtained:

$$\begin{aligned} \dot{V}_1 &= C_{12}\alpha|\dot{e}_1|^{\alpha-1}\left[S_1\Delta_{y_1} - \hat{M}_{11}|S_1| - \hat{M}_{12}S_1^2\right] - \tilde{M}_{11}\dot{\hat{M}}_{11} - \tilde{M}_{12}\dot{\hat{M}}_{12} \\ &\leq C_{12}\alpha|\dot{e}_1|^{\alpha-1}\left[(\Delta_{y_1} - M_{11})|S_1| - M_{12}S_1^2\right] + \tilde{M}_{11}\left[|\dot{e}_1|^{\alpha-1}|S_1| - \dot{\hat{M}}_{11}\right] \\ &\quad + \tilde{M}_{12}\left[|\dot{e}_1|^{\alpha-1}S_1^2 - \dot{\hat{M}}_{12}\right] \end{aligned} \quad (40)$$

Using Equation (38) and Assumption (1), Equation (40) can be represented as:

$$\dot{V}_1 \leq -C_{12}\alpha|\dot{e}_1|^{\alpha-1}M_{12}S_1^2 \leq 0 \quad (41)$$

□

3.2.2. Design of the ANFTSMC for the D-Component of the Stator Current

The adaptive control input for the y_2 dynamics is written as:

$$\begin{aligned} u_2 = u_{2eq} + u_{2r} &= \int_0^t (\dot{u}_{2eq}(\tau) + \dot{u}_{2r}(\tau))d\tau \\ &= g_2^{-1}\int_0^t\left[-\dot{f}_2 - \frac{|\dot{e}_2|^{2-\alpha}}{C_2\alpha} + \ddot{I}_{ds\_ref} - \hat{M}_{21}sign(S_2) - \hat{M}_{22}S_2\right]d\tau \end{aligned} \quad (42)$$

where $\hat{M}_{21}$ and $\hat{M}_{22}$ are the estimates of M_{21} and M_{22}, respectively. The following adaptive laws update the gains:

$$\begin{cases} \dot{\hat{M}}_{21} = \gamma_{21}|\dot{e}_2|^{\alpha-1}|S_2| \\ \dot{\hat{M}}_{22} = \gamma_{22}|\dot{e}_2|^{\alpha-1}S_2^2 \end{cases} \tag{43}$$

where γ_{21} and γ_{22} are positive constants.

Remark 2. *The adaptation gains γ_{21} and γ_{22} are adjusted on a systematical trial and error basis; then, they are kept constant when the desired responses are achieved.*

The main results of the ANFTSMC for the y_2 dynamics are summarized in the following theorem:

Theorem 4. *Suppose the information about the upper bounds of the lumped disturbances of Equation (18) is unavailable if the NFTSMC surface is selected as Equation (29); the adaptive controller is developed as (42), and the trajectory tracking error asymptotically converges to zero.*

Proof of Theorem 4. Equation (33) can be modified as follows:

$$V_2 = \frac{1}{2}S_2^2 + \frac{1}{2\gamma_{21}}\tilde{M}_{21}^2 + \frac{1}{2\gamma_{22}}\tilde{M}_{22}^2 \tag{44}$$

where $\tilde{M}_{21} = M_{21} - \hat{M}_{21}, \tilde{M}_{22} = M_{22} - \hat{M}_{22}$. Computing the time derivative of Equation (44) gives:

$$\dot{V}_2 = S_2\dot{S}_2 - \tilde{M}_{21}\dot{\hat{M}}_{21} - \tilde{M}_{22}\dot{\hat{M}}_{22} \tag{45}$$

The following relationship can be obtained after substituting Equations (30) and (42) into Equation (45):

$$\dot{V}_2 = C_2|\dot{e}_2|^{\alpha-1}\left[S_2\Delta_{y_2} - \hat{M}_{21}|S_2| - \hat{M}_{22}S_2^2\right] - \tilde{M}_{21}\dot{\hat{M}}_{21} - \tilde{M}_{22}\dot{\hat{M}}_{22}$$

$$\leq C_2\alpha|\dot{e}_2|^{\alpha-1}\left[(\Delta_{y_2} - M_{21})|S_2| - M_{22}S_2^2\right] + \tilde{M}_{21}\left[|S_1| - \frac{\dot{\hat{M}}_{21}}{\gamma_{21}}\right] + \tilde{M}_{22}\left[S_2^2 - \frac{\dot{\hat{M}}_{22}}{\gamma_{22}}\right] \tag{46}$$

Equation (46) can be modified using Equation (43) and Assumption (1) as:

$$\dot{V}_1 \leq -C_2\alpha|\dot{e}_2|^{\alpha-1}M_{22}S_2^2 \leq 0 \tag{47}$$

$\square$

Remark 3. *The chattering issue due to the discontinuous control component (sign(.) function) is solved by replacing it with the tanh(.) function [45].*

4. Simulation Results and Discussions

The simulation was performed in the MATLAB/SIMULINK 2020 Platform using a PC with an Intel(R) Core(TM) i7-10510U CPU @ 2.3 GHz and 8 GB RAM. The parameters of the WECS-PMSG were taken from [30]. The PMSG parametric variations of 40% were also taken into consideration in the simulation. The parameters of the proposed controller are given in Table 1. The initial conditions of the PMSG states and the adaptive laws were set as 0.01. To highlight the effectiveness of ANFTSMC in achieving the MPPT of the WECS-PMSG, a comparative study was executed with some existing control techniques such as FLC [27], passivity-based SMC (PSMC) [30], and adaptive STSMC (ASTSMC) [39] under four cases, e.g., the step change of the wind speed, the short-term random variation of the wind speed, the long-term random variation of the wind speed, and the real wind speed profile.

4.1. Step Change of the Wind Speed

In this case, it was assumed that the wind speed profile is a sequence of four-step changes, as shown in Figure 3. The performances of different controllers to achieve the MPPT of the WECS-PMSG are presented in Figures 4–7. Figure 4 shows that ANFTSMC was able to track the optimal rotor speed with greater accuracy than ASTSMC, PSMC, and FLC. The evolution of the maximum power coefficient is shown in Figure 5. From this figure, it can be seen that ANFTSMC was able to restore the power coefficient to the required value at a faster rate than ASTSMC, PSMC, and FLC whenever the wind speed in Figure 3 changed. The tracking responses of the optimal power under various control methods are depicted in Figure 6. From this figure, it is clear that ANFTSMC was able to follow the optimal power with greater accuracy than ASTSMC, PSMC, and FLC. The estimated parameters of the rotor speed and the d-component of the stator current controllers are presented in Figures 7 and 8, respectively. From these figures, it can be observed that the FLC controller gave the worst control performances in the presence of parametric uncertainties. Due to the robustness of PSMC, ASTSMC, and ANFTSMC, the uncertainties in the WECS-PMSG were mitigated, and better control performances were obtained. However, the WECS-PMSG under the proposed ANFTSMC attained the MPPT in a shorter time than FLC, ASTSMC, and FLC. Therefore, the effectiveness of the proposed ANFTSMC strategy under the step change of the wind speed was justified.

Table 1. Controller parameters.

Parameters	Values
α, β	3/2, 3
$C_{11}, C_{12}, C_{21}, C_{22}$	1, 1, 1, 1
$\gamma_{11}, \gamma_{12}, \gamma_{21}, \gamma_{22}$	1, 0.04, 10, 15

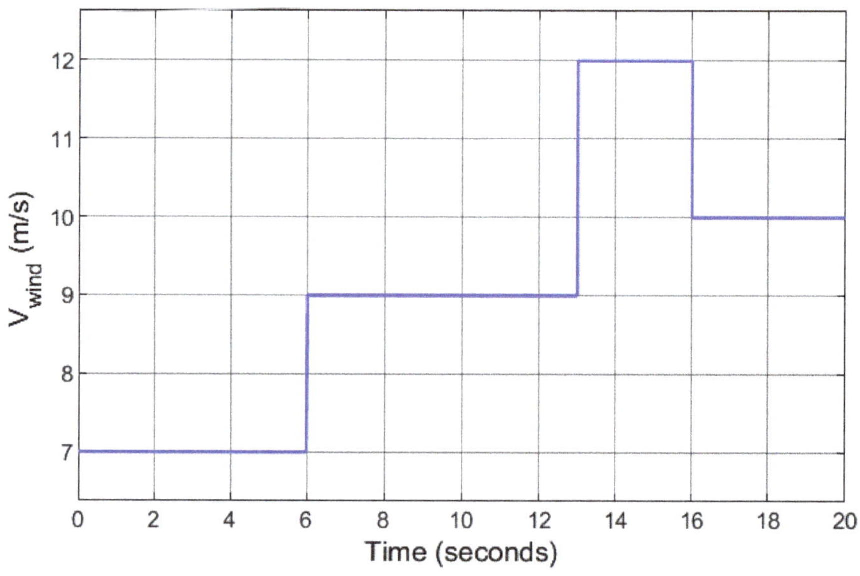

Figure 3. Step change of wind speed profile for the first case study.

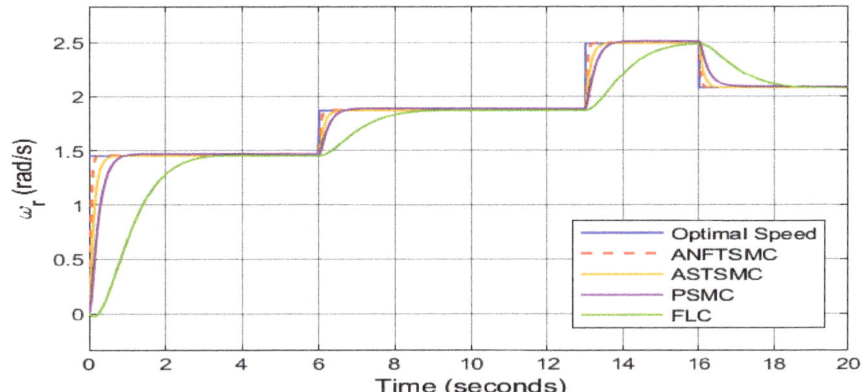

Figure 4. Tracking performance of the rotor speed under different control approaches for the first case study.

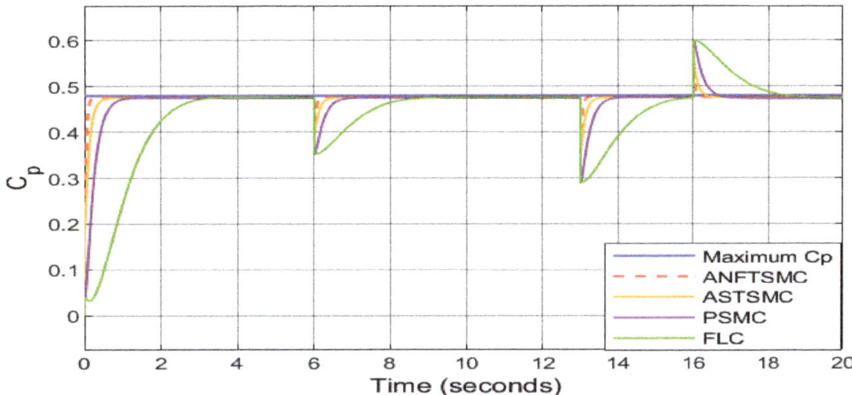

Figure 5. Power coefficient of the extracted power from the wind under different control approaches for the first case study.

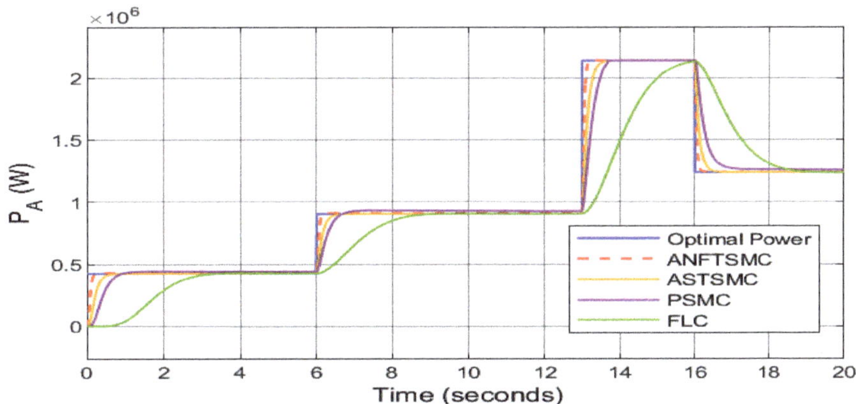

Figure 6. Extracted power from the wind under different control approaches for the first case study.

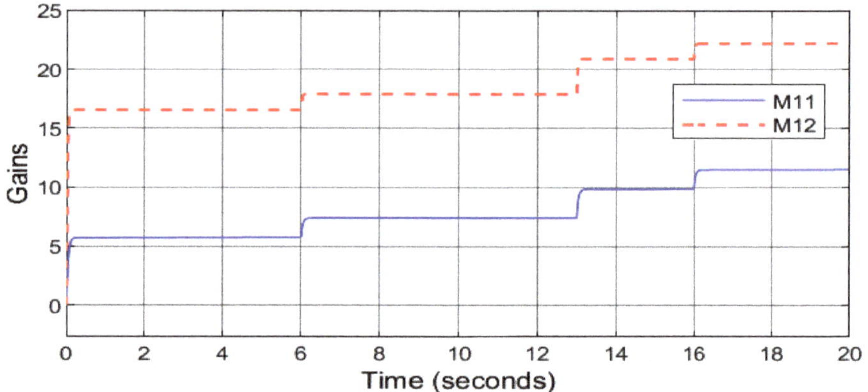

Figure 7. Adaptive parameters of the rotor speed controller for the first case study.

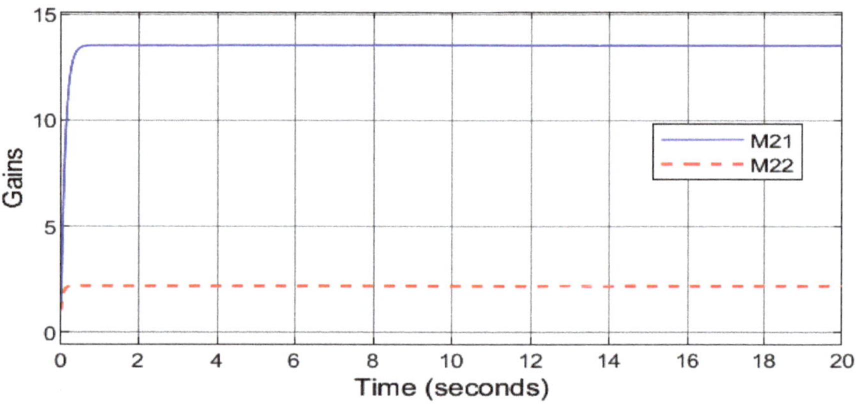

Figure 8. Adaptive parameters of the d-component of the rotor current controller for the first case study.

4.2. Random Variation of the Wind Speed

To further illustrate the effectiveness of the proposed ANFTSMC, a random wind speed profile with a mean value of 11 m/s, as shown in Figure 9, was applied to the WECS-PMSG. The control efforts of the four controllers are depicted in Figures 10–12. The rotor speed and the optimal rotor speed are presented in Figure 10. It can be observed from the figure that ANFTSMC showed the best optimal rotor speed tracking performance. Figure 11 illustrates the maximum power coefficient signals of the four control methods. Due to the random nature of the wind, the power coefficients under the control approaches fluctuated near the required maximum power coefficient. However, under the action of ANFTSMC, the power coefficient was closer to the required value than ASTSMC, PSMC, and FLC. The optimal power harnessed from the random wind is shown in Figure 12. It can be seen that ANFTSMC followed the fluctuating optimal wind power with more accuracy than ASTSMC, PSMC, and FLC. These figures show that FLC did not satisfactorily reach the MPPT of the WECS-PMSG as FLC requires an exact system modeling and is sensitive to the model uncertainties. On the other hand, ANFTSMC, ASTSMC, and PSMC were robust to the WECS-PMSG parametric uncertainties, and as such, they achieved the MPPT. Nevertheless, ANFTSMC exhibited more effectiveness as its responses were much closer to the MPPT under the random wind speed. Therefore, the efficacy of the proposed ANFTSMC strategy under the random variation of the wind speed was also justified.

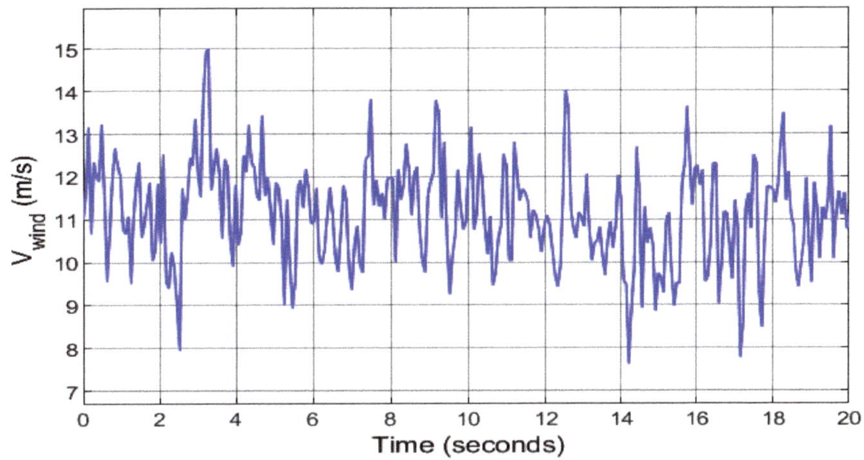

Figure 9. Random wind speed profile for the second case study.

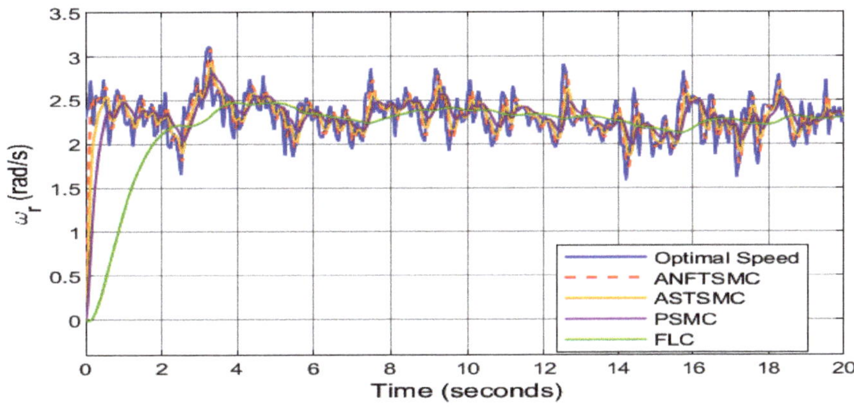

Figure 10. Tracking performance of the rotor speed under different control approaches for the second case study.

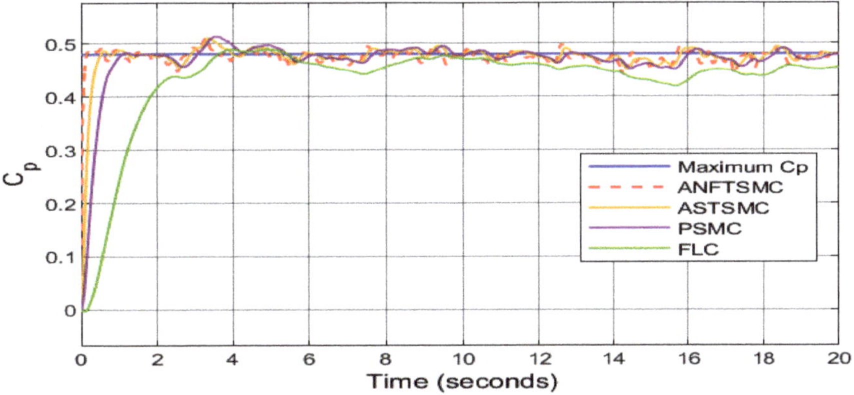

Figure 11. Power coefficient of the extracted power from the wind under different control approaches for the second case study.

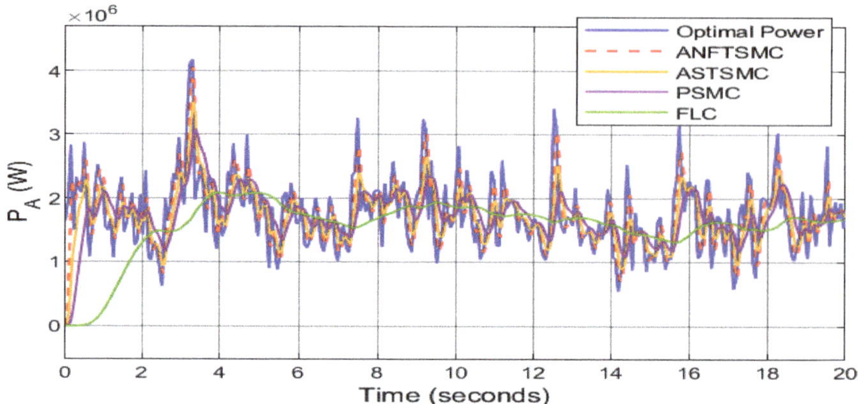

Figure 12. Extracted power from the wind under different control approaches for the second case study.

4.3. Long-Term Random Variation of the Wind Speed

The random wind speed profile in Figure 9 was extended for 1 h (3600 s), as shown in Figure 13, in order to highlight the performance of ANFTSMC over other strategies. From Figure 14, it is clear that the rotor speed was varied between 3.45 rad/s and 1.27 rad/s due to the random variation of the wind speed, and the ANFTSMC strategy tracked the variation more precisely (closer to the peaks and troughs of the rotor speed) than other strategies. The performance of the FLC strategy was the worst amongst the compared strategies as the responses were far away from the peak and trough values of the rotor speed. PMSC performed better than FLC, and ASTSMC performed better than PMSC. The power coefficient was varied rapidly with the rapid variation of the wind speed, as shown in Figure 15. The optimal power harnessed by the WECS-PMSG under the action of four different control approaches over a period of 1 h is depicted in Figure 16. This figure shows that ANFTSMC was able to follow the peaks and troughs of the optimal power with greater accuracy than ASTSMC, PSMC, and FLC. Therefore, the efficacy of the proposed ANFTSMC strategy under the long-term random variation of the wind speed was also justified.

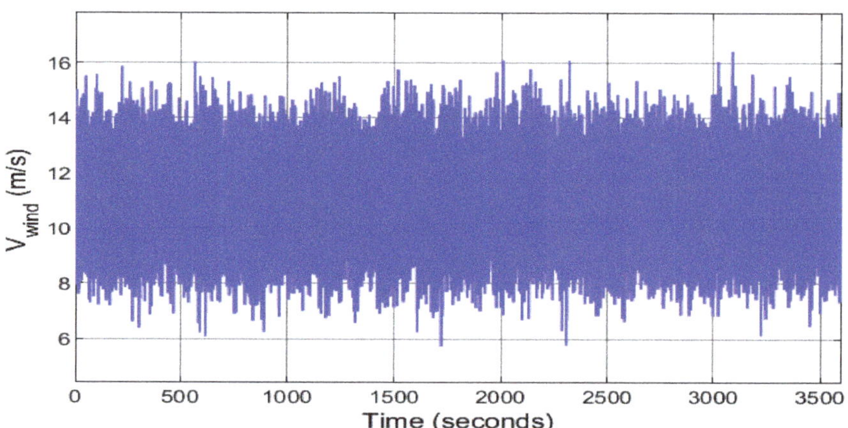

Figure 13. Random wind speed profile for the third case study.

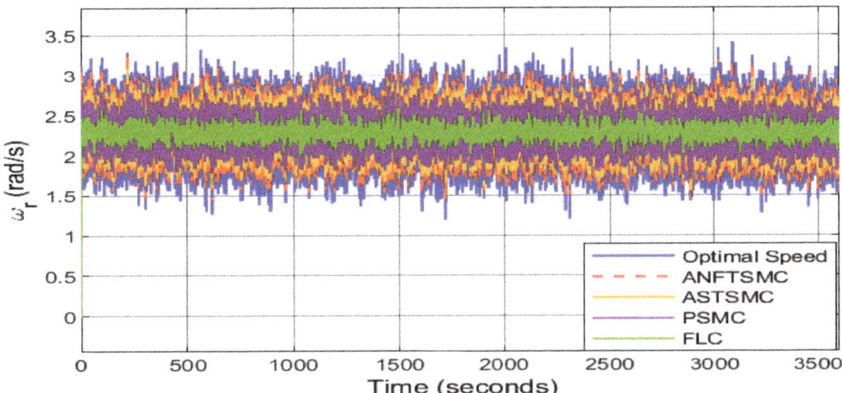

Figure 14. Tracking performance of the rotor speed under different control approaches for the third case study.

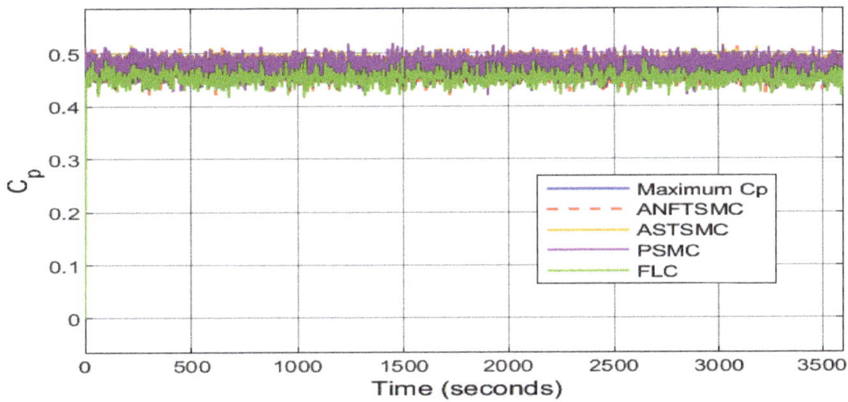

Figure 15. Power coefficient of the extracted power from the wind under different control approaches for the third case study.

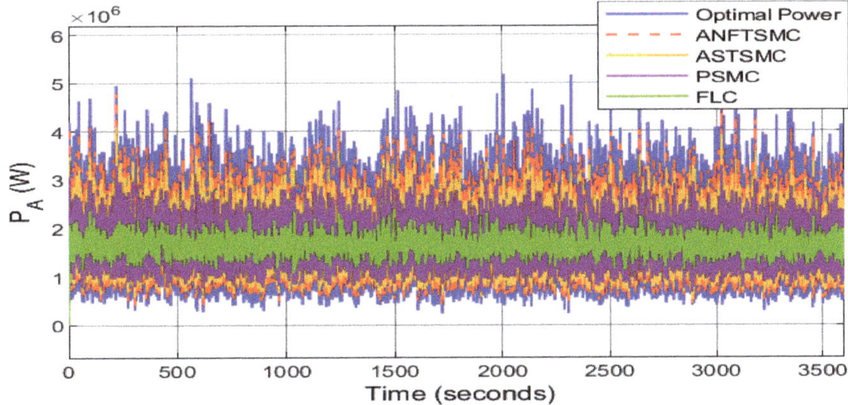

Figure 16. Extracted power from the wind under different control approaches for the third case study.

4.4. Historical Wind Speed Profile

The wind speed profile of Montreal on 31 March 2017 from 0:00 to 23:00 was applied to the WECS-PMSG to examine the performance of the proposed ANFTSMC on a real wind speed profile. Figure 17 depicts the 24 h wind speed profile of Montreal as collected from [46]. It can be seen from Figure 18 that the rotor speed varied between 0.61 rad/s and 2.87 rad/s due to the wind speed variation at the Montreal Weather Station. The ANFTSMC strategy tracked the variation more precisely (closer to the peaks and troughs of the rotor speed) than the other strategies. Similar to the previous cases, the responses of the FLC strategy were far away from the peak and trough values of the rotor speed. On the contrary, PMSC performed better than FLC, and ASTSMC performed better than PMSC. Figure 19 depicts the power coefficient variation of the extracted power. Figure 20 illustrates the optimal power harnessed by the WECS-PMSG for the real wind speed profile over a period of 24 h. It is observed from both Figures 19 and 20 that ANFTSMC maintained its superiority over other strategies. Therefore, the efficacy of the proposed strategy under the historical wind speed profile of Montreal was also justified.

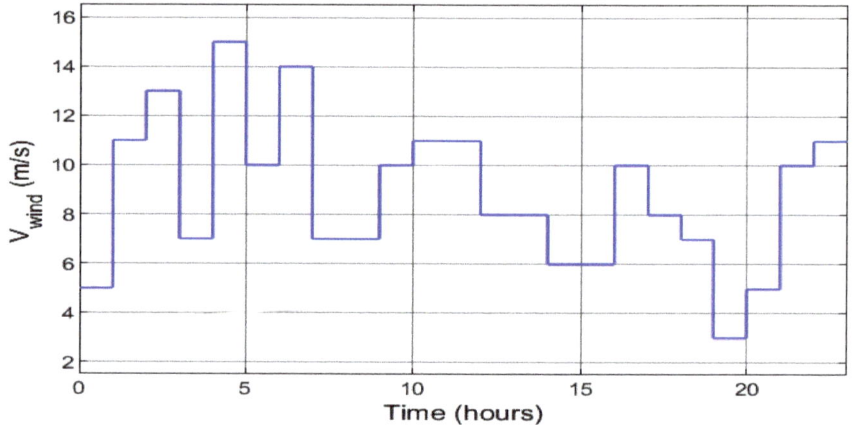

Figure 17. Montreal 24 h wind speed profile on 31 March 2017 for the fourth case study.

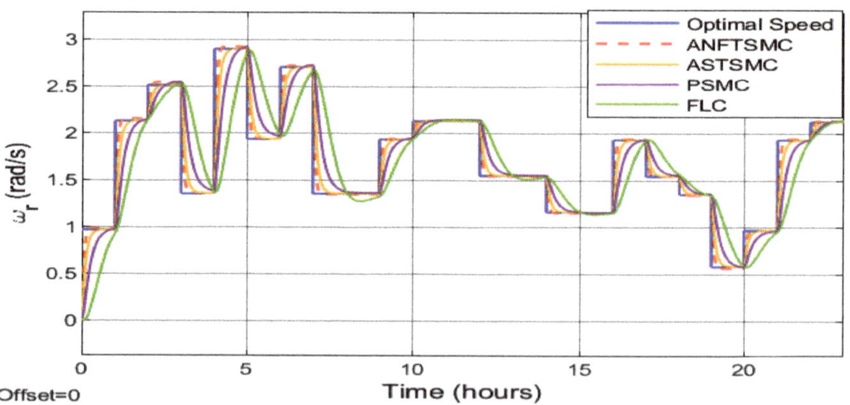

Figure 18. Tracking performance of the rotor speed under different control approaches for the fourth case study.

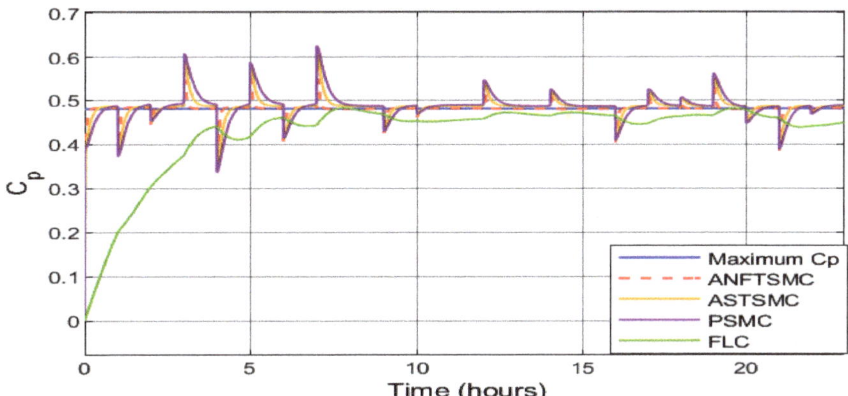

Figure 19. Power coefficient of the extracted power from the wind under different control approaches for the fourth case study.

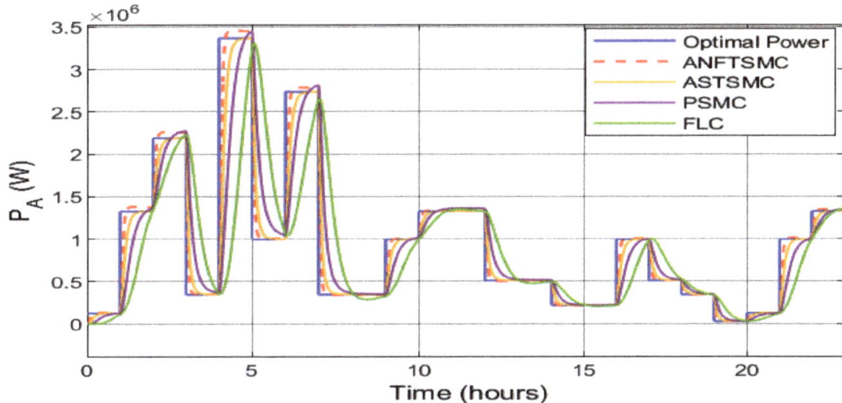

Figure 20. Extracted power from the wind under different control approaches for the fourth case study.

5. Conclusions

This paper presented an ANFTSMC strategy for the WECS-PSMG with model uncertainties to capture the maximum power. The proposed approach ensured singularity avoidance, robustness against unknown WECS-PSMG dynamic uncertainties, and a fast convergence rate to achieve the MPPT. Four case studies (step change of the wind speed, short-time random variation of the wind speed, long-term random variation of the wind speed, and a real wind speed profile) were considered to evaluate the efficient operation of the proposed strategy. In each case, the proposed method outperformed other techniques, including FLC, PSMC, and ASTSMC. As extensions of this work, a laboratory-scale experimental setup, very short-term wind forecasts, measurement uncertainty, and a system with energy storage can be considered.

Author Contributions: Conceptualization, M.M. and M.S.; methodology, M.M.; software, M.M.; validation, M.M., M.S., A.T.A.-A. and F.S.A.-I.; formal analysis, M.M., M.S. and A.T.A.-A.; investigation, M.M., M.S., A.T.A.-A. and F.S.A.-I.; resources, M.M. and M.S.; data curation, M.M., M.S. and F.S.A.-I.; writing—original draft preparation, M.M.; writing—review and editing, M.S., A.T.A.-A. and F.S.A.-I.; visualization, M.M., M.S., A.T.A.-A. and F.S.A.-I.; supervision, M.S., A.T.A.-A. and F.S.A.-I.; project administration, M.S., A.T.A.-A. and F.S.A.-I.; funding acquisition, A.T.A.-A.; All authors have read and agreed to the published version of the manuscript.

Funding: King Fahd University of Petroleum & Minerals (KFUPM): DF201006.

Institutional Review Board Statement: Not applicable.

Informed Consent Statement: Not applicable.

Data Availability Statement: Data will be made available on request.

Acknowledgments: The authors would like to acknowledge the support provided by the Deanship of Scientific Research (DSR) and Interdisciplinary Research Center for Smart Mobility and Logistics, King Fahd University of Petroleum & Minerals (KFUPM), for funding this work through Project No. DF201006.

Conflicts of Interest: The authors declare no conflict of interest.

References

1. Mínaz, M.R.; Akcan, E. An Effective Method for Detection of Demagnetization Fault in Axial Flux Coreless PMSG with Texture-Based Analysis. *IEEE Access* **2021**, *9*, 17438–17449. [CrossRef]
2. Maaruf, M.; Khan, K.A.; Khalid, M. Integrated Power Management and Nonlinear-Control for Hybrid Renewable Microgrid. In Proceedings of the 2021 IEEE Green Technologies Conference (GreenTech), Denver, CO, USA , 7–11 April 2021; pp. 176–180. [CrossRef]
3. Kaldellis, J.; Apostolou, D. Life cycle energy and carbon footprint of offshore wind energy. Comparison with onshore counterpart. *Renew. Energy* **2017**, *108*, 72–84. [CrossRef]
4. Ahmed, S.D.; Al-Ismail, F.S.; Shafiullah, M.; Al-Sulaiman, F.A.; El-Amin, I.M. Grid integration challenges of wind energy: A review. *IEEE Access* **2020**, *8*, 10857–10878. [CrossRef]
5. Feng, S.; Wang, K.; Lei, J.; Tang, Y. Influences of DC bus voltage dynamics in modulation algorithm on power oscillations in PMSG-based wind farms. *Int. J. Electr. Power Energy Syst.* **2021**, *124*, 106387. [CrossRef]
6. Shanmugam, L.; Joo, Y.H. Stabilization of permanent magnet synchronous generator-based wind turbine system via fuzzy-based sampled-data control approach. *Inf. Sci.* **2021**, *559*, 270–285. [CrossRef]
7. Zribi, M.; Alrifai, M.; Rayan, M. Sliding Mode Control of a Variable- Speed Wind Energy Conversion System Using a Squirrel Cage Induction Generator. *Energies* **2017**, *10*, 604. [CrossRef]
8. Tripathi, S.; Tiwari, A.; Singh, D. Grid-integrated permanent magnet synchronous generator based wind energy conversion systems: A technology review. *Renew. Sustain. Energy Rev.* **2015**, *51*, 1288–1305. [CrossRef]
9. Bonfiglio, A.; Delfino, F.; Gonzalez-Longatt, F.; Procopio, R. Steady-state assessments of PMSGs in wind generating units. *Int. J. Electr. Power Energy Syst.* **2017**, *90*, 87–93. [CrossRef]
10. Yang, B.; Yu, T.; Shu, H.; Zhang, X.; Qu, K.; Jiang, L. Democratic joint operations algorithm for optimal power extraction of PMSG based wind energy conversion system. *Energy Convers. Manag.* **2018**, *159*, 312–326. [CrossRef]
11. Abd El Hamied, M.A.; Amary, N.H.E. Permanent Magnet Synchronous Generator Stability Analysis and Control. *Procedia Comput. Sci.* **2016**, *95*, 507–515. [CrossRef]
12. Zarei, M.E.; Ramirez, D.; Prodanovic, M.; Medrano Arana, G. Model Predictive Control for PMSG-based Wind Turbines with Overmodulation and Adjustable Dynamic Response Time. *IEEE Trans. Ind. Electron.* **2021**, *69*, 1573–1585. [CrossRef]
13. Maaruf, M.; Ferik, S.E.; Mahmoud, M.S. Integral Sliding Mode Control With Power Exponential Reaching Law for DFIG. In Proceedings of the 2020 17th International Multi-Conference on Systems, Signals Devices (SSD), Monastir, Tunisia, 20–23 July 2020; pp. 1122–1127. [CrossRef]
14. Rhaili, S.; Abbou, A.; Ziouh, A.; Elidrissi, R. Comparative study between PI and FUZZY logic controller in vector controlled five-phase PMSG based variable-speed wind turbine. In Proceedings of the 2018 IEEE 12th International Conference on Compatibility, Power Electronics and Power Engineering (CPE-POWERENG 2018), Doha, Qatar, 10–12 April 2018; pp. 1–6.
15. Giraldo, E.; Garces, A. An Adaptive Control Strategy for a Wind Energy Conversion System Based on PWM-CSC and PMSG. *IEEE Trans. Power Syst.* **2014**, *29*, 1446–1453. [CrossRef]
16. Shafiullah, M.; Abido, M.A; Coelho, L.S. Design of robust PSS in multimachine power systems using backtracking search algorithm. In Proceedings of the 2015 18th International Conference on Intelligent System Application to Power Systems (ISAP), Porto, Portugal, 11–16 September 2015; IEEE: Piscataway, NJ, USA, 2015; pp. 1–6.
17. Saad, N.H.; El-Sattar, A.A.; Marei, M.E. Improved bacterial foraging optimization for grid connected wind energy conversion system based PMSG with matrix converter. *Ain Shams Eng. J.* **2018**, *9*, 2183–2193. [CrossRef]
18. Alam, M.S.; Shafiullah, M.; Hossain, M.I.; Hasan, M.N. Enhancement of power system damping employing TCSC with genetic algorithm based controller design. In Proceedings of the 2015 International Conference on Electrical Engineering and Information Communication Technology (ICEEICT), Savar, Bangladesh, 21–23 May 2015; IEEE: Piscataway, NJ, USA, 2015; pp. 1–5.
19. Haridy, A.L.; Abdelbasset, A.A.M.; Hemeida, M. Optimum Controller Design Using the Ant Lion Optimizer for PMSG Driven by Wind Energy. *J. Electr. Eng. Technol.* **2021**, *16*, 367–380. [CrossRef]

20. Shafiullah, M.; Rana, M.J.; Coelho, L.S.; Abido, M.A. Power system stability enhancement by designing optimal PSS employing backtracking search algorithm. In Proceedings of the 2017 6th International Conference on Clean Electrical Power (ICCEP), Santa Margherita Ligure, Italy, 27–29 June 2017; IEEE: Piscataway, NJ, USA, 2017; pp. 712–719.
21. Shafiullah, M.; Rana, M.J.; Shahriar M.S.; Zahir, M.H. Low-frequency oscillation damping in the electric network through the optimal design of UPFC coordinated PSS employing MGGP. *Measurement* **2019**, *138*, 118–131.
22. Qais, M.H.; Hasanien, H.M.; Alghuwainem, S.; Elgendy, M.A. Output Power Smoothing of Grid-Tied PMSG-Based Variable Speed Wind Turbine Using Optimal Controlled SMES. In Proceedings of the 2019 54th International Universities Power Engineering Conference (UPEC), Bucharest, Romania, 3–6 September 2019; pp. 1–6.
23. Qais, M.H.; Hasanien, H.M.; Alghuwainem, S. A novel LMSRE-based adaptive PI control scheme for grid-integrated PMSG-based variable-speed wind turbine. *Int. J. Electr. Power Energy Syst.* **2021**, *125*, 106505. [CrossRef]
24. Shafiullah, M.; Rana, M.J.; Shahriar, M.S.; Al-Sulaiman, F.A.; Ahmed, S.D.; Ali, A. Extreme learning machine for real-time damping of LFO in power system networks. *Electr. Eng.* **2021**, *103*, 279–292. [CrossRef]
25. Mahmoud, M.S.; Maaruf, M.; El-Ferik, S. Neuro-adaptive output feedback control of the continuous polymerization reactor subjected to parametric uncertainties and external disturbances. *ISA Trans.* **2021**, *112*, 1–11. [CrossRef]
26. Mahmud, M.A.; Roy, T.K.; Littras, K.; Islam, S.N.; Amanullah Oo, M.T. Nonlinear Partial Feedback Linearizing Controller Design for PMSG-Based Wind Farms to Enhance LVRT Capabilities. In Proceedings of the 2018 IEEE International Conference on Power Electronics, Drives and Energy Systems (PEDES), Chennai, India, 18–21 December 2018; pp. 1–6.
27. Soufi, Y.; Kahla, S.; Bechouat, M. Feedback linearization control based particle swarm optimization for maximum power point tracking of wind turbine equipped by PMSG connected to the grid. *Int. J. Hydrogen Energy* **2016**, *41*, 20950–20955. [CrossRef]
28. El Mourabit, Y.; Derouich, A.; El Ghzizal, A.; El Ouanjli, N.; Zamzoum, O. Nonlinear backstepping control for PMSG wind turbine used on the real wind profile of the Dakhla-Morocco city. *Int. Trans. Electr. Energy Syst.* **2020**, *30*, e12297. [CrossRef]
29. Saidi, Y.; Mezouar, A.; Miloud, Y.; Kerrouche, K.D.E.; Brahmi, B.; Benmahdjoub, M.A. Advanced non-linear backstepping control design for variable speed wind turbine power maximization based on tip-speed-ratio approach during partial load operation. *Int. J. Dynam. Control* **2020**, *30*, e12297. [CrossRef]
30. Yang, B.; Yu, T.; Shu, H.; Zhang, Y.; Chen, J.; Sang, Y.; Jiang, L. Passivity-based sliding-mode control design for optimal power extraction of a PMSG based variable speed wind turbine. *Renew. Energy* **2018**, *119*, 577–589. [CrossRef]
31. Xiong, L.; Li, P.; Ma, M.; Wang, Z.; Wang, J. Output power quality enhancement of PMSG with fractional order sliding mode control. *Int. J. Electr. Power Energy Syst.* **2020**, *115*, 105402. [CrossRef]
32. Sami, I.; Ullah, S.; Ullah, N.; Ro, J.S. Sensorless fractional order composite sliding mode control design for wind generation system. *ISA Trans.* **2021**, *111*, 275–289. [CrossRef]
33. Dursun, E.H.; Kulaksiz, A.A. Second-order sliding mode voltage-regulator for improving MPPT efficiency of PMSG-based WECS. *Int. J. Electr. Power Energy Syst.* **2020**, *121*, 106149. [CrossRef]
34. Nasiri, M.; Mobayen, S.; Zhu, Q.M. Super-Twisting Sliding Mode Control for Gearless PMSG-Based Wind Turbine. *Complexity* **2019**, *2019*, 6141607. [CrossRef]
35. İrfan Yazıcı.; Yaylacı, E.K. Discrete-time integral terminal sliding mode based maximum power point controller for the PMSG-based wind energy system. *IET Power Electron.* **2019**, *12*, 3688–3696. [CrossRef]
36. Dursun, E.H.; Kulaksiz, A.A. Second-Order Fast Terminal Sliding Mode Control for MPPT of PMSG-based Wind Energy Conversion System. *Elektron. Elektrotechnika* **2020**, *26*, 39–45. [CrossRef]
37. Sami, I.; Ullah, S.; Ali, Z.; Ullah, N.; Ro, J.S. A Super Twisting Fractional Order Terminal Sliding Mode Control for DFIG-Based Wind Energy Conversion System. *Energies* **2020**, *13*, 2158. [CrossRef]
38. Kord, H.; Barakati, S.M. Design an adaptive sliding mode controller for an advanced hybrid energy storage system in a wind dominated RAPS system based on PMSG. *Sustain. Energy Grids Netw.* **2020**, *21*, 100310. [CrossRef]
39. Tang, Y.; Zhang, Y.; Hasankhani, A.; VanZwieten, J. Adaptive Super-Twisting Sliding Mode Control for Ocean Current Turbine-Driven Permanent Magnet Synchronous Generator. In Proceedings of the 2020 American Control Conference (ACC), Denver, CO, USA, 1–3 July 2020; pp. 211–217.
40. Sun, K.; Li, D.; Zheng, B. Adaptive Global Fast Terminal Sliding Mode Control for MPPT of Direct-driven PMSG. In Proceedings of the 2018 13th World Congress on Intelligent Control and Automation (WCICA), Changsha, China, 4–8 July 2018; pp. 1658–1663.
41. Wang, J.; Bo, D.; Ma, X.; Zhang, Y.; Li, Z.; Miao, Q. Adaptive backstepping control for a permanent magnet synchronous generator wind energy conversion system. *Int. J. Hydrogen Energy* **2019**, *44*, 3240–3249. [CrossRef]
42. Wang, J.; Bo, D. Adaptive fixed-time sensorless maximum power point tracking control scheme for DFIG wind energy conversion system. *Int. J. Electr. Power Energy Syst.* **2022**, *135*, 107424. [CrossRef]
43. Mahmoud, M.S.; Maaruf, M. Robust Adaptive Multilevel Control of a Quadrotor. *IEEE Access* **2020**, *8*, 167684–167692. [CrossRef]
44. Yang, L.; Yang, J. Nonsingular fast terminal sliding-mode control for nonlinear dynamical systems. *Int. J. Robust Nonlinear Control* **2011**, *21*, 1865–1879. [CrossRef]
45. Labbadi, M.; Cherkaoui, M. Robust adaptive backstepping fast terminal sliding mode controller for uncertain quadrotor UAV. *Aerosp. Sci. Technol.* **2019**, *93*, 105306. [CrossRef]
46. Beniaguev, D. Historical Hourly Weather Data 2012–2017. Available online: https://www.kaggle.com/selfishgene/historical-hourly-weather-data (accessed on 31 October 2021).

 *sustainability*

Article

Water-Energy-Food Nexus Approach to Assess Crop Trading in Saudi Arabia

Mohammad Tamim Kashifi [1], Fahad Saleh Mohammed Al-Ismail [2,3], Shakhawat Chowdhury [1,4], Hassan M. Baaqeel [5], Md Shafiullah [6], Surya Prakash Tiwari [2] and Syed Masiur Rahman [2,*]

[1] Department of Civil and Environmental Engineering, King Fahd University of Petroleum & Minerals, Dhahran 31261, Saudi Arabia; tamim_motebahher@yahoo.com (M.T.K.); schowdhury@kfupm.edu.sa (S.C.)
[2] Applied Research Center for Environment & Marine Studies, King Fahd University of Petroleum & Minerals, Dhahran 31261, Saudi Arabia; fsalismail@kfupm.edu.sa (F.S.M.A.-I.); surya.tiwari@kfupm.edu.sa (S.P.T.)
[3] Electrical Engineering Department and K.A.CARE Energy Research & Innovation Center, King Fahd University of Petroleum & Minerals, Dhahran 31261, Saudi Arabia
[4] Interdisciplinary Research Center for Construction and Building Materials, King Fahd University of Petroleum & Minerals, Dhahran 31261, Saudi Arabia
[5] Department of Chemical Engineering, King Fahd University of Petroleum & Minerals, Dhahran 31261, Saudi Arabia; baaqeel@kfupm.edu.sa
[6] Interdisciplinary Research Center for Renewable Energy and Power Systems, Dahran 31261, Saudi Arabia; shafiullah@kfupm.edu.sa
* Correspondence: smrahman@kfupm.edu.sa

Citation: Kashifi, M.T.; Al-Ismail, F.S.M.; Chowdhury, S.; Baaqeel, H.M.; Shafiullah, M.; Tiwari, S.P.; Rahman, S.M. Water-Energy-Food Nexus Approach to Assess Crop Trading in Saudi Arabia. *Sustainability* **2022**, *14*, 3494. https://doi.org/10.3390/su14063494

Academic Editors: Lirong Liu and Baojie He

Received: 21 October 2021
Accepted: 7 February 2022
Published: 16 March 2022

Publisher's Note: MDPI stays neutral with regard to jurisdictional claims in published maps and institutional affiliations.

Abstract: Water scarcity is a global challenge, especially in arid regions, including Middle Eastern and North African countries. The distribution of water around the earth is not even. Trading water in the form of an embedded commodity, known as the water footprint (*WF*), from water-abundant regions to water-scarce regions, is a viable solution to water scarcity problems. Agricultural products account for approximately 85% of the earth's total *WF*, indicating that importing water-intense crops, such as cereal crops, can partially solve the local water scarcity problem. This study investigated water, energy, and food nexus dynamics for the trades of a few major crops, specifically considering Saudi Arabia. It analyzed the trade of crops and its impact on *WF*, energy, and carbon dioxide (CO_2) emission savings. The findings revealed that importing major cereal crops to Saudi Arabia could significantly reduce the local *WF*. The imports of wheat, maize, rice, and barley reduced approximately 24 billion m^3 per year of consumable *WF* (i.e., blue and green water footprint) in the global scale. Similarly, the trade of major crops had a significant impact on energy and CO_2 emission savings. The energy savings from the wheat, maize, and barley trades in Saudi Arabia was estimated to be approximately 9 billion kWh. It also saved about 7 million tons per year of CO_2 emissions. The trades of cereal crops in Saudi Arabia reduced water consumption, energy usage, and CO_2 emissions significantly.

Keywords: water footprint; agricultural product; energy footprint; carbon dioxide emission; water-energy-food nexus

1. Introduction

Freshwater is one of the critical global resources. Its availability for consumption is a global challenge [1,2], and this issue has been a concern for many years [3,4]. Day by day, the increase in the freshwater demand has imposed an elevated pressure on groundwater extraction. The exploitation of groundwater at a higher rate than the recharge may lead to the depletion of non-renewable groundwater aquifers. Groundwater withdrawals are expected to increase by half in developing countries and by one-fifth in developed countries by 2025 compared to 2011 demands [5].

Freshwater resources are not evenly distributed around the earth. Middle Eastern countries face higher water scarcity than many other countries due to the lack of renewable

Table 1. Data description.

Number	Data	Description/Value	Source
1	Wheat, barley, rice, maize	Import and production data	[34]
2	*WF*	*WF* data	[35]
3	Groundwater pumping energy requirement	0.764 Kwh/m^3	[33]
4	Wastewater treatment energy requirement	0.4 Kwh/m^3	[33]
5	Shipment Energy Intensity (SEI) for crop transportation (for an average ship speed of 20 knots)	0.015 Kwh/ton-km	International Maritime Organization (IMO) [36]
6	Emission intensity of energy production	0.73 Kg-CO$_2$/Kwh	[37]

2.2. Method

The analysis starts from crop trades and *WF* savings from trading (Figure 1). The *WF* of imports is based on a hypothetical *WF* estimation. It indicates the quantity of *WF* consumed if these crops were produced inside Saudi Arabia. On the other hand, the actual *WF* of imports will be calculated based on the *WF* of crops in the regions where these crops were cultivated. In this study, long-term global averages of the *WF* from recent years were used [35]. If the water requirement of crops in exporting countries is less than in importing countries, water trade can improve water efficiency [38]. This notion leads to the water-saving concept. The water-saving by water trade between two countries is estimated by multiplying the volume of the traded crops by the difference between *WF* per unit of the crops of the importing and exporting countries [11].

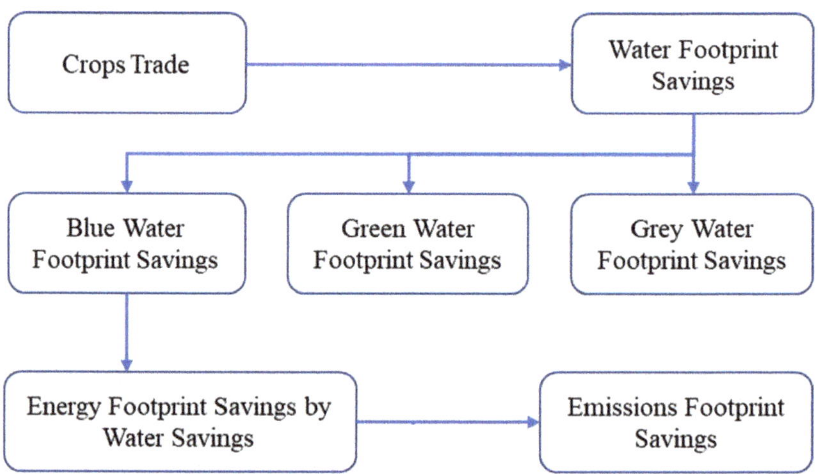

Figure 1. Methodology to estimate water and emission footprint savings through crop trading.

The energy savings associated with the *WF* were investigated. The green *WF* does not require any energy for applying in the crop field. The blue *WF* requires pumping from groundwater aquifers and transporting water to an irrigation area, and the grey *WF* requires wastewater treatment and transport. The grey *WF* savings are likely to be much lower than the blue WF savings because of its insignificant contribution to the agricultural sector to date. As such, grey water energy footprint savings is not considered in this study. The details of the methods are summarized below:

Step 1: *WF* savings

The total hypothetical *WF* of crops is calculated as follows:

$$HWF_n = \sum_i^m WFL_i \times W_i \tag{1}$$

where HWF_n is the total hypothetical WF of all types (n = 1, 2, 3; 1. Blue, 2. Green, and 3. Grey) in million m^3 (Mm^3) for m types of crops; WFL_i is the hypothetical *WF* of crop i in the local area (i.e., Saudi Arabia) in (m^3/ton), and W_i is the gross weight of imported crop i in million ton (MT). The total actual *WF* of crops for global average is calculated as follows:

$$AWF_n = \sum_i^m WFG_i \times W_i \tag{2}$$

where AWF_n is the total actual *WF* of all types in Mm^3, WFG_i is the global average of *WF* (m^3/ton), and W_i is the gross weight of imported crop i in MT. The total *WF* savings from crop trade is calculated as follows:

$$WFS_n = \sum_i^m (WFL_i - WFG_i) W_i \tag{3}$$

where WFS_n is the total *WF* savings of *WF* type n (Mm^3) by *WF* trade.

Step 2: Energy footprint savings

The energy requirements for water extraction in the case of local production is calculated as follows:

$$EC_n = \sum_j^l WF_n \times P_j \times EI_j \tag{4}$$

where EC_n is the energy consumption (in kWh) for water extraction and processing for WF type (n). The WF is shown as WF_n, and the energy intensity for irrigation is shown as EI_j, where j indicates the type of irrigation water collected from underground aquifers and wastewater treatment plants. The percentage of irrigation by groundwater or wastewater is shown as P_j (j = groundwater or wastewater). EI_j indicates energy intensity (kWh/m^3) for irrigation type j. The total energy requirement for water extraction and processing is calculated as follows.

$$EE = \sum_i^n WFS_n \times EC_n \tag{5}$$

where EE is the total energy consumption for three WFS_n in kWh.

The energy requirement for transportation of imported crops from outside Saudi Arabia is calculated as follows:

$$ET = SEI \times D \times W \tag{6}$$

where SEI is shipment energy intensity (Kwh/ton-km), D is distance (km), and W is the weight (ton) of the crops. The distance (D) is approximated as the average distance from four major grain exporters (i.e., Germany, Canada, Poland, and Lithuania) to Saudi Arabia. The total energy savings (Kwh) from crop trades can be calculated as follows:

$$ES = EE - ET \tag{7}$$

Step 3: Emission footprint savings

The emission savings through energy savings can be estimated as:

$$EFS = EFI \times ES \tag{8}$$

where EFS is the emission footprint savings in Kg-CO$_2$, EFI is emission footprint intensity (Kg-CO$_2$/Kwh).

3. Results and Discussion

The production and import of four major cereal crops in the recent years are presented in Figure 2. The wheat production was the highest in 2011 but gradually decreased until

2015. By 2015, Saudi Arabia banned wheat production for two years (2016 and 2017). Wheat production started again in 2018 on a limited scale to support the small-scale growers. On the other hand, wheat imports fluctuated between 2 to 4 million tons (MT)/year (Figure 2). Maize imports have been gradually increasing since 2011. However, the production is relatively constant and much lower than the imports. Barley and rice are mostly imported, and the local production of these crops is negligible.

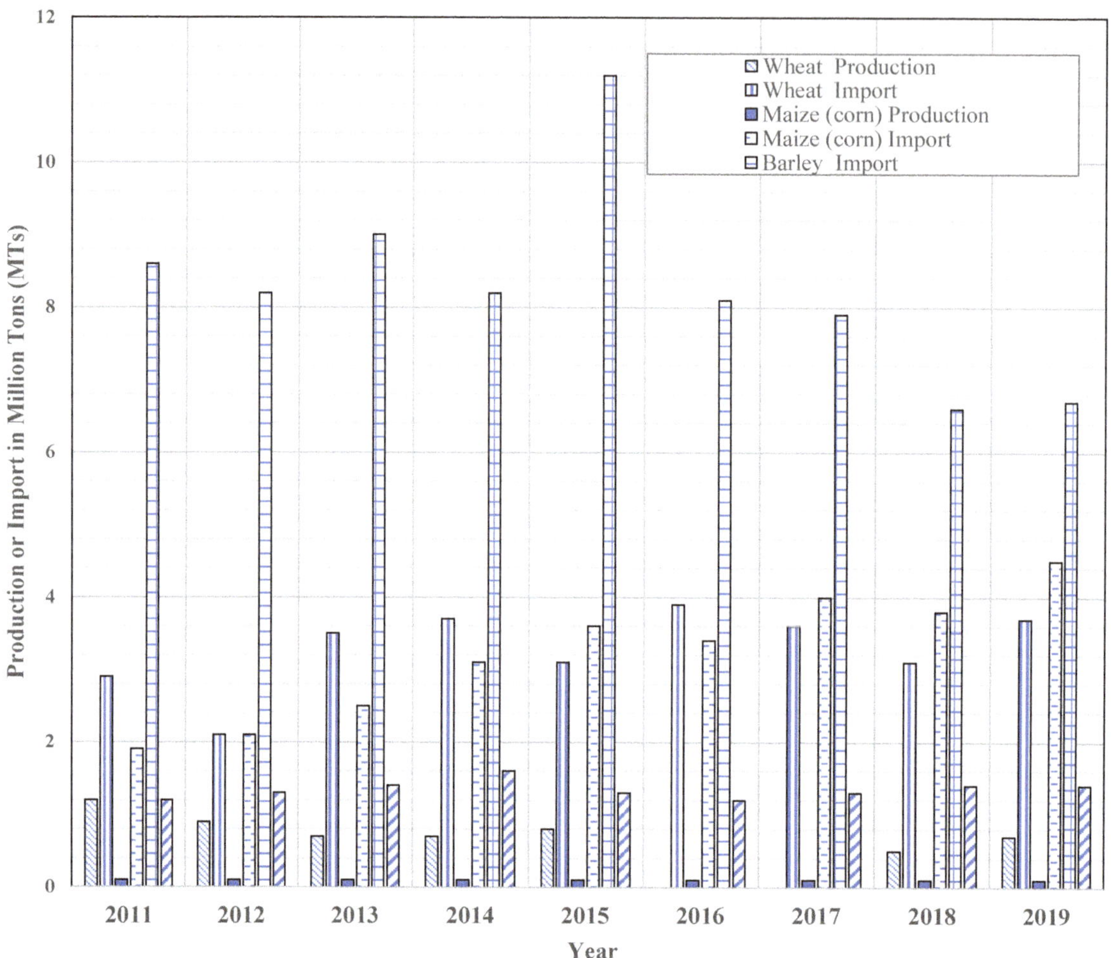

Figure 2. Major crop production and import for Saudi Arabia (source of data: International Grains Council, [34]).

The *WF* of four main crops in Saudi Arabia and global averages are provided in Figure 3. The local *WF* of rice is not available. The *WF* of crops in Saudi Arabia has a higher value for the blue *WF* and a lower value for the green *WF* due to lower annual rainfall. The hypothetical blue *WFs* of three major crops (i.e., wheat, maize, and barley) were higher than the green and grey *WFs*. Thus, importing these major crops can save groundwater extractions from aquifers.

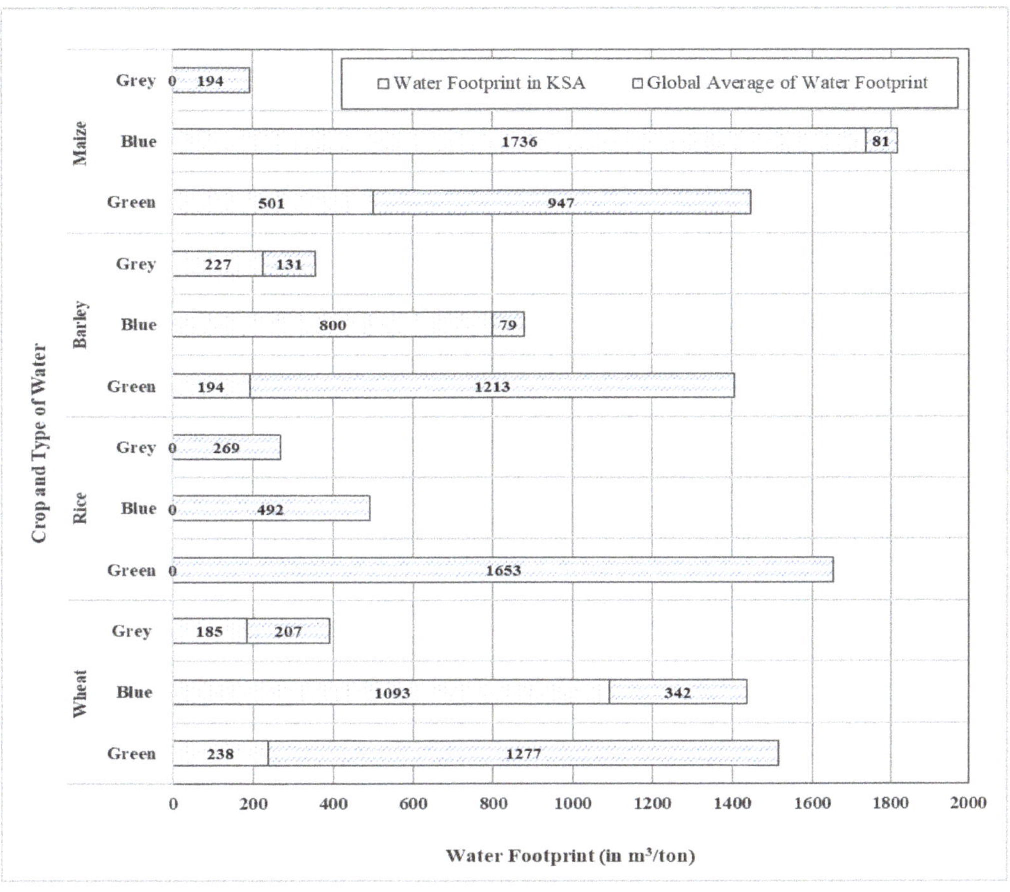

Figure 3. *WF* of Saudi Arabia and global average of *WF* (Hoekstra and Mekonnen 2010).

The total hypothetical *WF* for four major crops is calculated (Table 2). The wheat *WF* was highest in 2011 and experienced a continuous decrease until 2015. The wheat *WF* of production was very low during 2016 and 2017. The hypothetical *WF* of imported and produced major crops is calculated for between 2011 and 2020 (Table 2). The wheat import hypothetical *WFs* primarily rely on the blue type. The green and grey *WFs* comprise a relatively small portion of the total *WF*. The local *WF* rates are not available for rice, and the global averages are adopted, so the green *WF* is the dominant type in this case. The barley and maize import *WFs* are also mostly the blue *WF*. However, the maize does not have a grey *WF*. Analysis of a hypothetical *WF* indicated that they mostly consumed groundwater if the imported products were produced locally. Among the four major crops, only wheat and maize are produced locally to a considerable extent. The local production of wheat *WFs* varied significantly over the years. The *WFs* of wheat approached to zero during 2016 and 2017. On the other hand, the *WFs* of maize were stable over the years.

The green and blue *WFs* are consumable as these are not available after use. The consumable *WFs* for production and imports were calculated. The total consumable *WF* was lower in 2012 (19,852 Mm³) and 2018 (23,080 Mm³) compared to the years before and after these years, respectively (Figure 4). There was a significant rise in total consumable *WF* in 2015 (27,387 Mm³), attributed to the increased import of barley. The total consumable *WF* of the selected crops was dominated by imports (Figure 4).

Table 2. Hypothetical *WFs* for imports and production, in a million m^3 (Mm^3).

		Import								
Crop	Type of *WF*	2011	2012	2013	2014	2015	2016	2017	2018	2019
Wheat (Durum)	Green	691.3	500.6	834.3	882.0	739.0	929.7	858.2	739.0	882.0
	Blue	3170.3	2295.7	3826.2	4044.8	3388.9	4263.5	3935.5	3388.9	4044.8
	Grey	535.9	388.1	646.8	683.8	572.9	720.8	665.3	572.9	683.8
Rice	Green	1983.6	2148.9	2314.2	2644.8	2148.9	1983.6	2148.9	2314.2	2314.2
	Blue	590.1	639.3	688.4	786.8	639.3	590.1	639.3	688.4	688.4
	Grey	322.9	349.9	376.8	430.6	349.9	322.9	349.9	376.8	376.8
Barley	Green	1664.8	1587.4	1742.3	1587.4	2168.2	1568.1	1529.3	1277.7	1297.0
	Blue	6878.7	6558.8	7198.6	6558.8	8958.3	6478.8	6318.8	5279.0	5359.0
	Grey	1951.6	1860.8	2042.3	1860.8	2541.6	1838.1	1792.7	1497.7	1520.4
Maize (corn) starch	Green	952.3	1052.5	1253.0	1553.7	1804.3	1704.1	2004.8	1904.6	2255.4
	Blue	3299.2	3646.5	4341.1	5382.9	6251.1	5903.9	6945.7	6598.4	7813.9
	Grey	0.0	0.0	0.0	0.0	0.0	0.0	0.0	0.0	0.0
		Production								
Wheat (Durum)	Green	286.1	214.5	166.9	166.9	190.7	0.0	0.0	119.2	166.9
	Blue	1311.8	983.9	765.2	765.2	874.6	0.0	0.0	546.6	765.2
	Grey	221.8	166.3	129.4	129.4	147.8	0.0	0.0	92.4	129.4
Maize (corn) starch	Green	50.1	50.1	50.1	50.1	50.1	50.1	50.1	50.1	50.1
	Blue	173.6	173.6	173.6	173.6	173.6	173.6	173.6	173.6	173.6
	Grey	0.0	0.0	0.0	0.0	0.0	0.0	0.0	0.0	0.0

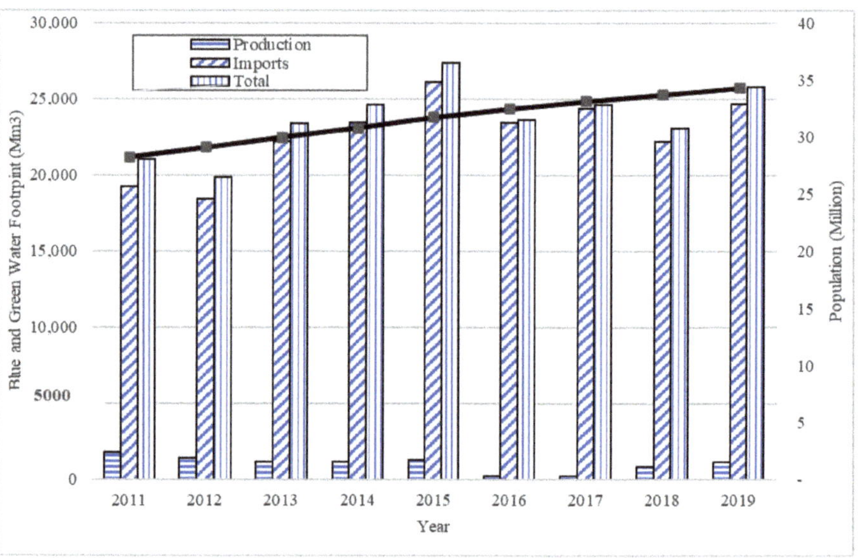

Figure 4. Trends of the consumable *WFs* of four major grains over the years.

Although green and blue *WFs* are precious and consumable, their values are not the same considering *WF* sustainability. The green *WF* is renewable (rainwater). The blue *WF* comes from scarce sources (non-renewable underground aquifers or surface water stored in the dams or shallow aquifers), and some of these are barely renewable. Therefore, extensive use of the blue *WF* may be a threat to sustainability. Crop production in Saudi Arabia requires a significantly higher blue *WF* than global averages due to the scarcity of rainfall and higher rates of evapotranspiration. The increased dependency on local crop production will require an additional blue *WF*. Therefore, Saudi Arabia can import crops without exerting additional pressure on non-renewable water resources.

This study estimated water savings from wheat, barley, and maize considering global perspectives. By importing crops, Saudi Arabia is saving the blue *WF* locally; however, the production of the imported crops will cost the *WF* in the exporting countries. The *WF* savings by crop imports in Saudi Arabia is presented in Figure 5.

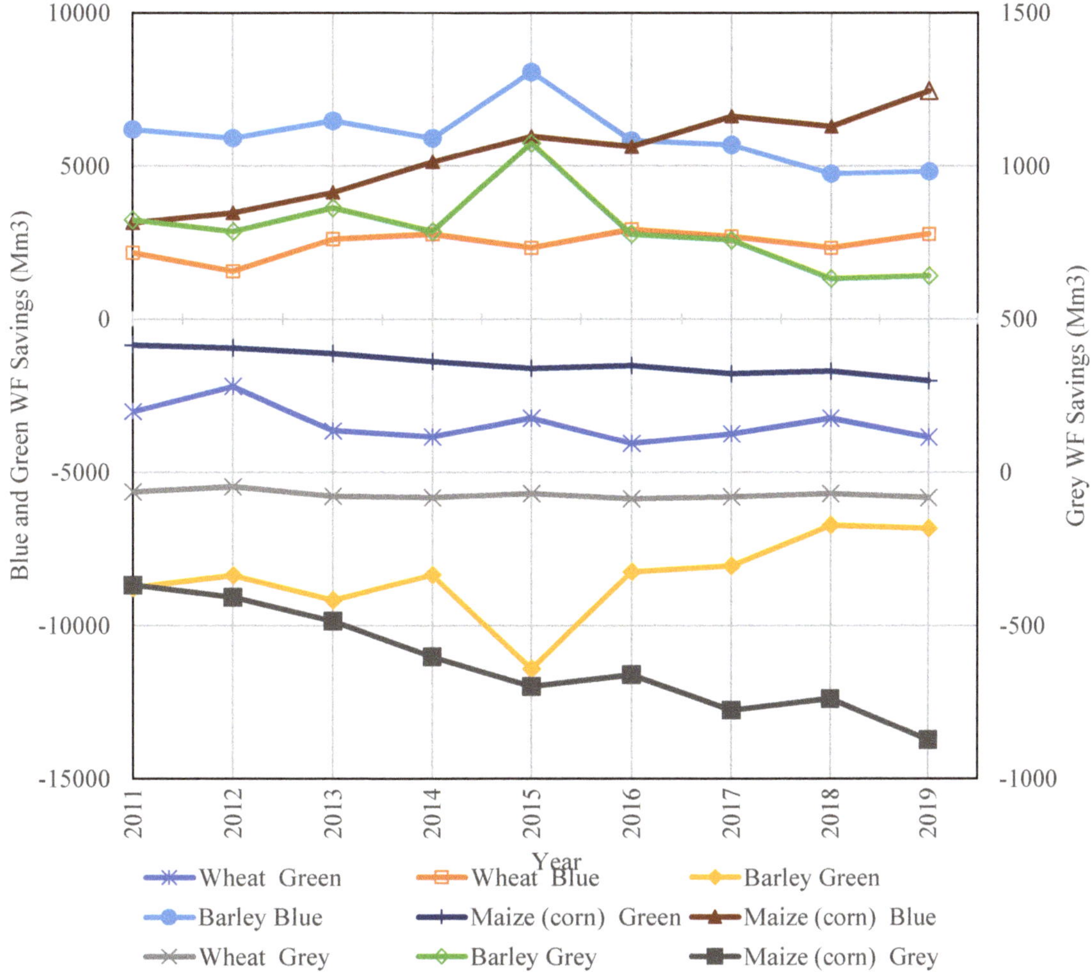

Figure 5. Saudi Arabia *WF* savings from the import of three major crops.

Overall, there is limited yearly *WF* savings in the greywater footprint compared to the blue and green *WFs* (Figure 5). Further, the barley grey *WF* savings is positive, whereas it is negative for wheat and maize. The reason for the barley grey *WF* being positive, unlike the wheat grey *WF* savings and maize grey *WF* savings, is that the grey *WF* of barley is lower than the global averages (Figure 3). This means when barley is produced locally it pollutes less water compared to global averages, unlike wheat and maize. However, there is a significant blue *WF* savings due to the import of the three main crops, and there is a loss of the green *WF* due to crop imports as the green *WF* is wasted. Therefore, the imports of cereal crops save the blue *WF* locally at the cost of the green *WF* in the global perspective (Figure 5). The energy and emission footprint savings due to the trading of crops are shown in Figure 6. These savings have resulted due to the replacement of the blue *WF* by the green *WF*. According to Figure 6, the energy savings due to the barley trade was highest among the three crops from 2011 up to 2015, whereas after that the maize trade resulted in the highest savings. Energy savings due to the wheat trade was the least among the three crops that ranged between 1–1.9 billion Kwh/year. The *WF* savings leads to energy and emission footprint savings. The emission intensity of 0.73 Kg-CO_2/kWh for Saudi Arabia was used to estimate carbon dioxide (CO_2) footprint savings from energy reduction [37]. The yearly CO_2 footprint reduction due to crop imports was in the range of 5.80–8.66 MT during 2011–2019 (Figure 6), which is around 1.5% of Saudi Arabia's total yearly emissions [39]. Similar to the energy savings, the emissions savings due to barley have been decreasing since 2015 while the opposite is true for maize. These fluctuations can be attributed to the import of amount of imports of these crops (Table 2). The emission footprint savings from wheat has a relatively stable trend between 2011 and 2019.

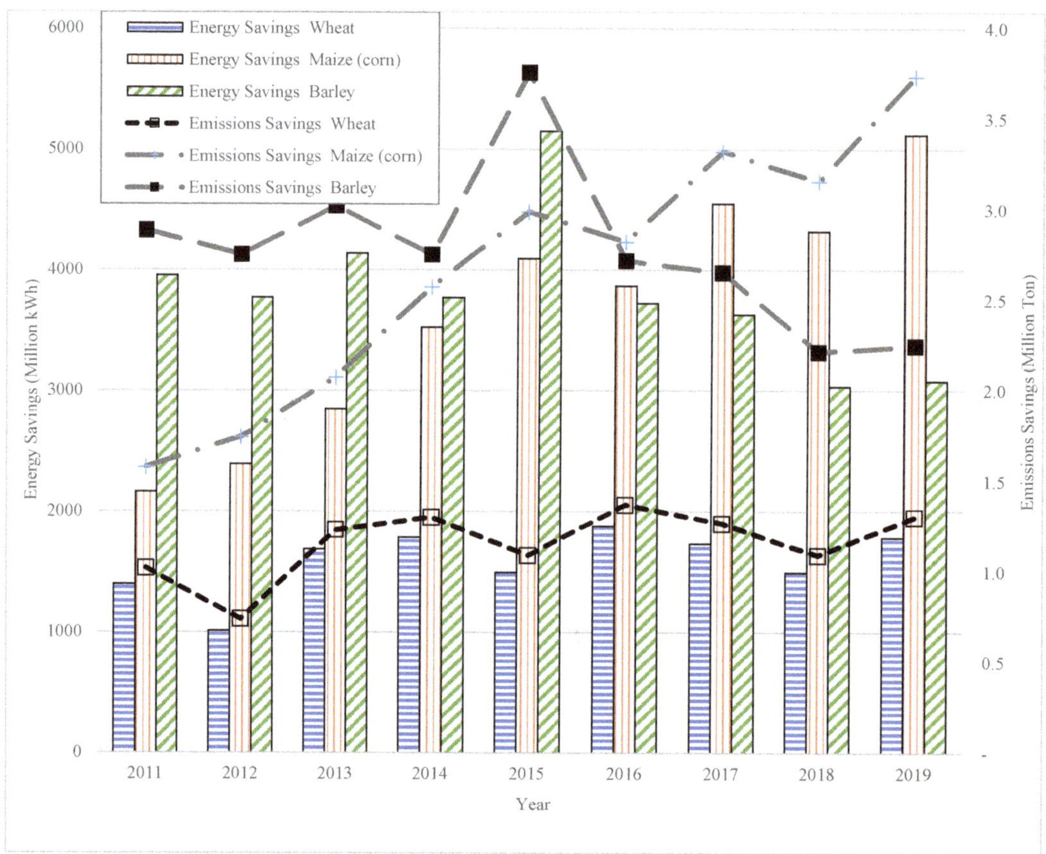

Figure 6. Energy and emissions footprint savings due to crop trade in Saudi Arabia.

The Kingdom has been increasing energy and emission savings during the last few years. However, water conservation and self-sufficiency have a trade-off. Typically, the country imports different crop items from more than 70 countries [12]. Due to the diversified supply chain, it is expected that the Kingdom's food security may not suffer significantly due to uncertainty in crop imports. However, it can only be verified through more detailed analysis focusing on location of importing countries, type, price, and quantity of crop, and mode of transportation. Recently, the Kingdom has been investigating diversified ways to meet water demand in a sustainable manner including artificial cloud seeding [40].

Therefore, the results of this study should be assessed with due consideration of other available opportunities in the Kingdom to reduce *WF*, energy, and emission savings.

WF trading has been growing among countries including Saudi Arabia. Water-intensive crop imports have reduced stress on the scarce groundwater resources of Saudi Arabia. Therefore, the country may continue *WF* trading for achieving water sustainability, by trading barley, wheat, maize, and rice. However, a nation's extensive exports of water-intense products will lead to water depletion and cause unsustainable solutions on a global scale. Based on the results of this study, the local crop trades have been benefitting Saudi Arabia without compromising global sustainability. On the other hand, being a water-scarce country with significant energy reservoirs, Saudi Arabia faces the risk of embodied water in energy export, water consumed by energy extraction, and transformation called embodied water on energy resources. As a result, it faces water scarcity and water embodied in energy lost due to energy export [41]. The crop-trading-related policies have a significant positive impact on the WEF nexus.

4. Conclusions

Proper management of essential resources such as water, energy, and food requires interaction among these resources. Policymakers can take advantage through incorporating the nexus in these resources to optimize the trade-off and synergies among the resources and protect environmental quality. In this study, water, energy, and food nexus dynamics for few major crop trading were investigated for Saudi Arabia. This study quantified the effects of crop trades on the savings of three types of *WFs*. Additionally, the *WF* savings that would lead to energy and emission savings was investigated. The recent trades of four major crops significantly improved *WF* savings, leading to energy and emission footprint savings.

In recent years, the trade of crops has saved 1100 to 16,000 Mm^3/year of blue *WF* at the cost of probable green *WF* in the exporting countries. The savings will reduce pressure on local groundwater resources. Further, the effects of crop trades on energy consumption footprint and emission footprint savings were estimated. The energy savings from trading three major crops (wheat, maize, and barley) in Saudi Arabia was around 9 billion kWh. This energy savings leads to emission savings of about 7 million tons of CO_2 yearly. However, these results should be evaluated with appropriate consideration of other available opportunities in the Kingdom to reduce *WF*, energy, and GHG emissions.

This study used the global averages to estimate *WF* indicators related to crop imports. Nevertheless, water productivity differs among countries due to different rates of rainfall and temperature. In the future, country-specific *WF* information can be used to accurately identify the global impact of Saudi Arabia's crop trades. Additionally, this study considered the impact of trading the major crops. The other crops should be included in future for exploring the impact of their trades on the *WF*, as well as energy savings and GHG reduction.

Author Contributions: Conceptualization, M.T.K. and S.M.R.; methodology, M.T.K.; software, M.T.K. and S.M.R.; validation, S.C., F.S.M.A.-I., S.P.T. and H.M.B.; formal analysis, M.T.K.; investigation, S.M.R.; resources, M.T.K. and S.M.R.; data curation, S.M.R. and M.T.K.; writing—original draft preparation, M.T.K.; writing—review and editing, S.M.R., M.S. and S.C.; visualization, M.T.K., S.M.R. and S.P.T.; supervision, S.M.R. and S.C.; project administration, F.S.M.A.-I. and S.M.R. All authors have read and agreed to the published version of the manuscript.

Funding: No funding was received for this paper.

Institutional Review Board Statement: Not applicable.

Informed Consent Statement: Not applicable.

Data Availability Statement: The data for this paper were received from publicly available sources and the sources were mentioned in the paper.

Acknowledgments: The authors gratefully acknowledge the support of King Fahd University of Petroleum and Minerals in conducting this research.

Conflicts of Interest: The authors declare that the research was conducted without any commercial or financial relationships that could be construed as potential conflicts of interest.

References

1.   Hoff, H. Global water resources and their management. *Curr. Opin. Environ. Sustain.* **2009**, *1*, 141–147. [CrossRef]
2.   Postel, S.L.; Daily, G.C.; Ehrlich, P.R. Human Appropriation of Renewable Fresh Water. *Science* **1996**, *271*, 785–788. [CrossRef]
3.   Postel, S.L. Entering an era of water scarcity: The challenges ahead. *Ecol. Appl.* **2000**, *10*, 941–948. [CrossRef]
4.   United Nations. *The United Nations World Water Development Report 2015: Water for a Sustainable World—UNESCO Digital Library*; United Nations Educational, Scientific and Cultural: Paris, France, 2015.
5.   Water Futures. Water Futures Report. Available online: http://www.sabmiller.com/docs/default-source/investor-documents/reports/2011/sustainability/water-futures-report-2011.pdf?sfvrsn=4 (accessed on 3 June 2021).
6.   Bob, M.; Rahman, N.A.; Elamin, A.; Taher, S. Rising Groundwater Levels Problem in Urban Areas: A Case Study from the Central Area of Madinah City, Saudi Arabia. *Arab. J. Sci. Eng.* **2016**, *41*, 1461–1472. [CrossRef]
7.   Brown, A.; Matlock, M.D.; Vörösmarty, C.J.; Douglas, E.M.; Green, P.A.; Revenga, C.; Ashraf, B.; Aghakouchak, A.; Alizadeh, A.; Mousavi, B.; et al. *Water Footprint Manual*; Water Footprint Network: Enschede, The Netherlands, 2009.
8.   Hoekstra, A.Y. Virtual Water Trade. In Proceedings of the Internacional Expert Meeting on Virtual Water Trade, Delft, The Netherlands, 12–13 December 2002.
9.   Daniels, P.L.; Lenzen, M.; Kenway, S. The ins and outs of water use—A review of multi-region input–output analysis and water footprints for regional sustainability analysis and policy. *Econ. Syst. Res.* **2011**, *23*, 353–370. [CrossRef]
10. Perry, C. Efficient irrigation; inefficient communication; flawed recommendations. *Irrig. Drain.* **2007**, *56*, 367–378. [CrossRef]
11. Mekonnen, M.M.; Hoekstra, A.Y. The green, blue and grey water footprint of production and consumption. *Ecol. Econ.* **2011**, *70*, 749–758. [CrossRef]
12. Chowdhury, S.; Kabir, F.; Chowdhury, I.R.; Papadopoulou, M.P. Importing and Exporting Agricultural Crop Products: An Assessment of Virtual Water Flow (VWF) in Saudi Arabia. *Arab. J. Sci. Eng.* **2019**, *44*, 4911–4920. [CrossRef]
13. Rasul, G. Food, water, and energy security in South Asia: A nexus perspective from the Hindu Kush Himalayan region☆. *Environ. Sci. Policy* **2014**, *39*, 35–48. [CrossRef]
14. Al-Saidi, M.; Saliba, S. Water, Energy and Food Supply Security in the Gulf Cooperation Council (GCC) Countries—A Risk Perspective. *Water* **2019**, *11*, 455. [CrossRef]
15. Chowdhury, S.; Ouda, O.K.M.; Papadopoulou, M.P. Virtual water content for meat and egg production through livestock farming in Saudi Arabia. *Appl. Water Sci.* **2017**, *7*, 4691–4703. [CrossRef]
16. Kotilaine, J.T. GCC Agriculture. Economic Research. Available online: https://www.farmlandgrab.org/wp-content/uploads/2010/05/20100301163254eGCC-Agriculture-Sector-March-2010.pdf (accessed on 3 June 2021).
17. Ji, L.; Huang, G.H.; Niu, D.X.; Cai, Y.P.; Yin, J.G. A Stochastic Optimization Model for Carbon-Emission Reduction Investment and Sustainable Energy Planning under Cost-Risk Control. *J. Environ. Inform.* **2020**, *36*, 107–118. [CrossRef]
18. Sharifzadeh, M.; Hien, R.K.T.; Shah, N. China's roadmap to low-carbon electricity and water: Disentangling greenhouse gas (GHG) emissions from electricity-water nexus via renewable wind and solar power generation, and carbon capture and storage. *Appl. Energy* **2019**, *235*, 31–42. [CrossRef]
19. Lee, M.; Keller, A.A.; Chiang, P.-C.; Den, W.; Wang, H.; Hou, C.-H.; Wu, J.; Wang, X.; Yan, J. Water-energy nexus for urban water systems: A comparative review on energy intensity and environmental impacts in relation to global water risks. *Appl. Energy* **2017**, *205*, 589–601. [CrossRef]
20. Lv, J.; Li, Y.; Huang, G.; Suo, C.; Mei, H. Quantifying the impact of water availability on China's energy system under uncertainties: A perceptive of energy-water nexus. *Renew. Sustain. Energy Rev.* **2020**, *134*, 110321. [CrossRef]
21. Li, X.; Liu, J.; Zheng, C.; Han, G.; Hoff, H. Energy for water utilization in China and policy implications for integrated planning. *Int. J. Water Resour. Dev.* **2016**, *32*, 477–494. [CrossRef]
22. Liao, X.; Hall, J.W.; Eyre, N. Water use in China's thermoelectric power sector. *Glob. Environ. Chang.* **2016**, *41*, 142–152. [CrossRef]
23. Liao, X.; Hall, J.W. Drivers of water use in China's electric power sector from 2000 to 2015. *Environ. Res. Lett.* **2018**, *13*, 094010. [CrossRef]
24. Ringler, C.; Bhaduri, A.; Lawford, R. The nexus across water, energy, land and food (WELF): Potential for improved resource use efficiency? *Curr. Opin. Environ. Sustain.* **2013**, *5*, 617–624. [CrossRef]
25. Karnib, A. Bridging Science and Policy in Water-Energy-Food Nexus: Using the Q-Nexus Model for Informing Policy Making. *Water Resour. Manag.* **2018**, *32*, 4895–4909. [CrossRef]
26. Radmehr, R.; Ghorbani, M.; Ziaei, A.N. Quantifying and managing the water-energy-food nexus in dry regions food insecurity: New methods and evidence. *Agric. Water Manag.* **2020**, *245*, 106588. [CrossRef]
27. Li, M.; Fu, Q.; Singh, V.P.; Ji, Y.; Liu, D.; Zhang, C.; Li, T. An optimal modelling approach for managing agricultural water-Tenergy-food nexus under uncertainty. *Sci. Total Environ.* **2019**, *651*, 1416–1434. [CrossRef] [PubMed]

28. Zhou, Y.; Li, H.; Wang, K.; Bi, J. China's energy-water nexus: Spillover effects of energy and water policy. *Glob. Environ. Chang.* **2016**, *40*, 92–100. [CrossRef]
29. Zhai, M.; Huang, G.; Liu, L.; Zheng, B.; Guan, Y. Inter-regional carbon flows embodied in electricity transmission: Network simulation for energy-carbon nexus. *Renew. Sustain. Energy Rev.* **2020**, *118*, 109511. [CrossRef]
30. Al-Zahrani, M.; Musa, A.; Chowdhury, S. Multi-objective optimization model for water resource management: A case study for Riyadh, Saudi Arabia. *Environ. Dev. Sustain.* **2015**, *18*, 777–798. [CrossRef]
31. MOEP (Ministry of Economy and Planning). *Ninth Development Plan, 1431–1435 H (2010–2014)*; Ministry of Economy and Planning: Riyadh, Saudi Arabia, 2010.
32. FAO. *Saudi Arabia Country Profile*; Food and Agriculture Organization of the United Nations (FAO): Rome, Italy, 2008.
33. Christopher, N.; García Téllez, B. *Energy for Water in Agriculture: A Partial Factor Productivity Analysis*; King Abdullah Petroleum Studies and Research Center (KAPSARC): Riyadh, Saudi Arabia; p. 2016.
34. Urueña, R. International Grains Council. In *Handbook of Transnational Economic Governance Regimes*; Brill Nijhoff: Leiden, The Netherlands, 2010; pp. 695–703.
35. Mekonnen, M.M.; Hoekstra, A.Y. The green, blue and grey water footprint of crops and derived crop products. *Hydrol. Earth Syst. Sci.* **2011**, *15*, 1577–1600. [CrossRef]
36. Von Knorring, H.J.; Karisson, R. *Air Pollution and Energy Efficiency*; International Maritime Organization (IMO): London, UK, 2016.
37. Climate Action Tracker (CAT). Saudi Arabia. PEI Power Engineering International (Vol. 13), 2016. Available online: https://climateactiontracker.org/countries/saudi-arabia/sources/ (accessed on 16 January 2021).
38. Oki, T.; Kanae, S. Virtual water trade and world water resources. *Water Sci. Technol.* **2004**, *49*, 203–209. [CrossRef]
39. Worldometer. Saudi Arabia CO_2 Emissions—Worldometer. Available online: https://www.worldometers.info/co2-emissions/saudi-arabia-co2-emissions/ (accessed on 16 January 2021).
40. Alam, T.; Khan, M.A.; Gharaibeh, N.K.; Gharaibeh, M.K. Big Data for Smart Cities: A Case Study of NEOM City, Saudi Arabia. In *Smart Cities: A Data Analytics Perspective*; Springer: Cham, The Netherlands, 2021; pp. 215–230. [CrossRef]
41. Zhang, J.C.; Zhong, R.; Zhao, P.; Zhang, H.W.; Wang, Y.; Mao, G.Z. International energy trade impacts on water resource crises: An embodied water flows perspective. *Environ. Res. Lett.* **2016**, *11*, 074023. [CrossRef]

 *sustainability*

Article

Enhancing Transient Response and Voltage Stability of Renewable Integrated Microgrids

Luay Elkhidir [1], Khalid Khan [1], Mohammad Al-Muhaini [1,2] and Muhammad Khalid [1,2,3,*]

[1] Electrical Engineering Department, King Fahd University of Petroleum and Minerals (KFUPM), Dhahran 31261, Saudi Arabia; g201707630@kfupm.edu.sa (L.E.); g201604320@kfupm.edu.sa (K.K.); muhaini@kfupm.edu.sa (M.A.-M.)
[2] Center for Renewable Energy and Power Systems, KFUPM, Dhahran 31261, Saudi Arabia
[3] K.A. CARE Energy Research & Innovation Center, Dhahran 31261, Saudi Arabia
* Correspondence: mkhalid@kfupm.edu.sa

Abstract: Integration of renewable generation coupled with an energy storage system (ESS) in a power system increases the complexity of networks' stability analysis and control. Therefore, an accurate stability assessment of power networks is expected to become a big challenge in the future. In this work, an effective approach to prevent power outage by controlling the source voltage of the power network is formulated to mitigate the effects of grid faults. Small signal stability studies are conducted on a renewable integrated IEEE 9 bus system as a case study with optimized size and allocation of ESS for reducing output power variability of renewables. An assessment is performed to study the effects of load-sharing devices on parallel generators under 6-cycle three-phase fault disturbances. The damping of the power network is increased at nominal and light loading conditions with 6-cycle three-phase fault disturbances through coordinated power system stabilizer (PSS) and static VAR compensator (SVC) at bus 9. The developed framework is extensively analyzed in steady-state conditions using a load flow program. Based on the results obtained, the proposed coordinated PSS-SVC device proves to possess comparatively better performance in terms of enhancing most of the system response rate under various load conditions with overall improved stability.

Keywords: grid fault restoration; renewable microgrid; power system stabilizer; voltage stability

Citation: Elkhidir, L.; Khan, K.; Al-Muhaini, M.; Khalid, M. Enhancing Transient Response and Voltage Stability of Renewable Integrated Microgrids. *Sustainability* **2022**, *14*, 3710. https://doi.org/10.3390/su14073710

Academic Editor: Pablo García Triviño

Received: 27 January 2022
Accepted: 9 March 2022
Published: 22 March 2022

Publisher's Note: MDPI stays neutral with regard to jurisdictional claims in published maps and institutional affiliations.

1. Introduction

Today, power system grids are more complicated and expansive, as electricity plays an important role in almost all aspects of humankind. Therefore, it is pertinent to mitigate the blackout probability and its period to increase the level of security and welfare. Small signal stability is defined as "the ability of a power system to maintain synchronism under small disturbance". The impact of power quality appears in dynamic system and electric power industry which can be significantly expensive [1]. Power quality is usually defined as the ability of the power system networks to transfer a stable, uninterruptible, and clean power supply with a pure noise-free sinusoidal waveform. Power system plants are frequently exposed and sustain disturbances as they are non-linear dynamic systems. These disturbances may lead to partial or blackout, which can produce severe consequences [2].

Nowadays, parallel standby power systems are used instead of single large generator units. These backup power systems play an increasingly significant role in ensuring an uninterrupted supply of power. Parallel operation of generator sets (parallel power systems) provides many benefits such as reliability, expandability, flexibility, ease of maintenance, and quality performance. Generally, the load shedding technique is investigated under these operations [3,4]. Decades ago, the series capacitive compensation technique was used for reactive power control and damping out the oscillations to improve the transmitted power [5,6]. Then the use of automatic controls like power system stabilizers (PSSs) in large power systems grids became essential to maintain stability. The power system stabilizer is used to provide supplementary feedback stabilizing control signals to the excitation system for mitigating the electro-mechanical oscillations [7].

The concept of microgrid formulation facilitates exclusive control over selected intensive problems associated with renewable integration [8–10]. Typically, a microgrid includes control theory to sustain a distributed generator, energy storage system, and local loads. The microgrid can be operated in an islanded as well as grid-connected mode [11]. This allows the formation of a deregulated power network that is pertinent considering the complexities of renewable integration. Hence, microgrid helps to increase renewable penetration in the energy sector with enhanced control over the grid elements maintaining the reliability and security of the supplied power [12–14]. Accordingly, further development in power electronics has led to the large-scale incorporation of flexible AC transmission system (FACTS) devices in electrical power plants [15]. This technique is one type of variable series compensation which is very effective for enhancing stability as well as controlling power flow in the transmission lines. The occurrence of electrical disturbances like faults and lightning are damped out by incorporation of static VAR compensator (SVC) in combination with PSS and automatic voltage regulators (AVR) in large power systems [16].

A variety of energy sources with different characteristics decreases the techno-economic significance of renewable energy sources (RES) primarily due to their time-varying energy capacity [17]. For instance, solar PV energy is available during the day, so at night other alternatives or energy storage support are pertinent. Similarly, wind energy systems also impose similar challenges and limitations usually attributed to their unpredictable variability. Such time-varying complexity of RESs makes the integration of energy storage systems (ESS) and dispatchable energy sources pertinent, especially for autonomous RES applications (Figure 1) for various applications, such as appropriate energy mix, ensuring reliability, and reducing operational costs of sustainable energy system [18–20].

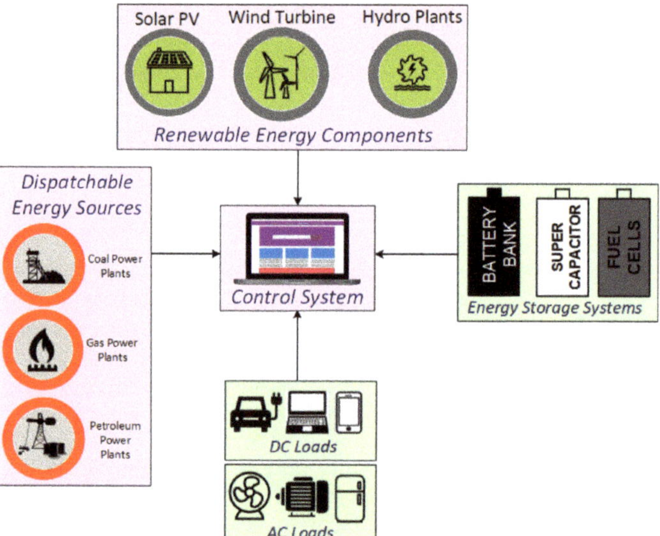

Figure 1. Standalone hybrid energy systems.

The benefits of ESSs are substantial and have long been recognized to be essential towards a coordinated and successful operation of utility grids. Power storage systems mostly include batteries, flywheels, pumped hydro-power storage, supercapacitors, and compressed air energy storage [21]. ESS improves RES integration flexibility through peak-shifting, mitigation of forecasting errors, providing frequency and voltage support among other operational services [22]. Furthermore, expensive grid improvements or outages due to unforeseen demand or any trip-off of any sources connected to the national grid network can be obviated [23–25]. Accordingly, the meager inertia, voltage, and frequency support introduced due to RES integration can be further facilitated through hybridized ESS [26–28]. In particular, the deteriorating power quality at the distribution level can be obviated through dedicated energy management algorithms to optimally integrate RESs in accordance with the requirements of the energy market, grid standards, and contingencies [29–31].

In this study, a coordinated PSS-SVC is developed to enhance the stability of RES integrated power networks with load sharing device that increases the damping by adding more power system stabilizer value to the system. The main objective of this paper is to enhance the transient stability of renewable integrated power networks. A modified IEEE 9 bus system is considered with solar and wind energy integration incorporated with an appropriate energy storage system that aims to mitigate renewable variability under nominal and light loading conditions. Moreover, a comparative study is also presented between the base case wherein the modified IEEE 9 bus system is incorporated with individual PSS and SVC.

The remainder of the paper is organized as follows. Section 2 presents the related work associated with PSS tuning using different algorithms. Section 3 formulates the equations for stability and load shedding investigation and presents the modeling of the modified IEEE 9 bus under study. Section 4 discusses the results obtained and presents numerous stability studies based on light and nominal loading conditions, and with system fault conditions, followed by the conclusion in Section 5.

2. Related Work

Tuning PSS parameters and input signal play an important role in small signal stability investigations of microgrids. The main function of PSS is to produce a torque in phase with the rotor speed deviation and compensate the generator terminal voltage by inserting additional signal [32]. Artificial intelligence (AI) techniques over the last years have been frequently used for PSS tuning. Artificial neural network (ANN) and their types are employed as they can robustly perform based on incomplete data tasks for complex problems while dealing with non-linear problems by easily learning from the historical data. Several network structures have been contemplated for PSS design that includes feed-forward neural networks, recurrent neural network [33], and pole shifting method [34–36].

In the last few years, optimization algorithms have also been developed and proposed to solve PSS designing problems. Tabu search and genetic algorithm (GA) methods are mentioned in [37] for designing PSS. They prove to be more advantageous as the resultant solutions generated are not trapped at the local optimum. Another technique like simulated annealing (SA) is illustrated in [36], for tuning the parameters of PSS. In similar terms, numerous evolutionary and heuristic algorithms have been proposed for parameter tuning of PSS, such as bacteria foraging (BF) process [38] and particle swarm optimization technique (PSO) [39]. A new optimization algorithm that mimics a whale's hunting behavior known as whale optimization algorithm (WOA) is illustrated in [40] in tuning PSS to shift the eigenvalues to a predefined stable zone. Most of the recent power system stability researches investigate new approaches to enhance transient stability effectively and efficiently [41].

However, there are certain limitations to these algorithms. The ANN technique consumes a long training time to choose the number of layers and neurons in each layer and exhaustive training is required [39]. The Pole shifting method imposes a memory storage problem and the computational algorithms are highly complex. The SA method may produce inaccurate results due to being trapped at the local optimum. The GA method may require a long-running time depending on the complexity of the system. The BF algorithm suffers from a delay in reaching the global solution because the algorithm depends on random search directions. PSO has some limitations like partial optimism that effects the speed and direction regulation. Moreover, the algorithm suffers from a weak ability to search locally and that may inadvertently lead to trapping in local minimum solutions. Although PSS and SVC techniques are mature and prominent, it is also vital to assess their implementation with renewable integration particularly coupled with energy storage systems. An effective operation of PSS and SVC with renewable and energy storage systems can play a significant role in appropriately outlining the potential stability of the system that can be achieved by the system operators. This paper presents quantifiable applicability of the PSS and SVC considering the system dynamics with the integration of renewable energy sources and energy storage systems that is not considered in the literature.

3. Problem Formulation and Proposed Framework

3.1. State Space Representation of the Power System Model

In control engineering, a state-space representation is a mathematical model of a physical system as a set of variables of input, output, and state connected by differential equations of the first order. "State space" refers to space whose axes are the state variables. The state of the system can be represented as a vector within that space [42]. A set of n first-order, nonlinear ordinary differential equations defined in (1) can describe the behavior of a dynamic power system.

$$\dot{x}_1 = f_1(x_1, x_2 \ldots, x_n; u_1, u_2, \ldots, u_r; t) \quad i = 1, 2, \ldots, n \tag{1}$$

where, n is the order of the system and r is the number of inputs. This can be written in the following form by using vector-matrix notation as described in (2):

$$\dot{x} = f(x, u, t) \tag{2}$$

The state equations of a power system with m number of power system stabilizers and n number of machines can be represented as:

$$\dot{x} = Ax + Bu \tag{3}$$

$$y = Cx + Du \tag{4}$$

$$A = \begin{bmatrix} \frac{\delta f_1}{\delta x_1} & \cdots & \frac{\delta f_1}{\delta x_n} \\ \vdots & \ddots & \vdots \\ \frac{\delta f_n}{\delta x_1} & \cdots & \frac{\delta f_n}{\delta x_n} \end{bmatrix} \quad B = \begin{bmatrix} \frac{\delta f_1}{\delta u_1} & \cdots & \frac{\delta f_1}{\delta u_r} \\ \vdots & \ddots & \vdots \\ \frac{\delta f_n}{\delta u_1} & \cdots & \frac{\delta f_n}{\delta u_r} \end{bmatrix}$$

$$C = \begin{bmatrix} \frac{\delta g_1}{\delta x_1} & \cdots & \frac{\delta g_1}{\delta x_n} \\ \vdots & \ddots & \vdots \\ \frac{\delta g_m}{\delta x_1} & \cdots & \frac{\delta g_m}{\delta x_n} \end{bmatrix} \quad D = \begin{bmatrix} \frac{\delta g_1}{\delta u_1} & \cdots & \frac{\delta g_1}{\delta u_r} \\ \vdots & \ddots & \vdots \\ \frac{\delta g_m}{\delta u_1} & \cdots & \frac{\delta g_m}{\delta u_r} \end{bmatrix} \tag{5}$$

where, A is the state matrix of size $n * n$, B is the input matrix of size $n * r$, C is the output matrix of size $m * n$ and D represents the feedforward matrix of size $m * r$ (5). The column vector u is the reference vector to the device. Furthermore, when the state variables derivatives are not explicit time functions, the system is said to be autonomous. In this case, (6) can be simplified to:

$$\dot{x} = f(x, u) \tag{6}$$

Similarly, the output variables (4) that can be observed in the system can be expressed in terms of the state variables and the input variables as:

$$y = g(x, u) \tag{7}$$

Therefore, the complex non-linear power systems and, hence, a set of non-linear differential equations can be defined:

$$x = \left[\delta, \omega, E'_q, E_{fd}, V_f\right] \tag{8}$$

where, δ is the rotor angle of the generator, ω is the synchronous speed of the generator, E'_q is the, E_{fd} represents the internal voltage of the generator, and V_f is the excitation voltage of the generator.

3.2. PSS Controller Structure

The PSS structure is represented in Figure 2. It consists of a gain constant, a washout filter to serve as a high pass filter, a dynamic compensator to compensate for the phase lag between the electric torque, and the excitation and limiter to prevent the excitation system from entering the saturation mode. The transfer function of the PSS is therefore expressed as:

$$\Delta U_i = k_i \frac{ST_w}{1 + ST_w}\left[\frac{1 + ST_{1i}}{1 + ST_{2i}}\right]\left[\frac{1 + ST_{3i}}{1 + ST_{4i}}\right]\Delta\omega_i \tag{9}$$

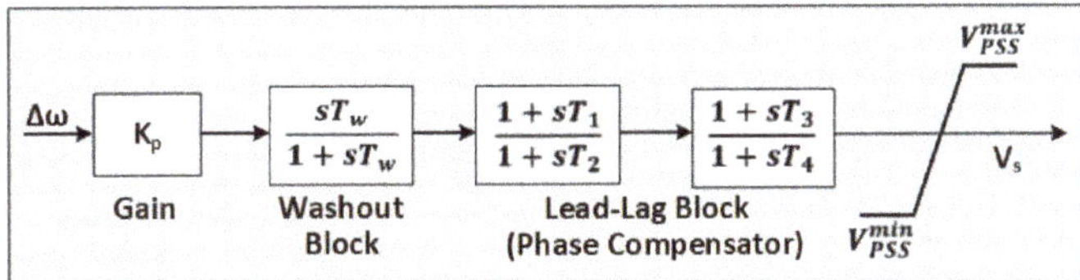

Figure 2. Structure of power system stabilizer.

3.3. Operation of IEEE 9 Bus System under Study

An IEEE 9 bus system is considered for this study. It consists of three generators and three loads as depicted in Figure 3. The tests are performed considering a time horizon of 24 h pertaining to its processed data [43]. Table 1 outlines the different loading conditions assessed for the analytical comparative study.

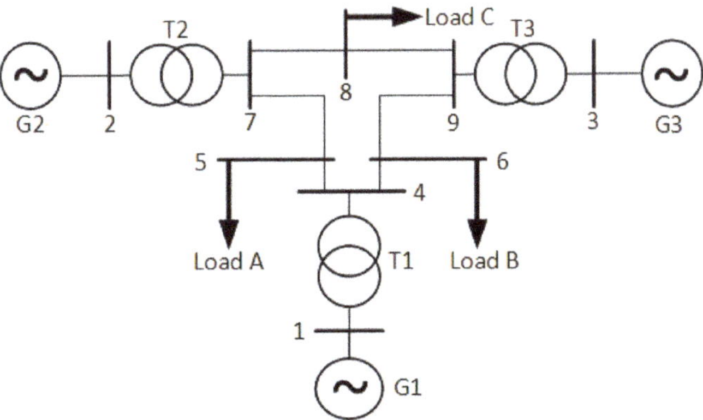

Figure 3. IEEE 9 Bus system without renewable generation.

Further, the load shedding is developed considering two renewable energy sources in the test system (wind turbines and PV cells) under AC power flow taking cost-minimizing as an objective function is illustrated in (10). Correspondingly, the ESS is incorporated in the network to mitigate the RES variability.

$$OF = \sum_{i,t} a_g \left(P_{i,t}^g \right)^2 + b_g \left(P_{i,t}^g \right) + C_g + \sum_{i,t} VOLL \left(P_{i,t}^{LS} \right) + VWC \left(P_{i,t}^{WC} \right) \tag{10}$$

where, $a_g, b_g,$ and c_g are the fuel cost coefficients of the thermal generation units VOLL is the value of the loss of load ($/MWh$), $P_{i,t}^g$ is the active power generated by the thermal unit, VMC is the renewable energy sources, $P_{i,t}^{LS}$ is the active load shedding at time t from bus i, and $P_{i,t}^{WC}$ denotes the curtailed power from the renewables at time t from bus i.

$$- P_{ij}^{max} \leq P_{ij}(t) \leq P_{ij}^{max} \tag{11}$$

$$P_g^{min} \leq P_g(t) \leq P_g^{max} \tag{12}$$

$$P_g(t) - P_g(t-1) \leq RU_g \tag{13}$$

$$- P_g(t-1) - P_g(t) \leq RD_g \tag{14}$$

$$SOC_i(t) = SOCi,(t-1) + (P_i^c(t)\eta_c - P_i^d / \eta_d)\Delta_t \tag{15}$$

$$P_{i,min}^c \leq P_i^c(t) \leq P_{i,max}^c \tag{16}$$

$$P_{i,min}^d \leq P_i^d(t) \leq P_{i,max}^d \tag{17}$$

$$SOC_{i,min} \leq SOC_{i,t} \leq SOC_{i,max} \tag{18}$$

The developed framework is optimized using GAMS to obtain an optimal size and allocation of ESS as presented in Table 2 for mitigating the impact of RES variability, and correspondingly the turn ratios of the distribution transformer are reduced to decrease the output voltage of the transformer. The optimization includes power balance equality constraints [44], transmission line constraint (11), generation constraint (12), the generation ramp up (13), and ramp down constraints (14). Furthermore, the constraints of the energy storage system include the charge/discharge characteristics, charge efficiency (η_c), discharge efficiency (η_d), and charge/discharge capability that is limited by their maximum power (15)–(18) [45]. The importance of the load shedding study is to avoid blackout points associated with large cost payment as shown in Figure 4 [46]. Therefore, an assessment is made after each optimization step to see the change in load and determine the value of the voltage corresponding to which the system experiences a brownout.

Table 1. Load conditions for the IEEE 9 bus microgrid (p.u.).

	Nominal Loading		**Light Loading**	
Generator	P	Q	P	Q
G1	1.7164	0.6205	0.9649	0.223
G2	1.630	0.0665	1.00	−0.1933
G3	0.85	−0.1086	0.45	−0.2668
Load	P	Q	P	Q
A	1.25	0.5	0.7	0.35
B	0.9	0.3	0.5	0.3
C	1.0	0.35	0.6	0.2

Table 2. Output of GAMS code of optimal size and allocation of ESS for IEEE 9 bus microgrid.

Bus	Optimal Size		Total Size	
	MW	MWh	MW	MWh
5	6.4	12.8	48.3	174.7
6	41.8	161.9	48.3	1744

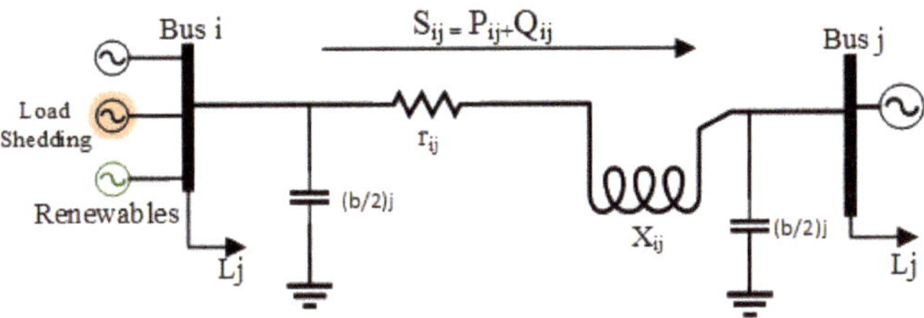

Figure 4. Input and output variables of AC power load flow.

3.4. Stabilization Paralleling and Load Sharing between Generators

If two or more generators are connected to a transmission line, assuming the frequency is constant, the models of the generators can be lumped into an equivalent that is powered by the sum of the individual mechanical torque output [47]. The block diagram representation of two parallel-connected generators in synchronous mode (Figure 5) depicts that separate feedback is required for every corresponding loop (here, ω_1 and ω_2).

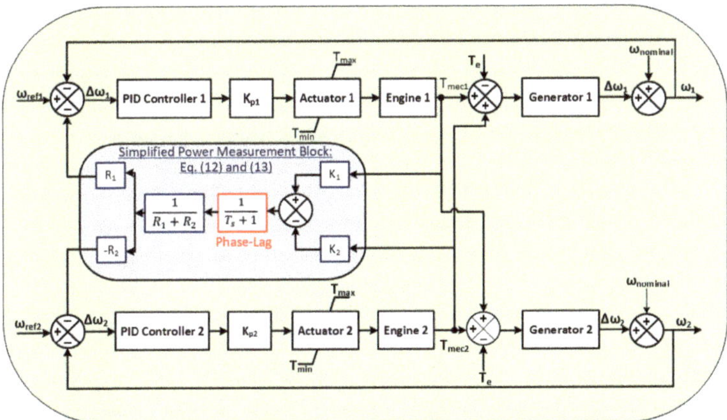

Figure 5. Block diagram of two generators connected in parallel in synchronous mode.

Accordingly, the average power of the two system generator sets with load sharing can be represented as an equivalent wattmeter as shown in Figure 6. Therefore, line-to-line voltage, line currents, and battery supply are considered as the inputs to the load sharing unit.

The load-sharing unit output is a DC voltage corresponding to the actual load. All parallel load-sharing units are connected via the parallel cable. To obtain the block diagram of the load sharing circuit, each power measuring circuit is modeled as a voltage source "battery" as shown in Figure 7. Hence, based on the circuit analysis, using Kirchoff's voltage law, on Figure 7, the voltage source representing the power measuring circuit has a value of the corresponding generator's electrical power (load) multiplied by a factor (k) expressed as:

$$V_1 = K_1 P_{L1}, \quad V_2 = K_1 P_{L1} \tag{19}$$

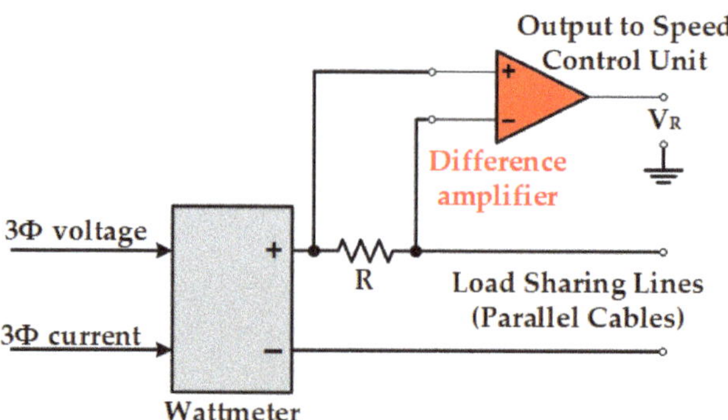

Figure 6. Equivalent load sharing unit.

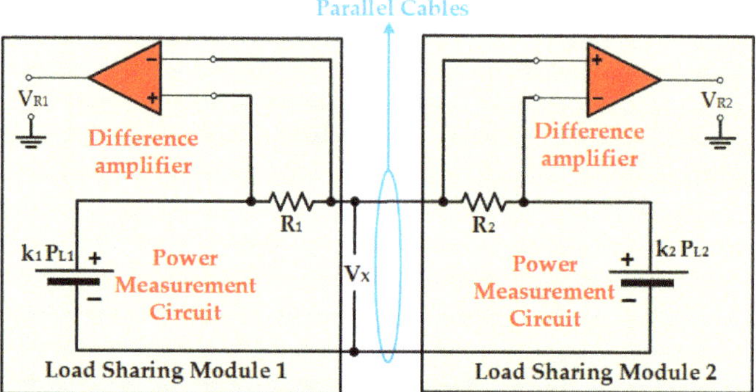

Figure 7. Equivalent load sharing circuit configuration.

The current going through the circuit is:

$$I_{ij,t} = \frac{bV_{i,t}}{2}\angle\left(\delta_{i,t} + \frac{\pi}{2}\right) + \frac{V_{i,t}\angle\delta_{i,t} - V_{j,t}\angle\delta_{j,t}}{Z_{ij}\angle\Theta_{ij}} \tag{20}$$

where, b denotes the line susceptance, $V_{i,j}$ is the voltage between the bus, $\delta_{i,j}$ represents the voltage angle, $Z_{i,j}$ is the line impedance, $\Theta_{i,j}$ is the phase angle difference between the current and voltage. The output of each difference amplifier (i.e., the voltage across each resistor) (21) and (22). Consequently, based on these formulations the simplified block diagram is developed as depicted in Figure 5. The resultant phase lag system for stabilizing paralleling and load sharing generators are developed and incorporated into the test system under study, that is systematically analyzed to observe the operational performance of the overall system framework and optimization. The apparent power (23) of the system is based on the current flow ($I_{ij,t}^*$) which is the complex conjugate of the current phasor flow between the modules. Accordingly, the active and reactive power flow of between the connecting modules is determined; wherein, $\theta_{i,j}$ is the angle between the active and reactive power the buses.

$$V_{R1} = I * R_1 = \frac{(K_1 P_{L2} - K_2 P_{L2})R1}{R_1 + R_2} \tag{21}$$

$$V_{R2} = -I * R_2 = \frac{-1}{R_1 + R_2}(K_1 P_{L2} - K_2 P_{L2})R_2 \tag{22}$$

$$S_{ij,t} = (V_{i,t}\angle\delta_{i,t})I_{ij,t}^* \tag{23}$$

$$P_{ij,t} = \frac{V_{i,t}^2}{Z_{ij}} cos(\theta_{ij}) - \frac{V_{i,t}V_{j,t}}{Z_{ij}} cos(\delta_{i,t} - \delta_{j,t} + \theta_{ij}) \tag{24}$$

$$Q_{ij,t} = \frac{V_{i,t}^2}{Z_{ij}} sin(\theta_{ij}) - \frac{V_{i,t}V_{j,t}}{Z_{ij}} sin(\delta_{i,t} - \delta_{j,t} + \theta_{ij}) - \frac{bV_{i,t}}{2} \tag{25}$$

4. Results and Discussion

The proposed optimization framework is tested on a modified IEEE 9 bus system (Figure 8). The average inputs values of renewable generation are 92 MW of PV array power at bus 6 and 69 MW of wind power located at bus 5. Furthermore, considering a constant impedance z corresponding to the varying frequency; $S = V^2/Z$ and $S = P + jQ$. Therefore, the load power is square of the system voltage at constant impedance. Based on this formulation the generator voltage is reduced by 1% in each step, by controlling the power across the load until the generation voltage ratio induces a system brownout and hence marks its critical voltage level. Based on the results obtained (Table 3), it was observed in the lower voltage bound that a system collapse is experienced when the reduction of voltage reached 2% of the nominal voltage of the corresponding bus (brownout voltage).

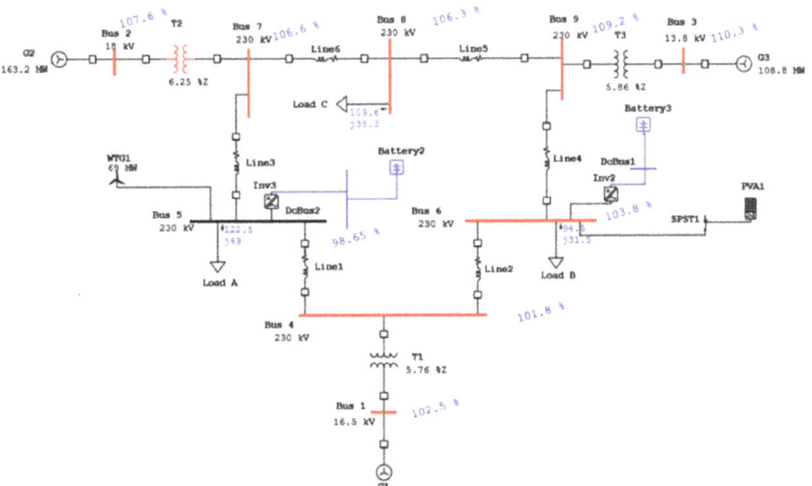

Figure 8. Circuit representation of the modified IEEE 9 bus microgrid under study.

Table 3. Characteristics of load power (MW and MVar) for each voltage ratio.

Voltage Ratio (kV)	Load A		Load B		Load C	
	MW	MVar	MW	MVar	MW	MVar
1	120.874	48.35	87.446	29.149	94.511	33.079
0.99	120.356	48.143	87.072	29.024	92.083	32.929
0.98	119.843	47.937	86.702	28.901	93.658	32.78
0.97	119.333	47.733	86.334	28.778	93.235	32.632
0.96	118.826	47.531	85.969	28.656	92.814	32.485
0.95	118.323	47.329	85.606	28.535	92.396	32.339
0.94	117.824	47.130	85.247	28.416	91.98	32.193
0.93	117.328	46.931	84.89	28.297	91.566	32.048
0.92	116.862	46.745	84.554	28.185	90.975	31.767
0.91	116.372	46.549	84.218	28.067	90.764	31.767
0.90	115.884	46.354	83.851	27.95	90.353	31.624

shape can be quickly determined from the right eigenvector components corresponding to state variables involved in the mode. Correspondingly, Table 9 shows the right eigenvector components for the weakly damped modes.

Table 8. Stability of G2 generation using phase-lag system.

Unstable Synchronous Mode	Unstable Synchronous and Load Sharing Mode	Stability with Phase-Lag System
-98.58	**$-100.09 + 415.54i$**	-129.92
112.39	**$-100.09 - 415.54i$**	-75.59
-111.12	-129.92	-20.41
-100	-75.59	-113.20
-25.97	-20.41	-97.67
$-0.03 + 0.98i$	$-5.55 + 4.81i$	$-5.55 + 4.81i$
$-0.03 - 0.98i$	$5.55 - 4.81i$	$-5.55 - 4.81i$
-26.04	-32.99	-25.896
-0.14×10^{-8}	-0.146	-2.5×10^{-14}
$3.28 + 2.15i$	-9.81×10^{-15}	-0.15
$3.28 - 2.15i$	-0.15	$-1.09 + 0.22i$
-	-	$1.09 - 0.22i$

In the case of mode 1, the sign of the real part of the right eigenvector component indicates that G1 swings against G2 yielding an inter-area oscillatory mode. For mode 2, the sign of the real part of right eigenvector components indicates that G1 and G2 swing coherently, yielding another inter-area oscillatory mode. Consequently, PSS is applied to G1 as it has the largest participation in mode 2 as indicated in Table 5. The resultant new system values after PSS insertion are shown in Table 10.

Table 9. Right eigenvector components associated with mode 1 and 2.

Mode	Machine Affected/Right Eigenvector	
1	G1/$0.0002 - 0.0069i$	G2/$-0.0006 - 0.0257i$
2	G1/$0.0004 - 0.0155i$	G2/$0.0005 - 0.0106i$

Table 10. IEEE 9 bus microgrid oscillation profile with integrated PSS in the microgrid.

Mode	Eigenvalues	Frequency (Hz)	Damping (%)
1	$-5.0135 \pm 17.8121i$	2.83	27.0938
2	$-0.4108 \pm 8.1018i$	1.28	5.0638
3	$-0.9165 \pm 1.9103i$	0.3	43.2572
4	$-1.0356 \pm 1.0097i$	0.16	71.6001
5	-52.8288	0	100
6	-9.0226	0	100
7	-0.2016	0	100

Therefore, the PSS adds some stability to the system by enhancing the poor damping modes observed in Table 4, i.e., for mode 1 from 2.51 to 27.09. Similarly, for mode 2 from 2.84 to 5.06. PSS added new modes to the system having no bad effect on the stability of the system. For comparison purposes, the system eigenvalues with and without the proposed PSS-SVC based controllers when applied individually and through coordinated design for two loading conditions (nominal and light) are determined in Table 11 and Table 12 respectively. The corresponding damping torque coefficient (K_d) versus the loading variations are shown in Figure 9. It can be observed that the damping characteristics of PSS outperforms SVC in terms at light loading conditions, whereas the coordinated PSS-SVC design facilitates better overall damping characteristics across the loading conditions highlighting comparatively better system stability. Conclusively, the microgrid damping is observably improved with effective coordination design with maximum estimated $K_d = 0.35$ (1/s) compared to 0.05 (1/s) for individual PSS (Figure 9). We observe that the microgrid plant does not help stabilize with individual SVC.

Table 11. System eigenvalues for nominal loading.

No Control	PSS	SVC	PSS &SVC
0.5255 ± 6.5919i	−4.88 ± 7.36i	−4.714 ± 6i	−7.61 ± 31.2i
−0.0795	0.550	0.6176	0.2369
10.694 ± 5.661i	−4.77 ± 7.5i	−4.72 ± 6.2i	−21.3625
5.6612i	7.51i	−20.223	−1.5361
-	−101.03	−2.5441	−1.4023
-	−0.400	−0.7052	−1.0797
-	−0.2	−0.2	−1.0065
-	-	-	−0.5401
-	-	-	−0.3733
-	-	-	−0.2002
-	-	-	−0.2000

Table 12. System eigenvalues for light loading.

No Control	PSS	SVC	PSS &SVC
0.0382 ± 0.3601i	−1.03 ± 6.58i	−0.66 ± 6.287i	−2.6 ± 2.8i
−0.006	0.15	0.1047	0.6801
−10.207 ± 6.385i	−8.96 + −7.08i	−9.37 ± 6.554i	−6.91 ± 16.7i
-	7.08i	6.5542i	−0.69 + −0.08i
-	−100.35	−20.08	−21.2253
-	−0.4	−1.3933	−0.2001
-	−0.2	−0.7989	−0.5226
-	-	−0.2	−0.3746
-	-	-	−0.2
-	-	-	-

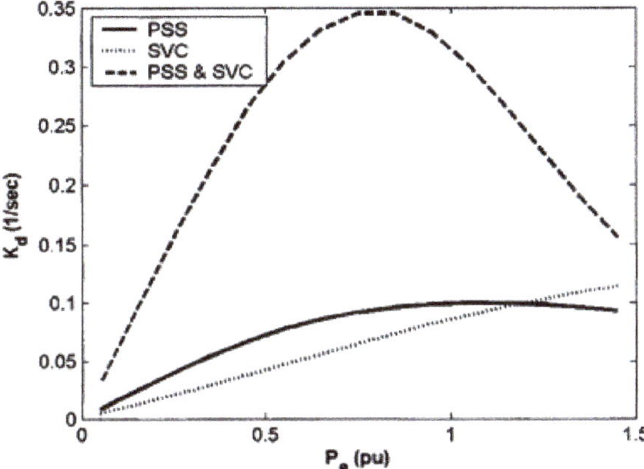

Figure 9. K_d with PSS-SVC based stabilizer.

4.2. Nonlinear Time-Domain Assessment for Coordinated PSS and SVC Design under System Fault Condition

To show the optimality and robustness of this coordinated design, the rotor angle (δ), speed deviation (ω), electrical power (P_e), and machine terminal voltage responses (V_t) are observed through an operational assessment carried out at the nominal and light loading condition specified in Table 1 under a 6-cycle three-phase fault induced in the system. As other generator parameters ($\Delta\omega, P, V_t$) are completely dependent on δ, the rotor

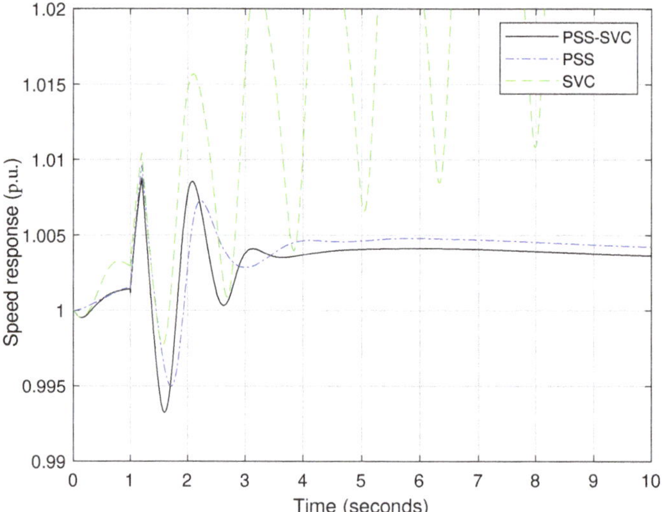

Figure 13. Speed response for 6-cycle fault with light loading.

In case of electrical power response, the individual SVC integrative support system response is unable to stabilize the system for both the nominal and light loading scenarios of the IEEE 9 bus system under 6-cycle fault disturbance. A better performance with the PSS-SVC electrical power response system is observed for setting time intervals that are at 4.1 s and 4 s for the nominal and the light loading conditions respectively (Figures 14 and 15). In comparison, the electrical power response of the individual PSS design respectively observes a settling time of 4.2 s and 4.8 s for both the loading scenarios respectively. However, the individual PSS outperforms in case of overshoot with the PSS-SVC incurring a 1.34 p.u. and 1.39 p.u. in comparison to the individual PSS with 1.28 p.u and 1.31 p.u. overshoot value respectively for the normal and light load system configurations.

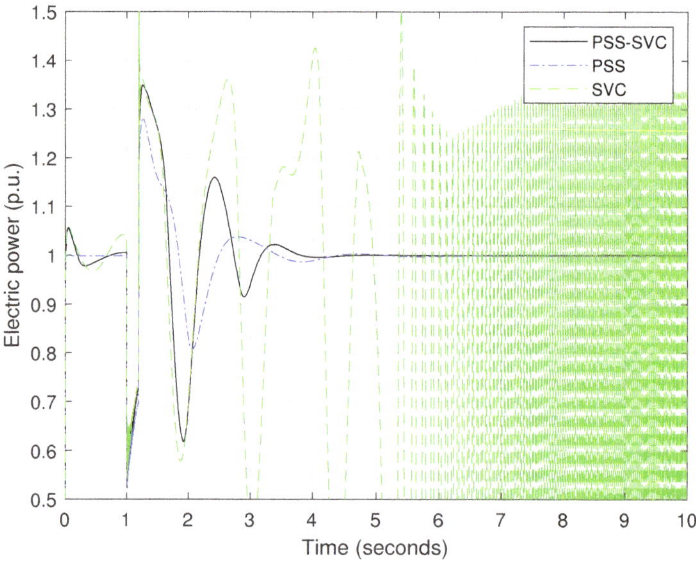

Figure 14. Electrical power response for 6-cycle fault with nominal loading.

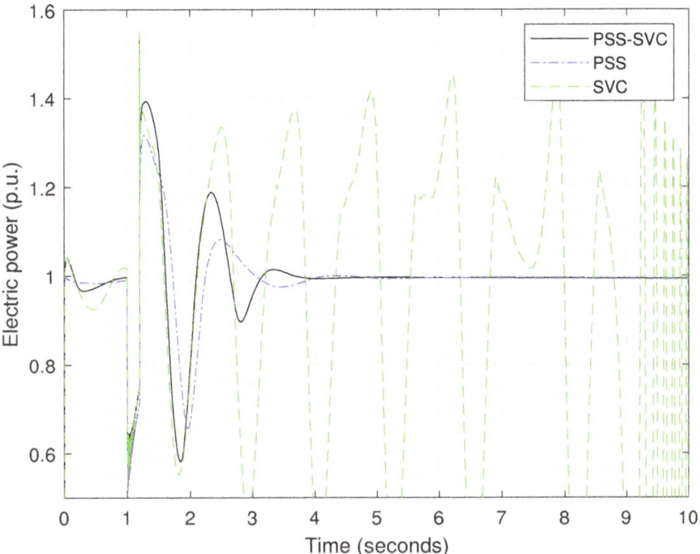

Figure 15. Electrical power response for 6-cycle fault with light loading.

Based on the results obtained pertaining to the terminal voltage as depicted in Figures 16 and 17, the individual PSS and PSS-SVC support response have an overshoot value of 1.22 and 1.04 p.u. during nominal loading, respectively. Accordingly, the terminal voltage response for both PSS-SVC and individual PSS are achieved at similar time interval of 4.2 s. Similarly, in case of light loading conditions of the IEEE 9 bus system, the PSS-SVC voltage response incurred an overshoot value of 1.07 p.u. as compared to the individual PSS terminal voltage response that reaches an overshoot of 1.21 p.u. with both having a settling time of 4.8 s. Furthermore, the individual SVC is observed to have the worst performance and is unable to clear the 6-cycle fault disturbance in both scenarios of the loading conditions.

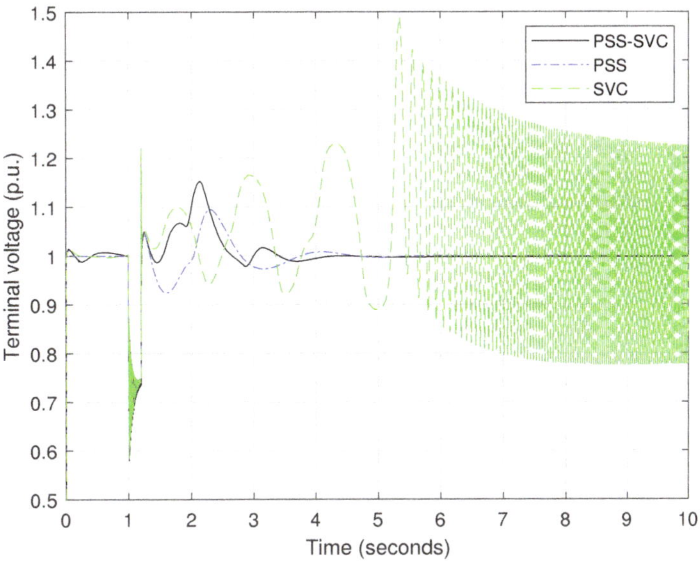

Figure 16. Terminal voltage response for 6-cycle fault with nominal loading.

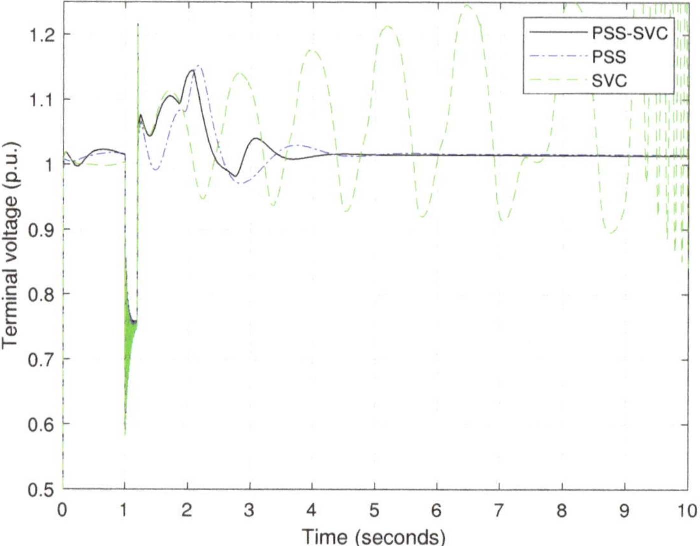

Figure 17. Terminal voltage response for 6-cycle fault with light loading.

Based on the calculated eigenvalues, the coordinated PSS-SVC is postulated to facilitate a better stability of the system with enhanced system responses. Furthermore, these designs are tested and validated on a renewable integrated IEEE 9 bus system with 6-cycle fault condition. During the fault occurrence, rotor angle, speed, electric power, and terminal voltage responses are evaluated for coordinated PSS-SVC, SVC, and PSS. While the coordinated PSS-SVC is observed to have comparatively overall better performance, PSS is observed to facilitate moderately better performance for speed response due to lower overshoot value and similar performance pertaining to the settling time of PSS-SVC for the terminal voltage response of the system.

5. Conclusions

This paper presents a coordinated PSS-SVC that was formulated to enhance the stability of hybrid energy system consisting of renewables and energy storage systems. The robustness of the proposed coordinated PSS-SVC design is verified under the most severe disturbance, wherein they facilitate appropriate damping characteristics to the network. The turn ratio of the distribution transformer was reduced to decrease the output voltage of the transformer. The voltage of the three generators was reduced by 1% in each step until blackout was reached and the brownout voltage was determined. The objective is to assess and compare the small signal stability of the IEEE 9 bus system, before and after the insertion of coordinated PSS-SVC design. The simulation results confirm the conclusion drawn for damping torque coefficient analysis that solves the problem of low effectiveness of the individual designs at light loading level. Furthermore, the systems were modeled and analyzed using the state-space method and these systems are two generators connected to a common load, each generator set with synchronous governing, and two generators are connected to a common load, with speed droop (using load sharing module). Therefore, based on an extensive comparative analysis performed with individual SVC and individual PSS, the proposed method improved the network quality in terms of eigenvalues, poles, and voltage profile.

Author Contributions: Conceptualization, L.E., K.K., and M.K.; Data curation, K.K. and M.A.-M.; Formal analysis, M.K.; Funding acquisition, M.A.-M. and M.K.; Investigation, L.E., K.K., and M.K.; Methodology, L.E. and M.A.-M.; Project administration, M.A.-M. and M.K.; Resources, M.A.-M. and M.K.; Software, L.E. and K.K.; Supervision, M.A.-M. and M.K.; Validation, L.E., K.K., M.A.-M., and M.K.; Visualization, L.E., M.A.-M. and M.K.; Writing—original draft preparation, L.E. and M.K.; Writing—review and editing, K.K. and M.K. All authors have read and agreed to the published version of the manuscript.

Funding: The authors would like to acknowledge the support provided by the Deanship of Research (DSR) at King Fahd University of Petroleum and Minerals (KFUPM) for funding this work through project No. DF191011. In addition, we would like to acknowledge the funding support provided by the King Abdullah City for Atomic and Renewable Energy (K.A. CARE).

Institutional Review Board Statement: Not applicable.

Informed Consent Statement: Not applicable.

Data Availability Statement: Not applicable.

Acknowledgments: The authors would like to acknowledge the support provided by the Deanship of Research (DSR) at King Fahd University of Petroleum and Minerals (KFUPM) for funding this work through project No. DF191011. In addition, we would like to acknowledge the funding support provided by the King Abdullah City for Atomic and Renewable Energy (K.A.CARE).

Conflicts of Interest: The authors declare no conflict of interest.

Abbreviations

The following abbreviations are used in this manuscript:

AI	Artificial Intelligence
ANN	Artificial neural network
AVR	Automatic voltage regulator
BF	Bacteria foraging
ESS	Energy storage system
FACTS	Flexible AC transmission system
GA	Genetic alogrithm
PSO	Particle swarm optimization
PSS	Power system stabilizer
RES	Renewable energy sources
SA	Simulated annealing
SVC	Static VAR compensator
WOA	Whale optimization algorithm
δ	Rotor angle
$\Delta\omega$	Deviation from the synchronous speed
V_t	Terminal voltage of generator
x	State variable vector
u, y	Input and output vector
g	Nonlinear function vector connected to u, y
A	State matrix
B	Input matrix
C	Output matrix
D	Feedforward matrix
a_g, b_g, c_g	Fuel cost coefficients of thermal unit g
OF	Total operating costs ($)
$P^g_{(i,t)}$	Active power generated by thermal unit g connected to bus i at time t (MW)

$VOLL$	Value of loss of load (\$/MW h)
$P_{(i,t)}^{LS}$	Active Load shedding in bus i at time t (MW)
VWC	Value of loss of wind (\$/MW h)
$P_{(i,t)}^{WC}$	Curtailed power of wind turbine connected to bus i at time t (MW)
ΔU_i	Transfer function of the PSS at bus i
T_w	Washout time constant
$T_{((1,2,3,4)i)}$	Compensation time constants
K_d	Damping torque coefficient

References

1. Lisin, E.; Shuvalova, D.; Volkova, I.; Strielkowski, W. Sustainable development of regional power systems and the consumption of electric energy. *Sustainability* **2018**, *10*, 1111. [CrossRef]
2. Mohammed, H. Power system transient stability enhancement by tuning of SSSC and PSS parameters using PSO technique. *J. Univ. Babylon* **2018**, *26*, 81–94. [CrossRef]
3. Gbadamosi, S.L.; Nwulu, N.I. Optimal power dispatch and reliability analysis of hybrid CHP-PV-wind systems in farming applications. *Sustainability* **2020**, *12*, 8199. [CrossRef]
4. Jethwa, U.K.; Bansal, R.K.; Date, N.; Vaishnav, R. Comprehensive load-shedding system. *IEEE Trans. Ind. Appl.* **2010**, *46*, 740–749. [CrossRef]
5. Mansori, M.; Zazi, M. Grid connected photovoltaic energy conversion system modeling and control using SMC-BSC approach. *Int. J. Renew. Energy Res. (IJRER)* **2019**, *9*, 1480–1491.
6. Nguyen, T.T.; Mohammadi, F. Optimal placement of TCSC for congestion management and power loss reduction using multi-objective genetic algorithm. *Sustainability* **2020**, *12*, 2813. [CrossRef]
7. Schäfer, B.; Yalcin, G.C. Dynamical modeling of cascading failures in the Turkish power grid. *Chaos Interdiscip. J. Nonlinear Sci.* **2019**, *29*, 093134. [CrossRef]
8. Dawood, F.; Shafiullah, G.; Anda, M. Stand-alone microgrid with 100% renewable energy: A case study with hybrid solar PV-battery-hydrogen. *Sustainability* **2020**, *12*, 2047. [CrossRef]
9. Khan, K.A.; Khalid, M. Hybrid energy storage system for voltage stability in a DC microgrid using a modified control strategy. In Proceedings of the IEEE Innovative Smart Grid Technologies-Asia (ISGT Asia), Chengdu, China, 21–24 May 2019; pp. 2760–2765.
10. Gao, K.; Yan, X.; Peng, R.; Xing, L. Economic design of a linear consecutively connected system considering cost and signal loss. *IEEE Trans. Syst. Man Cybern. Syst.* **2019**, *51*, 5116–5128. [CrossRef]
11. Nguyen, T.L.; Wang, Y.; Tran, Q.T.; Caire, R.; Xu, Y.; Gavriluta, C. A distributed hierarchical control framework in islanded microgrids and its agent-based design for cyber-physical implementations. *IEEE Trans. Ind. Electron.* **2020**, *68*, 9685–9695. [CrossRef]
12. Gao, K.; Wang, T.; Han, C.; Xie, J.; Ma, Y.; Peng, R. A review of optimization of microgrid operation. *Energies* **2021**, *14*, 2842. [CrossRef]
13. Khan, K.A.; Khalid, M. A reactive power compensation strategy in radial distribution network with high PV penetration. In Proceedings of the 8th International Conference on Renewable Energy Research and Applications (ICRERA), Brasov, Romania, 3–6 November 2019; pp. 434–438.
14. Kerdphol, T.; Rahman, F.S.; Mitani, Y.; Hongesombut, K.; Küfeoğlu, S. Virtual inertia control-based model predictive control for microgrid frequency stabilization considering high renewable energy integration. *Sustainability* **2017**, *9*, 773. [CrossRef]
15. Gandoman, F.H.; Ahmadi, A.; Sharaf, A.M.; Siano, P.; Pou, J.; Hredzak, B.; Agelidis, V.G. Review of FACTS technologies and applications for power quality in smart grids with renewable energy systems. *Renew. Sustain. Energy Rev.* **2018**, *82*, 502–514. [CrossRef]
16. Yari, S.; Khoshkhoo, H. A comprehensive assessment to propose an improved line stability index. *Int. Trans. Electr. Energy Syst.* **2019**, *29*, e2806. [CrossRef]
17. Alhumaid, Y.; Khan, K.; Alismail, F.; Khalid, M. Multi-Input nonlinear programming based deterministic optimization framework for evaluating microgrids with optimal renewable-storage energy mix. *Sustainability* **2021**, *13*, 5878. [CrossRef]
18. Tu, G.; Li, Y.; Xiang, J. Analysis, control and optimal placement of static synchronous compensator with/without battery energy storage. *Energies* **2019**, *12*, 4715. [CrossRef]
19. Movahedi, A.; Niasar, A.H.; Gharehpetian, G.B. LVRT improvement and transient stability enhancement of power systems based on renewable energy resources using the coordination of SSSC and PSSs controllers. *IET Renew. Power Gener.* **2019**, *13*, 1849–1860. [CrossRef]
20. Shah, R.; Mithulananthan, N.; Bansal, R. Damping performance analysis of battery energy storage system, ultracapacitor and shunt capacitor with large-scale photovoltaic plants. *Appl. Energy* **2012**, *96*, 235–244. [CrossRef]
21. Abo-Khalil, A.G. Grid connection control of DFIG in variable speed wind turbines under turbulent conditions. *Int. J. Renew. Energy Res.* **2019**, *9*, 1260–1271.
22. Wu, X.; Shen, X.; Cui, Q. Multi-objective flexible flow shop scheduling problem considering variable processing time due to renewable energy. *Sustainability* **2018**, *10*, 841. [CrossRef]

23. Ghazal, H.M.; Khan, K.A.; Alismail, F.; Khalid, M. Maximizing capacity credit in generation expansion planning for wind power generation and compressed air energy storage system. In Proceedings of the 2021 IEEE PES Innovative Smart Grid Technologies Europe (ISGT Europe), Espoo, Finland, 18–21 October 2021; pp. 1–5.
24. Singh, N.K.; Koley, C.; Gope, S.; Dawn, S.; Ustun, T.S. An economic risk analysis in wind and pumped hydro energy storage integrated power system using meta-heuristic algorithm. *Sustainability* **2021**, *13*, 13542. [CrossRef]
25. Al-Humaid, Y.M.; Khan, K.A.; Abdulgalil, M.A.; Khalid, M. Two-stage stochastic optimization of sodium-sulfur energy storage technology in hybrid renewable power systems. *IEEE Access* **2021**, *9*, 162962–162972. [CrossRef]
26. Luo, X.; Wang, J.; Dooner, M.; Clarke, J. Overview of current development in electrical energy storage technologies and the application potential in power system operation. *Appl. Energy* **2015**, *137*, 511–536. [CrossRef]
27. Akinyele, D.; Rayudu, R. Review of energy storage technologies for sustainable power networks. *Sustain. Energy Technol. Assess.* **2014**, *8*, 74–91. [CrossRef]
28. Khan, K.A.; Khalid, M. Improving the transient response of hybrid energy storage system for voltage stability in DC microgrids using an autonomous control strategy. *IEEE Access* **2021**, *9*, 10460–10472. [CrossRef]
29. Huang, H.; Li, F. Sensitivity analysis of load-damping characteristic in power system frequency regulation. *IEEE Trans. Power Syst.* **2012**, *28*, 1324–1335. [CrossRef]
30. Angalaeswari, S.; Sanjeevikumar, P.; Jamuna, K.; Leonowicz, Z. Hybrid PIPSO-SQP algorithm for real power loss minimization in radial distribution systems with optimal placement of distributed generation. *Sustainability* **2020**, *12*, 5787. [CrossRef]
31. Koohi-Fayegh, S.; Rosen, M. A review of energy storage types, applications and recent developments. *J. Energy Storage* **2020**, *27*, 101047. [CrossRef]
32. Machowski, J.; Bialek, J.; Bumby, J.R.; Bumby, J. *Power System Dynamics and Stability*; John Wiley & Sons: Hoboken, NJ, USA, 1997.
33. Kamalasadan, S.; Swann, G.D.; Yousefian, R. A novel system-centric intelligent adaptive control architecture for power system stabilizer based on adaptive neural networks. *IEEE Syst. J.* **2013**, *8*, 1074–1085. [CrossRef]
34. Georgilakis, P.S.; Hatziargyriou, N.D. Unified power flow controllers in smart power systems: Models, methods, and future research. *IET Smart Grid* **2019**, *2*, 2–10. [CrossRef]
35. Marinescu, B. Residue phase optimization for power oscillations damping control revisited. *Electr. Power Syst. Res.* **2019**, *168*, 200–209. [CrossRef]
36. Abido, M. Simulated annealing based approach to PSS and FACTS based stabilizer tuning. *Int. J. Electr. Power Energy Syst.* **2000**, *22*, 247–258. [CrossRef]
37. Majeed, A.; Ahmad, B.; Singh, Y. Power system stability, efficiency and controllability improvement using facts devices. *Int. J. Eng. Sci.* **2019**, *9*, 19590–19596.
38. Panwar, A.; Sharma, G.; Bansal, R.C. Optimal AGC design for a hybrid power system using hybrid bacteria foraging optimization algorithm. *Electr. Power Components Syst.* **2019**, *47*, 955–965. [CrossRef]
39. Abido, M. Optimal design of power-system stabilizers using particle swarm optimization. *IEEE Trans. Energy Convers.* **2002**, *17*, 406–413. [CrossRef]
40. Dasu, B.; Sivakumar, M.; Srinivasarao, R. Interconnected multi-machine power system stabilizer design using whale optimization algorithm. *Prot. Control Mod. Power Syst.* **2019**, *4*, 2. [CrossRef]
41. Elkhidir, L.; Hassan, A.; Khalid, M. SVC-based controller design via ant colony optimization algorithm. In Proceedings of the 8th International Conference on Renewable Energy Research and Applications (ICRERA), Brasov, Romania, 3–6 November 2019; pp. 301–308.
42. Grigsby, L.L. *Power System Stability and Control*; CRC Press: Boca Raton, FL, USA, 2012; Volume 5.
43. Wang, J.; He, Y. Study on model of interruptible load to participate in reserve market. In Proceedings of the 2009 Asia-Pacific Power and Energy Engineering Conference, Wuhan, China, 27–31 March 2009; pp. 1–4.
44. Mohamed, F.A.; Koivo, H.N. Online management genetic algorithms of microgrid for residential application. *Energy Convers. Manag.* **2012**, *64*, 562–568. [CrossRef]
45. Salman, U.; Khan, K.; Alismail, F.; Khalid, M. Techno-economic assessment and operational planning of wind-battery distributed renewable generation system. *Sustainability* **2021**, *13*, 6776. [CrossRef]
46. Soroudi, A. *Power System Optimization Modeling in GAMS*; Springer: Cham, Switzerland, 2017; Volume 78.
47. Nayak, N.; Pani, A. A short term forecasting of PV power generation using couple based particle swarm optimization pruned extreme learning machine. *Int. J. Renew. Energy Res.* **2019**, *9*. [CrossRef]
48. Ogata, K. *Discrete-Time Control Systems*; Prentice-Hall, Inc.: Hoboken, NJ, USA, 1995.

 *sustainability*

Article

Analysis of a Hybrid Wind/Photovoltaic Energy System Controlled by Brain Emotional Learning-Based Intelligent Controller

Hani Albalawi [1,2], **Mohamed E. El-Shimy** [1], **Hosam AbdelMeguid** [3], **Ahmed M. Kassem** [4] **and Sherif A. Zaid** [1,2,*]

[1] Electrical Engineering Department, Faculty of Engineering, University of Tabuk, Tabuk 47913, Saudi Arabia; halbala@ut.edu.sa (H.A.); moahmad@ut.edu.sa (M.E.E.-S.)

[2] Renewable Energy & Energy Efficiency Centre (REEEC), University of Tabuk, Tabuk 47913, Saudi Arabia

[3] Mechanical Engineering Department, Faculty of Engineering, University of Tabuk, Tabuk 47913, Saudi Arabia; hsaad73@ut.edu.sa

[4] Electrical Engineering Department, Faculty of Engineering, Sohag University, Sohag 82524, Egypt; kassem_ahmed53@hotmail.com

* Correspondence: shfaraj@ut.edu.sa

Abstract: Recently, hybrid wind/PV microgrids have gained great attention all over the world. It has the merits of being environmentally friendly, reliable, sustainable, and efficient compared to its counterparts. Though there has been great development in this issue, the control and energy management of these systems still face challenges. The source of those challenges is the intermittent nature of both wind and PV energy. On the other hand, a new intelligent control technique called Brain Emotional Learning-Based Intelligent Controller (BELBIC) has garnered more interest. This paper proposes the control and energy management of hybrid wind/PV microgrids using a BELBIC controller. To design the system, simple power and energy analyses were proposed. The proposed microgrid was modeled and simulated using MATLAB. The responses of the energy system were tested under two different types of disturbances, namely step and ramp disturbances. These disturbances are applied to the wind speed, the irradiation level of the PV, and the load power. The results indicate that the AC load voltage and frequency are steady with negligible transients against the previous disturbance. In addition, the performance is better than that of the classical PI controller. Also, energy management acts perfectly to compensate for the intermittence and disturbances of the wind and PV energies. On the other hand, the system robustness against model parameters uncertainties in the microgrid parameters are studied.

Keywords: BELBIC; photovoltaic; wind energy; maximum power point tracking

Citation: Albalawi, H.; El-Shimy, M.E.; AbdelMeguid, H.; Kassem, A.M.; Zaid, S.A. Analysis of a Hybrid Wind/Photovoltaic Energy System Controlled by Brain Emotional Learning-Based Intelligent Controller. *Sustainability* **2022**, *14*, 4775. https://doi.org/10.3390/su14084775

Academic Editor: Domenico Curto

Received: 17 March 2022
Accepted: 13 April 2022
Published: 15 April 2022

Publisher's Note: MDPI stays neutral with regard to jurisdictional claims in published maps and institutional affiliations.

1. Introduction

The beginning of this century was accompanied by worldwide industrial development and a growing population. These issues increased the world's electricity demand. However, the traditional sources of electricity are not sufficient and have many environmental problems [1]. Hence, renewable electricity resources (wind, solar, tidal, etc.) have gained great consideration. Renewable electricity resources have many environmental benefits. Nevertheless, it has a common disadvantage, namely intermittency [2]. Energy intermittency may not be a big problem when the system is connected to a large utility grid. However, isolated systems and small microgrids will suffer from this problem. One way to solve the intermittency problem is the integration of two or more renewable resources by introducing hybrid energy systems [3]. A common microgrid of such a type is the wind/PV microgrid [4].

Though the wind and PV energy resources are not steady, they may integrate to reduce the intermittency problem. Solar energy is available during the daytime. The availability of wind energy is not restricted to a certain time of the day. However, in some circumstances, the wind energy at night is greater than during day. Therefore, there is some form of integration between the two energy resources. Hence, for standalone applications, hybrid wind/PV systems are considered reliable and feasible alternatives to battery-coupled solar and wind-diesel systems [5].

Several research papers have been proposed for hybrid wind/PV systems [6–12]. Ref. [6] proposed a wind/PV system utilizing MPPT and fuzzy algorithms. The system's goal is to reduce storage requirements while also regulating load power. However, the controller was complex and expensive. Ref. [7] suggested a hybrid wind/PV system supplying an unbalanced load. The system has no storage, a simple controller, and extensive field tests. In ref. [8], a new hybrid wind/PV energy system was investigated for agricultural systems. Three different management algorithms were tested on the proposed power system. The results show that system efficiency was best in making the battery charging process have a priority over the system loads. A PWM rectifier is proposed by [9] to replace the boost converter of the conventional wind/PV energy system. Also, a composite sliding mode controller for load inverter was implemented for rural electrification applications. An implementation of a wind/PV microgrid operated in dual AC and DC modes was introduced by [10,11]. The control system adapted to the power exchange between AC and DC microgrids. The microgrid supplies dynamic and domestic loads. Ref. [12] has proposed a wind/PV microgrid with a distributed DC bus. The control system has implemented the MPPT of the PV array and the wind turbine. Step changes in the nonlinear load are applied to test the system performance.

Regarding wind/PV control systems, many control algorithms have been introduced in the literature [13–15]. Nevertheless, the intelligent control of nonlinear systems has gained great attention in the past decade [16]. Hence, widespread controllers of such types have been investigated, such as neural networks, and fuzzy and neuro-fuzzy controllers [17]. They have many merits, such as parameter linearization, good learning capabilities, built-in universal approximation, and model-free operation [18,19]. Therefore, it has too many applications in robust control, nonlinear control, adaptive control, robotics, and decision making [20–23].

A new controller called the brain emotional learning-based intelligent controller (BELBIC) was recently proposed [24,25]. The idea of this controller was derived from the computational model of the limbic system in the human brain [26]. It has various applications in space vehicles, electric power systems, and automotive systems [27–29]. The main advantages of the BELBIC controller are its good robustness, simplicity, effectiveness, and flexibility in selecting the emotional cues and sensory inputs for a certain application.

There are some recently published works in the proposed subject. Ref. [30] provides an intelligent energy management controller. It utilized a hybrid of fuzzy logic and fractional-order PID techniques. The proposed controller ensured continuous output power for both DC and AC loads. However, the harmonics of the load voltage and current are thought to be high. Also, the proposed microgrid has not been tested against ramp disturbances. Ref. [31] introduced a DC microgrid supplied by a hybrid wind-PV battery system. It used the classical PID controller and utilized the SEPIC converter. However, the system has a fair time response, and the load voltage has a steady-state error. Ref. [32] proposed a hybrid wind/PV energy system with an optimal MPPT controller. The controller provided energy management and tracked the peak power. Though the system was simple, its optimality was not ensured. As compared to the previous work, the proposed microgrid introduces a recently developed BELBIC controller to improve the energy management and time response of the wind/PV standalone microgrid under different disturbances in the insolation and wind speed as well as the load power. The disturbance types include the step and ramp form. Also, the load power quality is measured and compared to the standard values. It is thought that this is the first time the BELBIC controller was applied to the wind/PV standalone microgrid. The novelty items of this work include the energy analysis of the wind/PV microgrid, design of the system controllers (especially the

BELBIC controller), and simulation of the system response under step/ramp disturbances in the system load, wind speed, and solar radiation. The microgrid stability against model parameters uncertainties and variations in the microgrid parameters are also studied.

In this paper, a new simple analysis and design of a hybrid wind/PV energy system are proposed. A new simple analysis is investigated, generating closed-form design relations that are derived for the design purpose. The controller of the proposed system was designed based on the BELBIC control algorithm. The analysis and the design are verified by modeling and simulations. The introduced system contains a wind turbine, a PMSG, a rectifier, a PV array, two boost converters, a two-quadrant DC/DC converter, and an Energy Storage System (ESS). The introduced system and controllers were simulated using the MATLAB/Simulink platform. The research aims are:

1. Investigating simple energy and power analysis of the system. Hence, power and energy closed-form relations are derived. Also, equations for the size of the ESS are generated.
2. Designing the wind/PV microgrid for the BELBIC controller and other controllers.
3. Simulating and implementing the proposed system in the MATLAB platform. Then, the system performance is tested under step and ramp changes in the system load, wind speed, and solar radiation. Moreover, the system stability against model parameter uncertainties and variations in the microgrid parameters are discussed.

The paper structure is as follows: Section 2 explains the introduced system structure. Section 3 gives the analysis of the introduced wind/PV microgrid. Section 4 presents the power system design. The design of the controllers and BELBIC are presented in Section 5. Section 6 discusses the simulation results. The conclusions are presented in Section 7.

2. Explanation of the Proposed Microgrid

The proposed wind/PV standalone microgrid is presented in Figure 1. It has two renewable energy sources: wind and solar PV. Solar energy is available during the daytime. The availability of wind energy is not restricted to a certain time of the day. However, in some circumstances, the wind energy at night is greater than during day. Therefore, there is some form of integration between the two energy resources. Nevertheless, they do not generate steady energy due to the variations in the environmental state and solar irradiation. These issues give the wind and solar energies their intermittence nature. Hence, the utilization of the two resources increases the reliability and sustainability of the microgrid. Moreover, the size of the ESS system will be reduced.

The wind system includes the wind turbine coupled mechanically to a 3-φ Permanent Magnet Synchronous Generator (PMSG). The PMSG output is rectified through an uncontrolled rectifier, generating an unregulated DC voltage. This voltage is supplied to a boost converter. The function of the boost converter is to force the wind turbine towards the MPPT conditions. The output of the boost converter is attached to the DC bus of the microgrid.

The solar energy system consists of a PV array formed of three parallel stings. Each string includes modules. The PV output is supplied to another boost converter. Also, the boost converter is used to implement the MPPT conditions of the PV.

Due to the intermittent nature of the generated energy, ESS is usually utilized to compensate for the energy intermittency problem. The ESS consists of a group of lead-acid batteries connected in series and parallel to construct the required energy. These batteries are connected to the DC bus via a bidirectional converter. Generally, that converter is a DC/DC converter. Its function is to regulate the charge/discharge process of the ESS. Also, that converter represents the main adjustment actuator for the DC bus voltage and the microgrid energy balance.

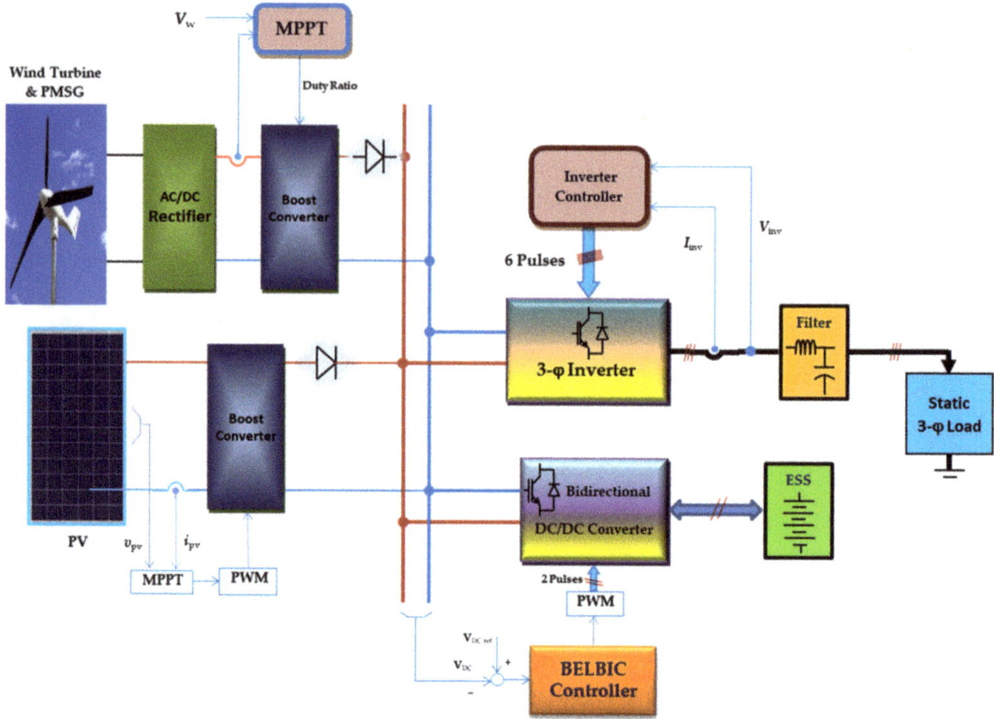

Figure 1. The proposed standalone Wind/PV microgrid.

3. Power Analysis of the Proposed Wind/PV Microgrid

The design of the system relies mainly on the power and energy relations of the system. Hence, deriving these relations will aid the design procedure. In this regard, it is assumed that the initial state of energy of the ESS (E_i), the load power of the microgrid (P_L), the swept area of the blades (A), the air density (ρ), and the average wind speed ($\overline{v}$) are given. The first step is the derivation of the average wind and solar power.

3.1. Average Wind Power

To get the annual average wind power (P_w) over a certain site:

$$P_w = \int_0^\infty p(v).f(v)dv \tag{1}$$

where ($p(v)$) is the wind power at the wind speed (v), and ($f(v)$) is the probability density function. Rayleigh is a common probability density function utilized for implementing the actual wind speed statistics; it is defined as [33]:

$$f(v) = \frac{\pi v}{2\overline{v}}e^{[-0.25\pi(v/\overline{v})^2]} \tag{2}$$

The wind power as a function of the wind speed is given by:

$$p(v) = 0.5\rho Av^3 \tag{3}$$

Substituting (2) and (3) in (1), and completing the integration, the formula becomes:

$$P_w = \frac{3}{\pi}\rho A\overline{v}^3 \tag{4}$$

The average wind speed can be determined by gathering site data for a long time. If the value of ($\overline{v}$) is determined, the average wind power is also determined.

3.2. Average Solar Power

Assume that the instantaneous PV power ($p_{pv}(t)$) of the array, as shown in Figure 2a, is given by:

$$p_{pv}(t) = P_m\left(1 - t^2/36\right) \tag{5}$$

where (P_m) is the maximum PV power and (t) is the time in hours. The solar energy is provided to start at 6:00 AM and has a duration of 12 h.

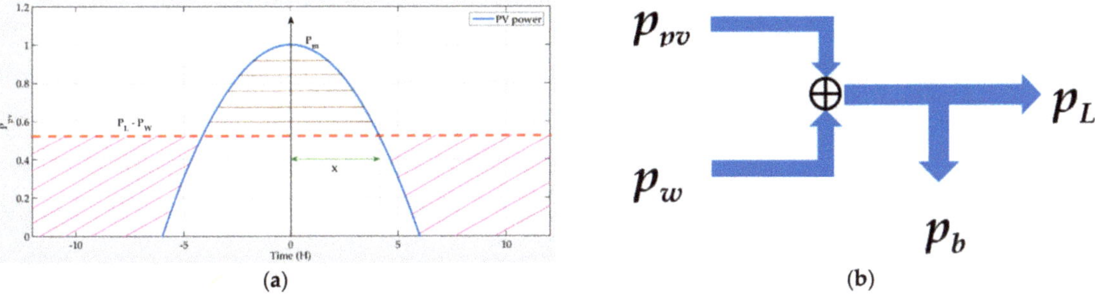

Figure 2. (**a**) The daily PV array power and (**b**) the system power flow diagram.

The average PV power may be calculated as:

$$P_{pv} = \frac{1}{24}\int_{-6}^{6} P_m\left(1 - t^2/36\right)dt = \frac{1}{3}P_m \tag{6}$$

The daily peak power (P_m) is determined from the statistics of the solar insolation at the specified site of the microgrid and averaged over the year.

The microgrid power flow diagram, presented in Figure 2b, generates the following instantaneous equation:

$$p_{pv}(t) + p_w(t) = p_L(t) + p_b(t) \tag{7}$$

where ($p_b(t)$) is the instantaneous ESS power. Take the daily average of Equation (7), which leads to:

$$P_{pv} + P_w = P_L + E_i/24 \tag{8}$$

As the average power of the ESS is supposed to be constant at ($E_i/24$).

3.3. Energy Analysis of the ESS

In this section, the instantaneous stored energy (E_b) can be determined by:

$$\int dE_b = \int p_b(t)dt \tag{9}$$

From (7), the instantaneous stored power is:

$$p_b(t) = \begin{cases} P_m\left(1 - t^2/36\right) + p_w(t) - p_L(t) & -6 \leq t \leq 6 \\ p_w(t) - p_L(t) & 6 \leq t \leq 18 \end{cases} \tag{10}$$

Assuming that the wind and the load power are constants at their average values:

$$p_w(t) - p_L(t) \cong P_w - P_L \tag{11}$$

Substituting (6), (8), and (10) into (9) and manipulating the integration:

$$E_b(t) = \begin{cases} \frac{1}{3}P_m\left(6 + 2t - \frac{t^3}{36}\right) + E_i\left(1.25 + \frac{t}{24}\right) - 6 \leq t \leq 6 \\ (E_i - 8P_m)t\frac{1}{24} + 6P_m + 0.25E_i\ 6 \leq t \leq 18 \end{cases} \tag{12}$$

Using traditional calculus, the maximum value of the stored energy takes place at:

$$t_{max} = \sqrt{24 + 1.5\frac{E_i}{P_m}} \tag{13}$$

Hence, the maximum stored energy is given by:

$$E_b|_{max} = E_b(t_{max}) \tag{14}$$

The rated energy storage can be determined using (14). From this analysis, if the required load power P_L is given, then Equations (6), (8), and (14) can help to determine the power rating of the system components. Based on the previous analysis, the PV power and the ESS size can be determined. Assume that the load power demand and the wind turbine power are given. Assume a suitable value for (E_i). Usually, the SOC of the ESS is from 20% to 95% [34]. Hence, from (6), (8), and (14), the PV power and the ESS energy will be determined.

4. The Control System Design

The proposed system controllers, shown in Figure 3, are the wind-PV MPPT, the ESS and DC link voltage controller, and the load inverter controller. The functions of the MPPT controllers are to extract the peak power from the wind turbine and PV array. They generate the required duty cycle signal to the boost converter, which in turn loads the wind turbine and the PV array with the MPPT load conditions. However, the ESS and DC link voltage controller regulate the DC link voltage and the charge/discharge process of the ESS. The third controller is used to regulate the load inverter voltage and frequency. The control design of them will be discussed in the following subsections.

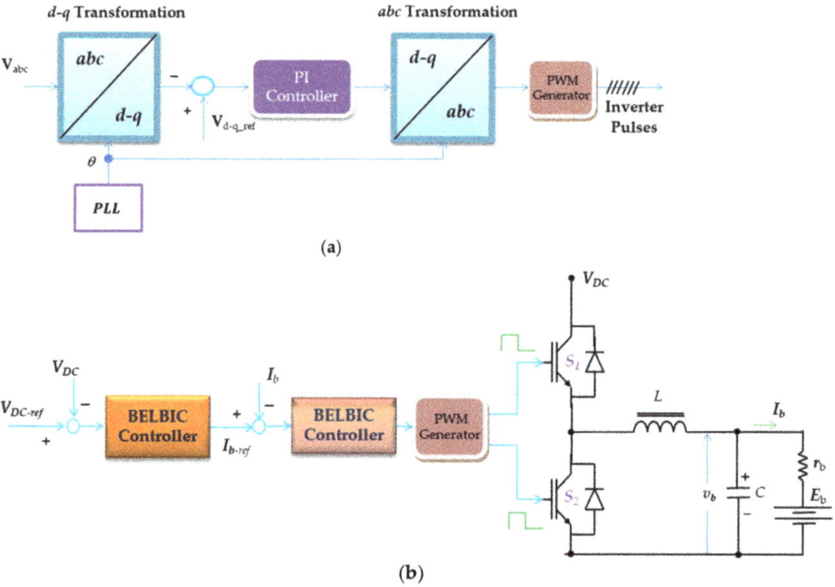

Figure 3. (**a**) The load inverter controller and (**b**) the storage and DC-link voltage controller.

4.1. The Wind and PV Array MPPT Controllers

These controllers are important for better utilization of wind and PV energy. In the last few years, many MPPT approaches for wind/PV hybrid systems have been introduced [35]. A common and frequently utilized approach is called "Perturb and Observe" (P&O). It has the merits of simple implementation and a straightforward algorithm. Two boost converters are used for this issue, one for the wind and another for the PV array. The output of each MPPT controller is the value of the duty cycle switch of the boost converter. The P&O algorithms for the two energy sources are identical. A flowchart for the P&O algorithm is presented in [34].

4.2. Load Inverter Controller

The objective of this controller is to supply the load with AC power at a regulated voltage and frequency. The control loop is shown in Figure 3a. The load 3-φ voltages are measured and transferred to the d-q frame with the help of Phase Locked Loop (PLL). The transferred d-q voltages are compared to their reference values. Then the resulting error is fed to a simple PI controller. Hence, the PI controller gains are tuned using the Ziegler–Nichols algorithm.

4.3. Storage and DC-Link Voltage Controller

Mainly, this controller aims to regulate the DC-link voltage that can be achieved by controlling the charging of the ESS. It includes two nested loops, as shown in Figure 3b. The outer loop adapts the DC-link voltage with the help of the BELBIC controller. However, the inner loop controls the ESS charging current with the help of another BELBIC controller. The output of the outer loop is the reference charging current of the inner loop. When the ESS is fully charged, the controller ends the charging process and the MPPT controllers stop.

5. The BELBIC Controller Design

The BELBIC controller emulates the process applied by the brain to process emotions. Its computational network includes the orbitofrontal cortex, amygdala, thalamus, and sensory input cortex [29]. The schematic diagram of the BELBIC controller is shown in Figure 4. The sensory input signals are handled and partially processed by the thalamus section. The output of the thalamus is the input of the sensory cortex. It helps in subclassing and favoritism of the thalamus output. The function of the orbitofrontal cortex is to prevent unstable performances from the amygdala. The amygdala section helps in following up the stimulus motion. Subtracting the amygdala and orbitofrontal cortex output signals produces the BELBIC controller output. Each sensory cortex output (S) has one node (A) in the amygdala. Another node is set to the thalamus output. Except for the thalamic node, each stimulus has one node (O). The outputs of the amygdala and orbitofrontal cortex are subtracted to form a common output node (MO).

Hence, the net output node MO is given by:

$$MO = \sum_k A_k - \sum_k O_k \tag{15}$$

The orbitofrontal part does not ban the thalamic signal. On the other hand, the other amygdala inputs are banned. Emotional learning, within the amygdala and the orbitofrontal cortex, is defined as:

$$\begin{bmatrix} \Delta G_{A_{ki}} \\ \Delta G_{O_k} \end{bmatrix} = \begin{bmatrix} \alpha S_{ki} \max(0,\ REW - \sum_{ki} A_{ki}) \\ \beta S_k R_o \end{bmatrix} \tag{16}$$

where

$$R_o = \begin{cases} \max(0,\ \sum_k A_k - REW) - \sum_k O_k & \forall REW \neq 0 \\ \max(0,\ \sum_k A_k - \sum_k O_k) & \forall REW = 0 \end{cases} \tag{17}$$

The two learning rules of (16) are similar. The node values are represented by:

$$\begin{bmatrix} A_k \\ O_k \end{bmatrix} = \begin{bmatrix} G_{A_k} S_k \\ G_{O_k} S_k \end{bmatrix} \tag{18}$$

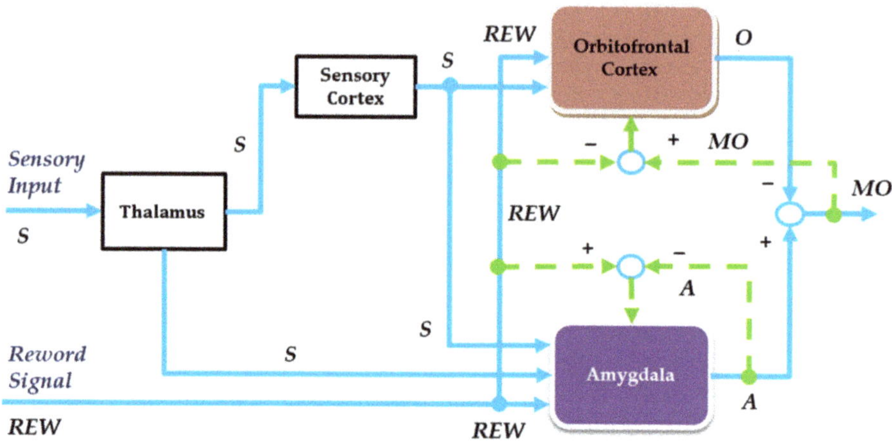

Figure 4. Scheme of the BELBIC structure.

The BELBIC controller operates in two ways. The first way is to learn the amygdaloid, then let it predict and respond to a certain REW. The second way is to direct the orbitofrontal to track diversions between REW and the system's predictions. Then it learned to ban the output corresponding to the diversions.

The *REW* signal is implemented based on the cost function used:

$$REW = J\left(e, y_p, S_k\right) \tag{19}$$

Also, the sensory inputs are functions of the system outputs:

$$S_k = f\left(y_p, u, \ r, e\right) \tag{20}$$

where (u) is the controller output, (r) is the reference input, (y_p) is the plant output, and (e) is an error signal.

The amygdala and the orbitofrontal have the continuous updating weights given by:

$$\begin{bmatrix} \frac{dG_{A_k}}{dt} \\ \frac{dG_{O_k}}{dt} \end{bmatrix} = \begin{bmatrix} \alpha S_k(REW - A_k) \\ \beta S_k(A_k - REW - O_k) \end{bmatrix} \tag{21}$$

6. The Simulation Results

The introduced hybrid wind/PV microgrid is simulated by the MATLAB/Simulink platform. The introduced parameters of the microgrid are presented in Table 1.

The proposed wind/PV microgrid is simulated using the MATLAB/Simulink platform. The simulation results of the proposed microgrid with the BILBIC controller, according to step changes in the solar insolation, the wind speed, and load power, are shown in Figure 5. Figure 5a shows the solar insolation level variations. It has 100% insolation during the first second and drops to zero during the remaining time. The wind speed of the wind turbine is presented in Figure 5b. It has step changes at the times 0.3 s, 0.6 s, and 1 s, respectively. The wind turbine response is shown in Figure 5c, where the torque is directly proportional to the wind speed. Figure 5d shows the state of charge of the ESS. The ESS is continuously charging during the first second, then discharges. As the PV and wind power are available and sufficient until 1 s, the SOC increases. However, after (1 s) the energy is not sufficient to supply the load. Hence, the ESS discharges to compensate for the energy drop. The PMSG speed is presented in Figure 5e. It is proportional to the wind or the turbine speed, except for some transients related to the turbine inertia. Figure 5f shows the ESS charging and discharging currents. For the period from 0 to 0.3 s, the charging current is 35 A, which is relatively high, as the PV energy is full and the wind energy corresponds to a 12 m/s wind speed. For the period (0.3 to 0.6 s), the charging current is 25 A, moderate as the PV energy is full and the wind energy drops. For the period (0.6 s to 0.8 s), the charging current is 50 A high as the PV energy is full and the wind energy is full, corresponding to a 14 m/s wind speed. For the period (0.8 to 1 s), the charging current drops as the load is increased. For the period from 0.8 to 1 s, both wind and PV energy are inhibited. Hence, the ESS will compensate for them during this period. The load voltage and current are shown in Figure 5g,h. They are sinusoidal with a stable frequency, and the voltage has a constant amplitude despite all the disturbances.

Table 1. Proposed Microgrid Parameters.

Item	Parameter	Value
Wind turbine	Rated power	10 KW
	Rated wind speed	12 m/s
	wind speed range	3.5–25 m/s
PV	SC current	21.2 A
	OC voltage	257.1 V
	Max. power	5.4 kW
Load	Voltage	110 V
	Frequency	50 Hz

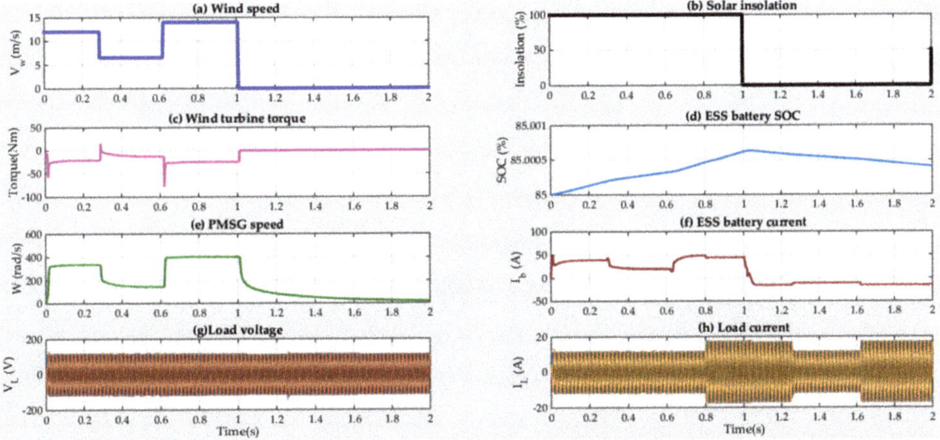

Figure 5. Simulation results of the proposed microgrid with the BILBIC controller (**a**) wind speed, (**b**) PV irradiation level, (**c**) wind turbine torque, (**d**) ESS battery SOC, (**e**) PMSG speed, (**f**) ESS battery current, (**g**) load voltage, and (**h**) load current.

Figure 6 compares the DC bus voltage responses for the BELBIC and PI controllers for the same microgrid. It tracks well with the reference voltage (300 V) for both controllers. However, the response of the BELBIC is excellent. It has no overshoot and smaller settling times.

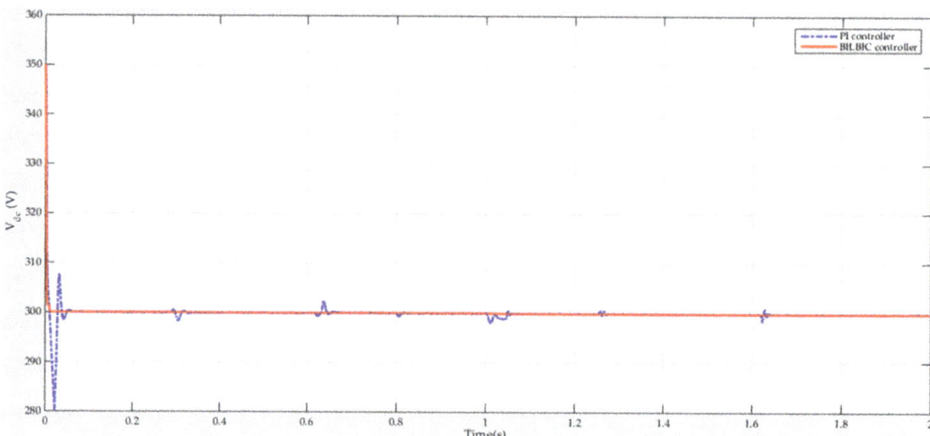

Figure 6. The DC link voltage response of the classical PI and BELBIC controllers.

To ensure robust stability against model parameter uncertainties, variations in the microgrid parameters are altered. Where the temperature of the PV is increased by 10%, the PV series resistance is increased by 10%, and the boost inductor of the wind MPPT is decreased by 10%. Figure 7 shows the proposed microgrid response with the BILBIC controller according to the previous step variations and under parameters uncertainties. It is indicated in the figure that the proposed controller can stabilize the load voltage and frequency with high accuracy, despite the modeling errors.

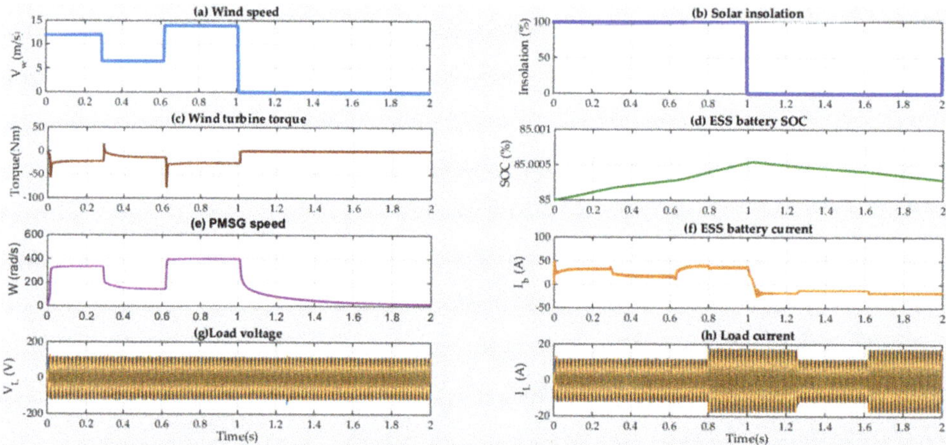

Figure 7. Simulation results of the proposed microgrid with the BILBIC controller under parameters uncertainty (**a**) wind speed, (**b**) PV irradiation level, (**c**) wind turbine torque, (**d**) ESS battery SOC, (**e**) PMSG speed, (**f**) ESS battery current, (**g**) load voltage, and (**h**) load current.

Figure 8 shows the spectrum analysis of the load current with the BELBIC and PI controllers. The load current THD in the case of the BELBIC controller is 2.22%. However, it is 3.68% in the case of the PI controller. The load current THD of both cases is lower than the standards specified in [36]. Hence, the load current quality is better in the case of the BELBIC controller than the PI controller.

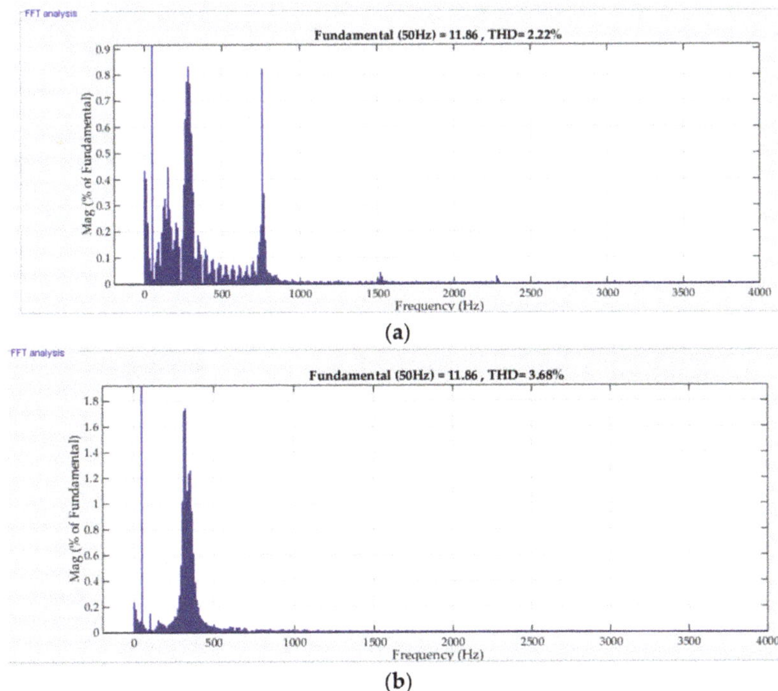

Figure 8. Spectrum analysis of the load current with the (**a**) BELBIC controller and (**b**) PI controller.

The simulation results of the proposed microgrid with the BILBIC controller according to ramp variations in the solar insolation and the wind speed are shown in Figure 9. Also, step load changes at 0.8 s, 1.25 s, and 1.63 s are presented. The wind speed of the wind turbine has the ramp changes indicated in Figure 9a. Figure 9b shows the solar insolation level ramp variations. It has a ramp increase of the insolation during the first 0.3 s, however, the wind speed has a constant value (12 m/s) during this time. The ESS charging current is increasing during this period, shown in Figure 9c, as the wind energy increases. During the period (0.3 s < t < 0.8 s), the PV energy decays, and the wind energy increases. As the energy rate of change is different, there is a drop in the generated power and the charging current. During the period 0.8 s < t < 1.25 s, the PV energy is at 100% insolation, the wind energy is very low, and load power is increased by 50%. The charging current drops during this period. During the remaining period, the solar energy decreases and the wind energy increases. However, the net generation is not sufficient to supply the load. Hence, the ESS discharges to compensate for the energy drop and the charging current is negative.

Figure 9d shows the state of charge of the ESS. The ESS is tracking the charging current. It is the integration of the charging current. Hence, when the charging current is positive, the SOC increases and vice versa. Also, the load voltage and current, shown in Figure 9e,f, have sinusoidal waveforms with stable frequency during all the disturbances.

Table 2 shows a comparative analysis of the extracted results with that of ref. [30]. It can be noticed that the proposed system has the best performance over the others. The disturbance function used in [30] was a simple one-step change in the wind speed. However, complex multi-step disturbances in the wind speed and solar insolation are applied to the proposed system. Also, the parameter uncertainties were not studied in [30].

21. Miranian, A.; Abdollahzade, M. Developing a local least-squares support vector machines-based neuro-fuzzy model for nonlinear and chaotic time series prediction. *IEEE Trans. Neural Netw. Learn. Syst.* **2013**, *24*, 207–218. [CrossRef]
22. Azadeh, A.; Gaeini, Z.; Haghighi, S.M.; Nasirian, B. A unique adaptive neuro fuzzy inference system for Optimum decision making process in a natural gas transmission unit. *J. Nat. Gas. Sci. Eng.* **2016**, *34*, 472–485. [CrossRef]
23. Petković, D.; Shamshirband, S.; Anuar, N.B.; Sabri, A.Q.M.; Rahman, Z.B.A.; Pavlovic, N.D. Input displacement neuro-fuzzy control and object recognition by compliant multi-fingered passively adaptive robotic gripper. *J. Intell. Robot. Syst.* **2016**, *82*, 177–187. [CrossRef]
24. Lotfi, E.; Rezaee, A.A. *Generalized BELBIC*; Neural Computing and Applications; Springer: Berlin, Germany, 2018.
25. Sharma, P. Design of novel BELBIC controlled semi-active suspension and comparative analysis with passive and PID controlled suspension. *Walailak J. Sci. Technol.* **2021**, *18*, 8989. [CrossRef]
26. Ershadi, M.H.; Shojaeian, S.; Keramat, R. A comparison of fuzzy and brain emotional learning-based intelligent control approaches for a full bridge DC-DC converter. *Int. J. Ind. Electron. Control Optim.* **2019**, *2*, 197–206.
27. Zirkohi, M.M. Stability analysis of brain emotional intelligent controller with application to electrically driven robot manipulators. *IET Sci. Meas. Technol.* **2020**, *14*, 182–187. [CrossRef]
28. Sharma, P.; Kumar, V. Design and analysis of novel bio inspired BELBIC and PSOBELBIC controlled semi active suspension. *Int. J. Veh. Perform.* **2020**, *6*, 399–424. [CrossRef]
29. Abd El-Gawad, A.; Elden, A.N.; Bahgat, M.E.; Ghany, A.A. BELBIC Load Frequency Controller Design for a Hydro-Thermal Power System. In Proceedings of the 2019 21st International Middle East Power Systems Conference (MEPCON), Cairo, Egypt, 17–19 December 2019.
30. Al Alahmadi, A.A.; Belkhier, Y.; Ullah, N.; Abeida, H.; Soliman, M.S.; Khraisat, Y.S.H.; Alharbi, Y.M. Hybrid wind/PV/battery energy management-based intelligent non-integer control for smart DC-microgrid of smart university. *IEEE Access* **2021**, *9*, 98948–98961. [CrossRef]
31. Karan, D.; Harish, V.S.K.V. Analysis of a wind-PV battery hybrid renewable energy system for a dc microgrid. *Mater. Today Proc.* **2021**, *46*, 5451–5457.
32. Kumar, G.B. Optimal power point tracking of solar and wind energy in a hybrid wind solar energy system. *Int. J. Energy Environ. Eng.* **2021**, 1–27. [CrossRef]
33. Masters, G.M. *Renewable and Efficient Electric Power Systems*, 2nd ed.; Wiley-IEEE Press: New York, NY, USA, 2013.
34. Atawi, I.E.; Hendawi, E.; Zaid, S.A. Analysis and design of a standalone electric vehicle charging station supplied by photovoltaic energy. *Processes* **2021**, *9*, 1246. [CrossRef]
35. Saad, S.S.; Zainuri, M.A.A.M.; Hussain, A. Implementation of Maximum Power Point Tracking Techniques for PV-Wind Hybrid Energy System: A Review. In Proceedings of the 2021 International Conference on Electrical Engineering and Informatics (ICEEI), Kuala Terengganu, Malaysia, 12–13 October 2021; pp. 1–6. [CrossRef]
36. *IEEE-519*; IEEE Recommended Practices and Requirements for Harmonic Control in Electric Power Systems. IEEE: New York, NY, USA, 1992.

Article

Robust Control for Optimized Islanded and Grid-Connected Operation of Solar/Wind/Battery Hybrid Energy

Muhammad Maaruf [1,2], **Khalid Khan** [3] **and Muhammad Khalid** [2,3,4,*]

[1] Systems Engineering Department, King Fahd University of Petroleum and Minerals (KFUPM), Dhahran 31261, Saudi Arabia; g201705070@kfupm.edu.sa

[2] K.A. CARE Energy Research & Innovation Center, King Fahd University of Petroleum and Minerals (KFUPM), Dhahran 31261, Saudi Arabia

[3] Electrical Engineering Department, King Fahd University of Petroleum and Minerals (KFUPM), Dhahran 31261, Saudi Arabia; g201604320@kfupm.edu.sa

[4] Center for Renewable Energy and Power Systems, King Fahd University of Petroleum and Minerals (KFUPM), Dhahran 31261, Saudi Arabia

* Correspondence: mkhalid@kfupm.edu.sa

Abstract: Wind and solar energy systems are among the most promising renewable energy technologies for electric power generations. Hybrid renewable energy systems (HRES) enable the incorporation of more than one renewable technology, allowing increased reliability and efficiency. Nevertheless, the introduction of variable generation sources in concurrence with the existing system load demand necessitates maintaining the power balance between the components of the HRES. Additionally, the efficiency of the hybrid power supply system is drastically affected by the number of converters interfacing its components. Therefore, to improve the performance of the HRES, this paper proposes a robust sliding mode control strategy for both standalone and grid-connected operation. The control strategy achieves maximum power point tracking for both the renewable energy sources and stabilizes the DC-bus and load voltages irrespective of the disturbances, change in load demand, variations of irradiance level, temperature, and wind speed ensuring an efficient energy management. Furthermore, the solar PV system is directly linked to the DC-bus obviating the need for redundant interfacing boost converters with decreased costs and reduced system losses. Lyapunov candidate function is used to prove the asymptotic stability and the convergence of the entire system. The robustness of the proposed control strategy is tested and validated under various conditions of HRES, demonstrating its efficacy and performance under various conditions of the HRES.

Keywords: energy storage system; hybrid microgrid; nonlinear control; power management; solar PV generation; wind power generation

Citation: Maaruf, M.; Khan, K.; Khalid, M. Robust Control for Optimized Islanded and Grid-Connected Operation of Solar/Wind/Battery Hybrid Energy. *Sustainability* **2022**, *14*, 5673. https://doi.org/10.3390/su14095673

Academic Editors: Akbar Maleki and Alberto-Jesus Perea-Moreno

Received: 17 February 2022
Accepted: 13 April 2022
Published: 8 May 2022

Publisher's Note: MDPI stays neutral with regard to jurisdictional claims in published maps and institutional affiliations.

1. Introduction

The need of renewable energy sources (RESs) in energy sector has progressively increased due to the global concern over environmental preservation and ever-increasing electric power demand. Therefore, RES technologies such as solar photovoltaics (PV), wind, hydro, geothermal, etc., are progressively utilized in electric power generation as they prove to be a more efficient and cheaper solution than conventional fuel-based generators, especially solar and wind energy sources, in terms of Levelized cost of energy [1,2]. Solar and wind energy sources are among the prominent RES technologies [3], attributable to their low cost, availability, modularity, and technological maturity [4]. In addition, recent advancements in power electronics technology enable a more flexible as well as desirable operational control and integration of RES to the power grid with reduced cost [5]. Nevertheless, large-scale RES integration is significantly limited due to their intermittent nature and geographical dependency, that is, they are highly dependent on ambient conditions such as wind speed, temperature, and degree of irradiance [6]. Such dependencies have a severely detrimental impact on the reliability and power quality of the system [7–9].

Realization of hybrid renewable energy system (HRES) includes incorporation of two or more power generation technologies [10]. Such HRES mitigates the intermittency of individual RESs enhancing the overall operational efficiency [11], and optimizing the capital investments through appropriate utilization of the available natural resources [12]. Wind and solar based HRESs have the dynamic capability to support the utility grid due to the availability of moment of inertia in the wind generation system [13], and the reliability of the power supply is increased due to the availability of multiple energy sources [14]. Besides, the complementary relationship between PV and wind energy sources ensures a high probability of continuous power supply, i.e., the output from the PV panel is high and the wind turbine power generation is low during the day. Accordingly, the PV production is negligible, whereas the wind turbine output increases at night [15].

The PV-wind HRES system deals with the intermittency of both energy sources. Concurrently, maximum power point tracking (MPPT) is vital for harnessing peak energy [16]. Energy storage systems (ESSs) store/supply the excess/deficient power generated by the RESs [17]. This enables a degree of controllability over the intermittency introduced by the variable RES and the loads. Seemingly, ESSs are proved to facilitate a multi-faced solution in RES-based power systems in the form of bulk energy services, transmission infrastructure services, customer energy management services, ancillary services, and off-grid operations [18–20]. However, HRESs that are designed to operate in both standalone and grid-connected modes needs to be meticulously operated to enable harnessing of maximum power from the RESs while maintaining acceptable power quality standards in terms of mitigating the impact of system uncertainties and maintaining acceptable voltage levels across the grid [21–23].

This paper presents a non-linear, multi-input–multi-output (MIMO) robust sliding mode control (SMC) for HRES consisting of wind/solar/battery. The control design facilitates a unified single controller for a safe, reliable and seamless operation for both standalone and grid-connected microgrid operation. A cost-efficient methodology is also achieved by connecting the PV array directly to the DC-bus without the interfacing DC-DC boost converter [24]. This approach facilities insights on the integration of hybrid renewable energy sources into the grid for power system designers and operators to not only minimize the cost of installation, but also to improve the efficiency of the PV output power in terms of practical applications. Besides, the proposed approach does not require an islanding detection system [25,26]. Thus, the drawbacks of islanding detection such as deviation in current and voltage due to mismatch in frequency, phase, and amplitude [27–29], during the switching between the islanded and grid modes are avoided. The outline and contribution of this paper is summarized as follows:

1. The development of a unified non-linear sliding mode MIMO controller ensuring a compliant, efficient, reliable, with low complexity, and safe operation of the components of the HRES both in standalone and grid-connected modes of the microgrid.
2. Ensuring a continuous power supply through the DC-DC buck/boost integrated ESS that allows power into and out of the battery with controlled charging and discharging operation.
3. Obviation of redundant converter incorporation with the integration of wind/PV hybrid RES using a back-to-back (B2B) converter topology and direct interconnection of solar PV to the DC-bus, hence facilitating higher efficiency and reducing power losses.
4. Formulation of autonomous MPPT operation for the solar and wind energy sources that is operable on the rotor side converter (RSC) and grid side converter (GSC) configuration of the B2B converter.
5. Investigation and evaluation of the proposed control architecture to perform following function: (i) stabilize the DC-bus and load voltages under the fluctuations of the generated RES power; (ii) achieving MPPT operation from solar and wind energy

sources; and (iii) maintaining the power balance of the HRES during both on-grid and off-grid operations.

The remainder of the paper is organized as follows: Section 2 presents the literature review. The mathematical models of the HRES components are derived in Section 3. The proposed power management architecture and MIMO sliding mode control scheme are presented in Sections 4 and 5, respectively. The simulation results are discussed in Section 6, followed by the conclusion in Section 7.

2. Related Works

The integration of HRES imposes numerous technical challenges on the utility grid such as voltage regulation, management of active and reactive power flow, and introduction of harmonics due to the integration of power electronic devices [30]. The variability of renewable energy requires innovative solutions that lead to the incorporation of auxiliary support systems such as energy storage systems. The power generated by the solar and wind energy sources are sporadically higher/lower to the load demand, and curtailment/injection of power is required inevitably to maintain the power balance of the grid [31]. Accordingly, a coordinated control framework is pertinent not only to mitigate the impact of but to assure a unified operation between RES, grid, and auxiliary systems while maintaining the power quality of the grid [32].

A novel and cost-effective technique based on fuzzy-PI methodology for energy management in HRES is presented in [24]. The wind energy conversion system (WECS) is integrated using a RSC and GSC architecture, and the PV is incorporated into the DC-bus via a DC-DC converter. The rotor and stator currents are regulated through the proposed fuzzy logic control (FLC). The corresponding PI gains are auto-tuned by the FLC, and the DC-DC link is controlled to maintain the active power flow during normal operating conditions and regulate the DC-bus during grid fault conditions. Therefore, the efficacy of the proposed framework is validated to achieve minimized rotor over-currents, enhance converter performance, protect wind/PV HRES during voltage disturbance, and minimize torque and rotor variations.

Furthermore, a vector control method is presented in [33], for wind-PV grid-connected B2B converters. This study proposes a separate MPPT algorithm for PV and WECS through the RSC and GSC, respectively. The control vector approach utilizes the GSC to regulate the DC-bus voltage under different operational modes. An adaptive least mean mixed norm control technique [34], is proposed to reduce the impact of the stochastic components in the HRES. The proposed control enables exclusive MPPT operation for PV/wind and regulates the RSC and GSC to reduce the impact of the varying solar irradiance, wind speed, and loads. The DC-bus voltage is regulated using a PI controller. This study presents an experimental validation that improves the power quality by reducing the disturbances and harmonic content. Nonetheless, the above studies did not utilize and collaborate their control theories considering ESSs.

A dynamic modeling and operational control strategy for wind/solar RES is presented in [35]. A multi-input current-source-interface DC-DC converter topology is proposed for a sustainable power network that mitigates the impact of parametric variations of solar irradiance and wind speed. The MPP for the wind and solar are achieved through variable speed control and incremental conductance control methods, respectively. The ESS is locally utilized as an energy buffer to reduce the effect of RES intermittency and support the islanded microgrid operations in extreme grid conditions of blackouts and natural disasters. The research in [36] developed an optimal fuel consumption technique for HRES that consists of solar/wind/battery/diesel generation systems. The control framework proposes a modified P&O MPPT technique for PV and ESS control that is incorporated through the DC-DC converter. The RSC is designed to extract maximum power from the WECS using field-oriented vector control [37]. The GSC is designed to perform load compensation, reactive power compensation, harmonics compensation, and optimal utilization and control of the diesel generators.

An MPPT algorithm is developed based on a sensorless approach for HRES consisting of a doubly-fed induction motor (DFIG) and solar PV [38]. This proposition posits the utilization of B2B converters to interconnect the standalone HRES. The system architecture and control logic presented in the study obviates the need for additional sensing devices ensuring an enhanced operation of PV-DFIG hybrid system with minimal errors that is almost equal to zero. The P&O method is widely used for MPPT due to its simplicity and ease of implementation [39–41]. Furthermore, the P&O technique can be sensorless that additionally reduces the complexity and error, but requires intensive expertise of the system parameters [42].

In [43], a robust fractional-order SMC is proposed for a variable speed wind turbine to attain MPPT. A non-linear control approach is used to develop the SMC algorithm. In comparison to the conventional SMC control algorithm, the authors postulated an improved performance through the suppression of external disturbances and reduction of overshoot. Similarly, an adaptive integral derivative SMC control theory is proposed for MPPT operation for PV system [44]. The authors combined the traditional perturb and observe (P&O) MPPT method with an SMC framework. The presented MPPT control theory is demonstrated to successfully obviate the overshoot during abrupt fluctuation of solar irradiance and reduce steady-state variations. Moreover, the controller gains are adjusted using an adaptive mechanism to ensure appropriate operation under numerous different irradiation levels.

In recent years, few results considering energy systems for grid-connected/on-grid and standalone/off-grid operations have been published. A review of optimization techniques for power generating systems operating standalone and grid-connected is presented in [45]. The on-grid and off-grid operation of permanent magnet synchronous generator (PMSG) driven wind turbine system is presented in [46]. The study posits a centralized control strategy for RGC-GSC of the WECS and the bidirectional DC-DC converter of ESS. Accordingly, a PI-fuzzy based power management scheme for wind/solar/battery HRES is presented in [47]. The control unit operated based on supervisory control theory ensures an enhanced efficiency performance both in off-grid and on-grid operations.

3. Mathematical Modeling

The hybrid renewable microgrid system consists of a PMSG wind turbine, solar PV, battery energy storage system, and load. The microgrid is connected to the utility grid through switch S_1 that is placed between the load and the utility network as shown in Figure 1. When the switch S_1 between the load and the utility grid is opened, the system is operating in standalone mode and is supplying to the load only. When the switch is closed, the system is in grid-connected mode. The PMSG wind turbine is integrated into the microgrid through the RSC of the B2B converter. The solar PV is directly connected to the DC-bus. A diode is connected in series with the PV array to avert destruction as a result of the reverse flow of current. Similarly, the battery energy storage system is tied to the system at the DC bus and controlled through a DC-DC buck–boost converter.

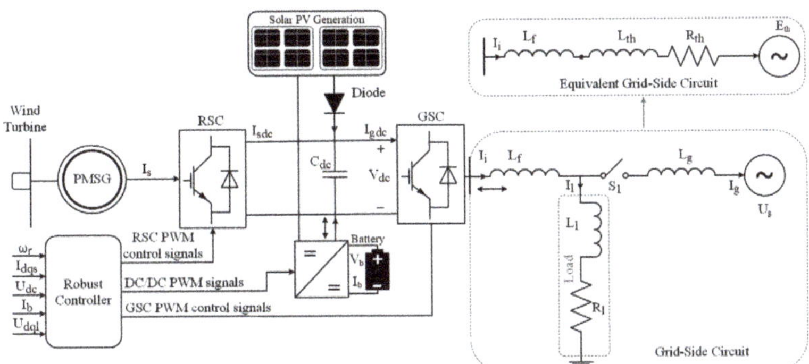

Figure 1. PV/wind/battery hybrid renewable energy system.

3.1. Wind Turbine Model

The mechanical power (P_w) captured by the wind turbine from the wind is given by [48]:

$$P_w = \frac{1}{2}\rho\pi R^2 C_p(\lambda,\beta)V_w^3 \tag{1}$$

$$C_p(\lambda,\beta) = 0.5176(\frac{116}{\lambda_i} - 0.4\beta - 5)e^{\frac{-21}{\lambda_i}} + 0.0068\lambda \tag{2}$$

$$\frac{1}{\lambda_i} = \frac{1}{\lambda + 0.08\beta} - \frac{0.035}{\beta^3 + 1} \tag{3}$$

$$\lambda = \frac{R\omega_r}{V_w} \tag{4}$$

$$\frac{d\omega_r}{dt} = J^{-1}P[T_m - T_e] \tag{5}$$

where, V_w is the wind speed, C_p is the power coefficient, ρ is the air density, R is the radius of the wind turbine, ω_r is the angular speed of the wind turbine, the tip-speed ratio is represented by λ, β is the pitch angle, J is the inertia of the mechanical shaft, T_e and T_m are the electrical and mechanical torque, respectively.

3.2. PMSG Model

The AC signals from the wind generation system are converted to DC using the RSC. The model of the PMSG is formulated in the d-q reference frame considering the dynamics of the RSC as follows [49]:

$$V_{ds} = L_d\frac{dI_{ds}}{dt} - \omega_r L_q I_{qs} + R_s I_{ds} \tag{6}$$

$$V_{qs} = L_q\frac{dI_{qs}}{dt} + \omega_r L_d I_{ds} + \omega_r \Lambda_r + R_s I_{qs} \tag{7}$$

$$T_e = \frac{3P}{2}[(L_d - L_q)I_{ds}I_{qs} + \Lambda_r I_{qs}] \tag{8}$$

where, V_{ds} is the d-axis stator voltage, Λ_r is the rotor flux, R_s is the stator resistance, V_{qs} is the q-axis stator voltage, L_d is the d-axis self-inducatnce, I_{ds} is the q-axis stator current, I_{qs} is the q-axis stator current, L_q is the q-axis self inductance, and R_s is the stator resistance. For nonsalient PMSG, $L_d = L_q$.

3.3. Modeling of Solar PV Module

The series and parallel connected PV cells formulate the solar PV module that utilizes the solar radiation to generate the DC voltage. In this respect, the equivalent circuit of the

PV cell comprises of a series resistor, parallel resistor, diode, and current source. The current output of the solar PV is expressed as [33,50]:

$$I_{pv} = N_p \left(I_{ph} - I_s \left[exp \left(\frac{q V_D}{N_s A K_B T} \right) - 1 \right] - \frac{V_D}{R_{sh}} \right) \tag{9}$$

$$V_D = \frac{V_{pv} + I_{pv} \frac{N_s}{N_p} R_{se}}{N_s} \tag{10}$$

$$I_{ph} = \left[I_{sc} + \psi_i (T - T_r) \right] \frac{S}{1000} \tag{11}$$

$$I_s = I_{rs} \left(\frac{T}{T_r} \right)^3 exp \left(\frac{q E_g}{A K_B T} \left[\frac{1}{T_r} - \frac{1}{T} \right] \right) \tag{12}$$

where, I_{ph} is the photo-generated current, I_{pv} is the PV output current, the leakage or reverse saturated current of the diode is denoted by I_s, I_{rs} is the saturated current at the operating temperature of the PV module, N_s and N_p are the number of series and parallel connected PV cells, q is the electron charge, V_D is the diode voltage, A is the p-n junction factor, E_g is the band gap energy of the semiconductor material used in the cell, K_B represents the Boltzmann constant, I_{sc} is short circuit current of the PV module, T is the ambient temperature, ψ_i is the temperature coefficient, S is the solar irradiance level, T_r is the operating temperature of the PV module, R_{se} and R_{sh} are the equivalent series resistance and shunt resistance of the PV cell, respectively.

Remark 1. *A number of PV modules are used to obtain a considerable power. The desired reference voltage of the DC-bus (V_{dc}^*), is obtained from the MPP voltage (V_{pv}^{max}) of the PV system. When the irradiance is not available, the V_{pv}^{max} is replaced by the nominal value of the DC-bus voltage.*

3.4. DC-DC Converter and Battery Modelling

In this study, a Lithium-ion (li-ion) battery is considered as the ESS component in the hybrid renewable microgrid. The ESS is incorporated into the DC-bus of the HRES via a buck/boost converter, as depicted in Figure 2. The converter allows bi-directional operation of the ESS, i.e., during ESS charging it operates as a buck converter and as a boost converter during ESS discharging (13). The converter operates as a buck converter during charging and as boost converter during discharging of the battery. Mathematically, the converter dynamics during the charging mode of the battery is formulated as [51]:

$$B_{mode} = \begin{cases} 0, & \text{if } I_b < 0 \text{ (buck)} \\ 1, & \text{if } I_b > 0 \text{ (boost)} \end{cases} \tag{13}$$

$$L_b \frac{dI_b}{dt} = V_b - I_b R_b - (1 - D_1) V_{dc} \tag{14}$$

$$C_{dc} \frac{dV_{dc}}{dt} = (1 - D_1) I_b - I_{gdc} \tag{15}$$

where, L_b is the battery inductance, I_b is the battery current, V_b is the battery voltage, R_b is the internal resistance, D_1 is the generated control signal during the charging mode of the battery, V_{dc} is the DC-bus voltage, C_{dc} is the capacitance of the DC-bus, and I_{gdc} is the DC current of the GSC converter. Similarly, converter dynamics during the discharging (boost) mode of the battery is expressed as:

$$L_b \frac{dI_b}{dt} = V_b - I_b R_b - D_2 V_{dc} \tag{16}$$

$$C_{dc} \frac{dV_{dc}}{dt} = D_2 I_b - I_{gdc} \tag{17}$$

where, D_2 is the generated control signal during the discharging mode of the battery. Accordingly, standardization can be made to reduce the complexity of the battery model to achieve general formulation using a virtual control (18). Hence, the overall generalized model of the of the battery model (14)–(17) is achieved, that is expressed as:

$$D_{12} = [B_{mode}(1 - D_1) + (1 - B_{mode})D_2] \tag{18}$$

$$L_b \frac{dI_b}{dt} = V_b - I_b R_b - D_{12} V_{dc} \tag{19}$$

$$C_{dc} \frac{dV_{dc}}{dt} = D_{12} I_b - I_{gdc} \tag{20}$$

$$\frac{d(SoC)}{dt} = -\frac{\eta_b}{Q_b} I_b \tag{21}$$

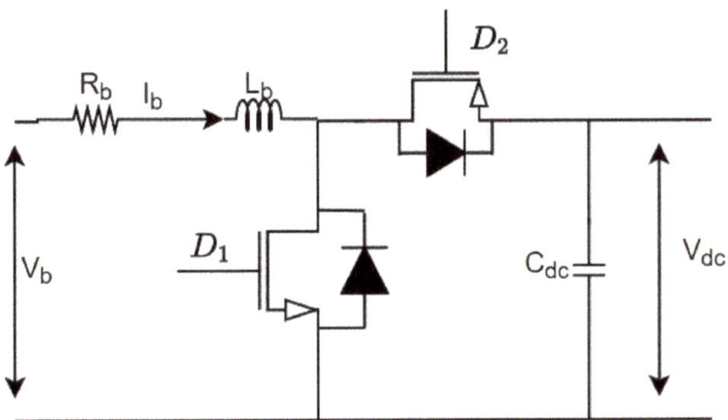

Figure 2. Buck/boost converter associated with ESS.

The state-of-charge (SoC) of the battery (21) is derived using the battery capacity (Q_b) and the battery efficiency (η_b) [52]. The SoC of the battery is constrained by the lower (SoC^{min}) and upper (SoC^{max}) limit of the battery capacity, as follows:

$$SoC^{min} < SoC < SoC^{max} \tag{22}$$

Remark 2. *The solution of* (21) *is* $SoC(t) = SoC(0) - \frac{\eta_b \int I_b dt}{Q_b}$, *where* $SoC(0)$ *is the initial charge of the battery. When the battery is charging,* I_b *is negative, and the* $SoC(t)$ *is increasing. On the other hand, when the battery is discharging,* I_b *is positive, and* $SoC(t)$ *is decreasing.*

3.5. GSC Modeling

The GSC is utilized to convert the DC signals to three-phase AC signals. Accordingly, GSC facilitates the power flow between the ESS and grid enabling the controllability over the load voltage, that is formulated in the d-q reference frame as [53]:

$$V_{di} = U_{dl} - L_f \omega_g I_{qi} + L_f \frac{dI_{di}}{dt} \tag{23}$$

$$V_{qi} = U_{ql} + L_f \omega_g I_{di} + L_f \frac{dI_{qi}}{dt} \tag{24}$$

where, V_{di} and V_{qi} denotes the d-axis and q-axis output voltage of the GSC, respectively, L_f and ω_g are the line filter inductance and grid electrical angular speed, respectively, U_{dl} and U_{ql} are the d-axis and q-axis load voltages, respectively, I_{di} and U_{qi} are the d-axis and q-axis GSC ouput currents.

3.6. Grid Side Modeling

The grid circuit consists of L_f, line inductance (L_g), RL load, switch (S_1), and the grid voltage. The leakage inductance of the transformer is included in L_g to simplify the circuit. In order to obtain the expression of the load voltage in terms of the grid voltage, the grid-side circuit is converted to the Thevenin's equivalent [46].

If the load impedance (Z_l) and the grid voltage (U_g) are given by $Z_l = R_l + j\omega_g L_l$ and $U_g = U_{dg} + jU_{qg}$, respectively, then the Thevenin's impedance (Z_{th}) and voltage (E_{th}) are, respectively, expressed as:

$$Z_{th} = \frac{j\omega_g L_g (R_l + j\omega_g L_l)}{R_l + j\omega_g (L_g + L_l)} \tag{25}$$

$$E_{th} = \frac{(U_{dg} + jU_{qg})(R_l + j\omega_g L_l)}{R_l + j\omega_g (L_g + L_l)} \tag{26}$$

where, U_{dg} and U_{qg} are the d-axis and q-axis grid voltages R_l is the load resistance, and L_l is the load inductance. The R_l and L_l are calculated from the power supplied to the load as follows:

$$R_l = \frac{3}{2} \frac{U_{ll}^2 P_l}{P_l^2 + Q_l^2}; \quad L_l = \frac{3}{2\omega_g} \frac{U_{ll}^2 Q_l}{P_l^2 + Q_l^2}; \tag{27}$$

where, U_{ll} is the line-to-line root-mean-square voltage. The active power (P_l) and the reactive power (Q_l) of the load are computed using its relationship with the d-axis load voltage (U_{dl}) and current (I_{dl}) as well as the q-axis load voltage (U_{ql}) and current (I_{ql}), as:

$$P_l = \frac{3}{2}(U_{dl} I_{dl} + U_{ql} I_{ql}); \quad Q_l = \frac{3}{2}(U_{ql} I_{dl} - U_{dl} I_{ql}) \tag{28}$$

The following equations are the d-q-axis representation of the grid side load voltages [46].

$$U_{dl} = R_{th} I_{di} - \omega_g L_{th} I_{qi} + E_{thd} \tag{29}$$

$$U_{ql} = R_{th} I_{qi} + \omega_g L_{th} I_{di} + E_{thq} \tag{30}$$

where, R_{th} represents the Thevenin's resistance, L_{th} is the Thevenin's inductance, E_{thd} and E_{thq} are the Thevenin's voltage in the d-axis and q-axis, respectively.

Remark 3. *When the hybrid microgrid is in standalone mode, $E_{th} = 0$ and $Z_{th} = Z_L$. A robust control law is essential to stabilized the load voltage when the hybrid microgrid is switching between the standalone mode and the grid-connected mode.*

3.7. Overall Model of the Hybrid Microgrid

To design a controller that can operate in both standalone and grid-connected modes, the dynamic equations of the components of the hybrid microgrid are expressed in state-space form. The system to be controlled is described by (5)–(7), (19)–(21), (23) and (24). It is an eight-order nonlinear MIMO system that has six control inputs and six controlled outputs. The state variables' vector, the inputs' vector, and the controlled outputs' vector are, respectively, defined as $x = [\omega_r,\ I_{qs},\ I_{ds},\ I_{di},\ I_{qi},\ I_b,\ V_{dc},\ SoC]^T$, $V = [V_{qs},\ V_{ds},\ V_{di},\ V_{qi},\ D_{12},\ I_{gdc}]^T$, and $y = h(x) = [\omega_r,\ I_{ds},\ U_{dl},\ U_{ql},\ I_b,\ V_{dc}]^T$. The dynamic equations of the hybrid microgrid system can be transformed to state-space model as follows [54]:

$$\dot{x} = f(x) + g(x)V \tag{31}$$
$$y = h(x)$$

where

$$f(x) = \begin{bmatrix} J^{-1}P[T_m - T_e] \\ [\omega_r L_q I_{qs} - R_s I_{ds}]/L_d \\ [-\omega_r L_d I_{ds} - \omega_r \Lambda_r - R_s I_{qs}]/L_q \\ U_{dl}/L_f - \omega_g I_{qi} \\ U_{ql}/L_f + \omega_g I_{di} \\ V_b/L_b \\ D_{12} I_b/C_{dc} \\ \eta_b/Q_b \end{bmatrix}$$

$$g(x) = \begin{bmatrix} 0 & 0 & 0 & 0 & 0 & 0 \\ \frac{1}{L_d} & 0 & 0 & 0 & 0 & 0 \\ 0 & \frac{1}{L_q} & 0 & 0 & 0 & 0 \\ 0 & 0 & \frac{1}{L_f} & 0 & 0 & 0 \\ 0 & 0 & 0 & \frac{1}{L_f} & 0 & 0 \\ 0 & 0 & 0 & 0 & -\frac{V_{dc}}{L_b} & 0 \\ 0 & 0 & 0 & 0 & 0 & -\frac{1}{C_{dc}} \\ 0 & 0 & 0 & 0 & 0 & 0 \end{bmatrix}$$

The input–output dynamics of the system is obtained by differentiating each output element of y with respect to time until at least one control input emerges. It is worth noting that ω_r has been differentiated twice before the input appears, while each of the remaining controlled outputs has been differentiated once. The input–output dynamics is thus:

$$\begin{bmatrix} \ddot{\omega}_r \\ \dot{I}_{ds} \\ \ddot{U}_{dl} \\ \dot{U}_{ql} \\ \dot{I}_b \\ \dot{V}_{dc} \end{bmatrix} = \begin{bmatrix} F_1 \\ F_2 \\ F_3 \\ F_4 \\ F_5 \\ F_6 \end{bmatrix} + \begin{bmatrix} \frac{3P^2\Lambda_r}{2JL_q} & 0 & 0 & 0 & 0 & 0 \\ 0 & \frac{1}{L_d} & 0 & 0 & 0 & 0 \\ 0 & 0 & \frac{R_{th}}{L_f} & -\frac{\omega_g L_{th}}{L_f} & 0 & 0 \\ 0 & 0 & \frac{\omega_g L_{th}}{L_f} & \frac{R_{th}}{L_f} & 0 & 0 \\ 0 & 0 & 0 & 0 & -\frac{V_{dc}}{L_b} & 0 \\ 0 & 0 & 0 & 0 & 0 & -\frac{1}{C_{dc}} \end{bmatrix} \begin{bmatrix} V_{qs} \\ V_{ds} \\ V_{di} \\ V_{qi} \\ D_{12} \\ I_{gdc} \end{bmatrix}$$

$$= F(x) + G(x)V \tag{32}$$

where,

$$F_1 = -\frac{3P^2\Lambda_r}{2J}[\omega_r L_q I_{qs} - R_s I_{ds}]/L_d \tag{33}$$

$$F_2 = [-\omega_r L_d I_{ds} - \omega_r \Lambda_r - R_s I_{qs}]/L_q \tag{34}$$

$$F_3 = R_{th}(\omega_g I_{qi} - \frac{U_{dl}}{L_f}) - \omega_g L_{th}(-\omega_g I_{di} - \frac{U_{ql}}{L_f}) + \dot{E}_{thd} \tag{35}$$

$$F_4 = R_{th}(-\omega_g I_{di} - \frac{U_{ql}}{L_f}) - \omega_g L_{th}(\omega_g I_{qi} - \frac{U_{dl}}{L_f}) + \dot{E}_{thq} \tag{36}$$

$$F_5 = \frac{D_{12} V_b}{L_b} \tag{37}$$

$$F_6 = \frac{I_b}{C_{dc}} \tag{38}$$

The number of differentiation of each output is the relative degree of the output with respect to its input. Therefore, the total relative degree of (32) is 8.

3.8. MPPT Derivation

3.8.1. Wind Turbine MPPT

A P & O scheme is employed for the MPPT operation of the wind turbine. The maximum power point of the wind turbine (P_{pw}) can be computed as follows [55]:

$$\frac{dP_w}{d\omega_r} = 0.5\rho V_w^3 \frac{dC_p(\lambda, \beta)}{d\omega_r} = 0 \tag{39}$$

By setting $\beta = 0$, C_p becomes a function of λ only. Therefore, $\frac{dC_p(\lambda,0)}{d\omega_r}$ is obtained as:

$$\frac{dC_p}{d\omega_r} = \frac{dC_p}{d\lambda_i} \times \frac{d\lambda_i}{d\omega_r} \tag{40}$$

Equation (39) can be rewritten as:

$$\frac{dC_p}{d\omega_r} = 0.5\rho V_w^3 \left(\frac{1260}{\lambda_i^3} - \frac{114.39}{\lambda_i^2}\right) e^{\frac{-21}{\lambda_i}} \times \frac{V_w R}{(V_w - 0.035R\omega_r)^2} \tag{41}$$

From (41), the condition for maximum power is $(V_w - 0.035R\omega_r) \neq 0$, then the optimal value of the power coefficient (C_p^{max}) and optimum tip speed ratio (λ_{opt}) are 0.48 and 8.1, respectively.

3.8.2. PV MPPT

A P & O scheme is employed for the MPPT operation of the PV module. The output power from the PV module (P_{pv}) is expressed as:

$$P_{pv} = I_{pv} V_{pv} \tag{42}$$

At MPP, we have [44]:

$$\frac{dP_{pv}}{dV_{pv}} = I_{pv} + V_{pv} \frac{dI_{pv}}{dV_{pv}} = 0 \tag{43}$$

Maximum power is extracted from the PV using the formulated MPP, that determines the corresponding DC-link voltage of the PVs. Similarly, the current enhanced centralized power converters enable direct integration of PVs ensuring power quality standard of operation [33].

4. Power Management

To prevent power shortage and damage of the microgrid components due to excess power, an energy management system is designed to coordinate the power flow between the grid power (P_b), P_{pv}, P_w, battery power (P_b), and the active load demand (P_l). The power balance equation for the hybrid microgrid is written as:

$$Grid\ connected: \ P_b + P_w + P_{pv} = P_l + P_g \tag{44}$$

$$Stand\ alone: \ P_b + P_w + P_{pv} = P_l \tag{45}$$

It is worth noting that

$$P_g = \begin{cases} P_g^- < 0 & when\ receiving\ power \\ P_g^+ < 0 & when\ sending\ power \end{cases} \tag{46}$$

The net power in the system (P_{net}) can be computed as:

$$P_{net} = P_l + P_g - (P_w + P_{pv}) \tag{47}$$

The charging and discharging modes of the battery depend on P_{net}. When $P_{net} < 0$, the excess power generated is transferred to the battery (charging mode) provided that $SoC < SoC^{max}$. When $P_{net} > 0$, the power shortage is compensated by discharging battery power to the load provided that $SoC > SoC^{min}$, otherwise load shedding is needed to maintain power balance.

5. Control Design

In this section, the nonlinear MIMO robust control of the hybrid microgrid system is designed. The presented control scheme works satisfactorily even under changing solar irradiation and varying wind speed. The control objectives are outlined as follows:

1. Harnessing the maximum power from the wind by optimally regulating the rotor speed, ω_r, to track the wind speed variations.
2. Achieving a unity power factor operation at the PMSG stator terminals by controlling I_{ds}.
3. MPPT operation of the PV module by controlling V_{pv}.
4. Meet the load voltage requirement by controlling the U_{dl} and U_{ql}.
5. Ensuring a smooth power management between the renewable energy sources, storage system, load and grid by controlling I_b.
6. Regulating the DC-bus voltage by controlling V_{dc}.

Calculation of the Reference Signals

The reference variables for ω_r, I_{ds}, U_{dl}, U_{ql}, I_b, and V_{dc} are set as ω_r^*, I_{ds}^*, U_{dl}^*, U_{ql}^*, I_b^*, and V_{dc}^*, respectively. The reference values are calculated as follows [55]:

1. The ω_r^* is computed as follows:

$$\omega_r^* = \frac{\lambda_{opt} V_w}{R} \tag{48}$$

2. The I_{ds}^* can be generated as follows: The stator's power factor angle (Θ_s) must remain zero in order to obtain unity power factor. The PMSG's stator current angle (Θ_I) and voltage phase angle (Θ_V) are expressed by the following equations [56]:

$$\Theta_I = tan^{-1}\left(\frac{I_{qs}}{I_{ds}}\right) \tag{49}$$

$$\Theta_V = tan^{-1}\left(\frac{V_{qs}}{V_{ds}}\right) = tan^{-1}\frac{\omega_r \Lambda_r - \omega_r L_d I_{ds}}{\omega_r L_q I_{qs}} \tag{50}$$

Subsequently, I_{ds}^* is computed such that the following condition is satisfied.

$$\Theta_s = \Theta_V - \Theta_I = 0 \tag{51}$$

The value of I_{ds}^* is thus:

$$I_{ds}^* = \frac{\Lambda_r - \sqrt{\Lambda_r^2 - 4L_d L_q I_{qs}^2}}{2L_d} \tag{52}$$

3. U_{dl}^* is selected to be equal to the grid voltage $(U_{dl}^* = |U_g|)$ so that the grid can easily synchronize with the microgrid at the point of common coupling.
4. U_{ql}^* is selected such that the reactive power is very close to zero. It is calculated as follows:

Assuming the GSC is ideal, then the active power along the two sides of the GSC are equal.

$$I_{gdc} V_{dc} = P_l + P_g^-$$
$$= U_{dl}^* I_{dl} + U_{ql}^* I_{ql} + P_g^- \tag{53}$$

Remark 4. *Note that P_g^- is the power received by the grid from the GSC as explained in* (46) *and I_{gdc} is the control input of V_{dc}.*

From (53), U_{ql}^* can be derived as follows:

$$U_{ql}^* = \frac{I_{gdc}V_{dc} - U_{dl}^* I_{dl} - P_g^-}{I_{ql}} \tag{54}$$

5. I_b^* is calculated by dividing P_{net} in (47) with V_b as follows:

$$I_b^* = \frac{P_{net}}{V_b} \tag{55}$$

6. V_{dc}^* is set as the MPPT voltage of the PV module ($V_{dc}^* = V_{pv}^{max}$). However, when the solar irradiance is low, V_{dc}^* is set as the nominal voltage of the DC-bus. The nominal value of the DC-bus voltage is calculated as [57]:

$$V_{dc}^* \geq \frac{1.6\sqrt{2}U_{ll}}{\sqrt{3}m_i} \tag{56}$$

where m_i is the modulation index.

The tracking errors are given as follows:

$$e_1 = \omega_r - \omega_r^* \tag{57}$$

$$e_2 = I_{ds} - I_{ds}^* \tag{58}$$

$$e_3 = U_{dl} - U_{dl}^* \tag{59}$$

$$e_4 = U_{ql} - U_{ql}^* \tag{60}$$

$$e_5 = I_b - I_b^* \tag{61}$$

$$e_6 = V_{dc} - V_{dc}^* \tag{62}$$

The sliding mode surfaces are defined as:

$$\begin{cases} \zeta_1 = \dot{e}_1 + k_1 \int e_1 dt + \gamma e_1 \\ \zeta_i = e_i + k_i \int e_i dt, \quad i = 2,3,4,5,6 \end{cases} \tag{63}$$

where k_i, α_i ($i = 1,2,\ldots,6$), and γ are positive constants. The time derivative of (63) yields:

$$\begin{cases} \dot{\zeta}_1 = \ddot{e}_1 + k_1 e_1 dt + \gamma \dot{e}_1 \\ \dot{\zeta}_i = \dot{e}_i + k_i e_i dt, \quad i = 2,3,4,5,6 \end{cases} \tag{64}$$

Define the vector $\zeta = [\zeta_1, \zeta_2, \zeta_3, \zeta_4, \zeta_5, \zeta_6]^T$. Then, (64) can be evaluated as follows:

$$\dot{\zeta} = F + GV + \begin{bmatrix} -\ddot{\omega}_r^* + k_1 e_1 + \gamma \dot{e}_1 \\ -\dot{I}_{ds}^* + k_2 e_2 \\ -\dot{U}_{dl}^* + k_3 e_3 \\ -\dot{U}_{ql}^* + k_4 e_4 \\ -\dot{I}_b^* + k_5 e_5 \\ -\dot{V}_{dc}^* + k_6 e_6 \end{bmatrix} \tag{65}$$

The output variables will converge toward their respective sliding mode surfaces and provide the desired steady-state performance by staying on the surfaces provided that

$\zeta_i = \dot{\zeta}_i = 0 \ (i = 1, 2, \ldots, 6)$. The equivalent control input vector (V_{eqv}) can be obtained by canceling the terms on the right-hand side of (65).

$$V_{eqv} = -G^{-1}F - G^{-1}\begin{bmatrix} -\ddot{\omega}_r^* + k_1 e_1 + \gamma \dot{e}_1 \\ -\dot{i}_{ds}^* + k_2 e_2 \\ -\dot{U}_{dl}^* + k_3 e_3 \\ -\dot{U}_{ql}^* + k_4 e_4 \\ -\dot{i}_b^* + k_5 e_5 \\ -\dot{V}_{dc}^* + k_6 e_6 \end{bmatrix} \tag{66}$$

Since $det(G(x)) \neq 0$, (66) is well defined. In order to compensate the external disturbances and parametric uncertainties, a switching control input vector is given by:

$$V_{sw} = -G^{-1}\alpha \, Sign(\alpha) \tag{67}$$

where $\alpha = diag(\alpha_1, \alpha_2, \alpha_3, \alpha_4, \alpha_5, \alpha_6)$ is a positive definite diagonal matrix, and $Sign(\zeta) = \left[sign(\zeta_1), \, sign(\zeta_2), \, sign(\zeta_3), \, sign(\zeta_4), \, sign(\zeta_5), \, sign(\zeta_6) \right]^T$.

The robust control input vector is given by:

$$\begin{aligned} V &= V_{eqv} + V_{sw} \\ &= V_{eqv} - G^{-1}\alpha \, Sign(\alpha) \end{aligned} \tag{68}$$

Figure 3 illustrates the RSC control for MPPT performance of the wind turbine and wind power transfer with unity power factor. The GSC control scheme has a cascaded control structure as shown in Figure 4. It consists of an outer DC-bus voltage controller and an inner loop U_{ql} controller to provide the pulse width modulation signals to the GSC. Additionally, U_{dl} is controlled through the GSC. Figure 5 depicts the control scheme for the ESS buck/boost converter. The controller adjusts the duty cycle of the buck/boost converter for charging/discharging operation of the ESS to balance the power.

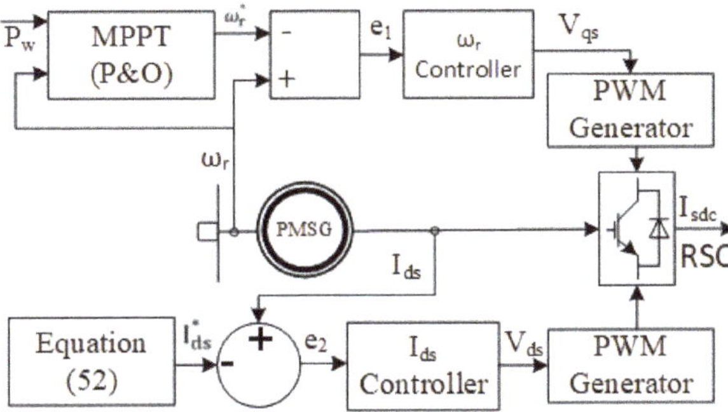

Figure 3. Control diagram of the RSC.

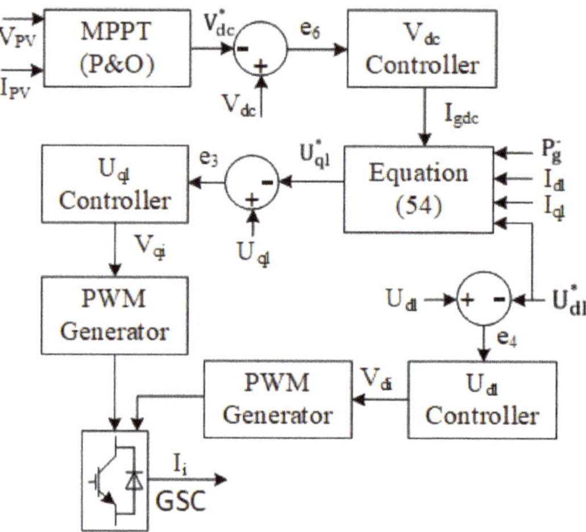

Figure 4. Control diagram of the GSC.

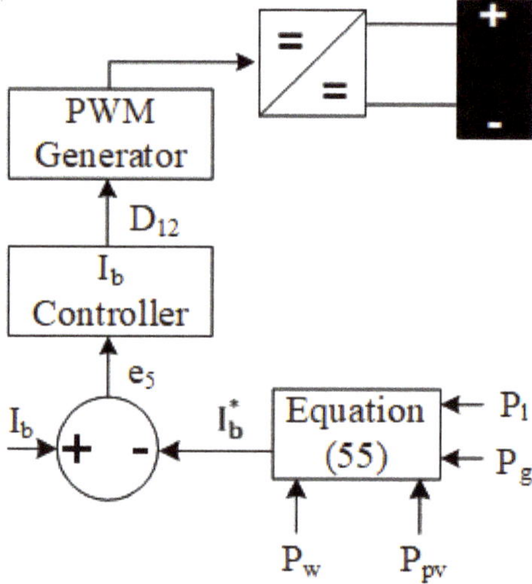

Figure 5. Control diagram of the buck/boost converter.

The state trajectories of the hybrid microgrid system achieves asymptotic convergence based on the sliding mode surfaces (63) and the robust control inputs (68).

Proof. Consider the following candidate Lyapunov function:

$$L = \frac{1}{2}\zeta^T\zeta \qquad (69)$$

The time derivative of L yields:

$$\dot{L} = \zeta^T\dot{\zeta} \qquad (70)$$

By substituting (65) and (68) into (70), the following equation is obtained.

$$\dot{L} = -\zeta^T\alpha Sign(\zeta) \qquad (71)$$

Taking the norm of (71) gives:

$$\dot{L} \leq -\|\alpha\|.\|\zeta\| \tag{72}$$

$\square$

Therefore, the closed-loop system is asymptotically stable.

6. Simulation Results

The hybrid microgrid system was developed on Matlab/Simulink environment using a time-domain simulation model. The harmonic effects of the converter have not been considered in this study. The system comprises of a number of PV module arranged in series and parallel to obtain a considerable output power of 1.2 MW, a nonsalient pole variable speed PMSG with rated power of 2.45 MW is deployed as the generator, and a li-ion battery is employed as the ESS. A discharging constraint of $10\% < SoC$ and charging constraint of $SoC < 90\%$ is employed to restrict ESS degradation [58,59]. The parameters of the renewable generators and ESS are given in Tables 1 and 2, respectively.

Table 1. Parameters of the renewable generation system [44,60].

Wind Turbine Generator			Solar PV Generator (KC200GH-2p)		
Parameter	Symbol	Value	Parameter	Symbol	Value
Air denisty (kg/m^3)	ρ	1.25	Ambient Temperature (°C)	T	25
Radius of wind turbine (m)	R	28.2	Maximum power at MPP (W)	P_{max}^{MPP}	200
			Maximum Voltage at MPP (V)	V_{max}^{MPP}	26.3
d-axis stator current (mH)	L_{ds}	9.8	P-N junction factor	A	1.8
			Temperature coefficient ($mA/°C$)	ψ_i	4.79
q-axis stator current (mH)	L_{qs}	9.8	Equivalent shunt resistor (Ω)	R_{sh}	313.33
Rotor flux (Wb)	Λ_r	28	Equivalent series resistor (Ω)	R_{se}	0.193
Inertia of Mechanical Shaft (kg·m^2)	J	4000	Short circuit current (A)	I_{sc}	8.21
			Maximum current at MPP (A)	I_{max}^{MPP}	7.61
Number of pole pairs	P	8	Number of parallel modules	N_s	68
Optimum tip speed ratio	λ_{opt}	8.1	Number of series modules	N_p	95
Power coefficient	C_p^{max}	0.48	Open-circuit voltage (V)	V_{oc}	32.9

The performance of the proposed controller is evaluated based on external and internal disturbances experienced in the HRES. The variations in renewable power and load demand are considered as the external disturbances and the internal disturbances includes a parametric uncertainty of $\pm 40\%$ that is introduced into the nominal parameters of the HRES components namely, on J, Λ_r, R_b, C_{dc}, and L_f. The controller gains are chosen as

$$K_1 = K_2 = 10, \ K_3 = 8, \ K_4 = 20, \ K_5 = 12 \text{ and } K_6 = 50, \ \alpha_1 = 2, \ \alpha_2 = 3, \ \alpha_3 = \alpha_4 = 5,$$
and $\alpha_6 = 15$.

Table 2. Grid and energy storage parameters.

Battery Energy Storage System			Grid Parameter		
Parameter	**Symbol**	**Value**	**Parameter**	**Symbol**	**Value**
Battery efficiency	η_b	0.9	Filter inductance (mH)	L_f	16.9
Battery capacity (AH)	Q_b	100	Line inductance (mH)	L_g	1.69
Battery power (MW)	P_b	1	Load demand (MW)	P_l	2
Battery voltage (V)	V_b	500	Line-to-line voltage (V)	U_{ll}	4000
Upper SoC limit (%)	SoC^{max}	90	DC-bus	C_{dc}	1670
Lower SoC limit (%)	SoC^{min}	10	capacitance (µF)		

Two case studies have been investigated. In case 1, a random wind speed profile, a constant solar irradiation level and load variation are considered. In case 2, step changing wind speed profile, varying solar irradiation, and load variation are considered.

6.1. Case 1: Random Wind and Fixed Solar with Varying Load

The wind speed comprises of a random profile (Figure 6), 1000 W/m² solar irradiance is fixed (Figure 7). Figure 8 depicts the optimal tracking performance of the rotor speed. It can be seen that the proposed controller closely tracks the optimal rotor speed calculated by the MPPT so that maximum power is produced by the PMSG. The response of the *d*-axis stator current for unity power factor operation is depicted in Figure 9. The DC-bus is receiving contribution from the solar PV, the wind power generator, and the ESS. The response of the DC-bus voltage together with the reference value which is the same as the MPPT voltage of the solar PV is shown in Figure 10. The controller can keep the DC-bus voltage stable and very close to the reference value despite the variation of the wind power, the ESS power, and the load demand.

As shown in Figure 11, an increment in the load demand at $t = 10$ s is experienced that rises from 2 MW to 2.8 MW, and further decreases to 1.9 MW at $t = 20$ s. The grid is receiving and sending power at $0 \text{ s} < t \leq 9 \text{ s}$ and $30 \text{ s} < t \leq 40 \text{ s}$, respectively, and the hybrid microgrid is off-grid at $9 \text{ s} < t \leq 30 \text{ s}$. In order to maintain the power flow, the battery is charging to store the excess power at $0 < t \leq 10 \text{ s}$ and $20 \text{ s} < t \leq 30 \text{ s}$. Similarly, the battery is discharging to the load at $10 \text{ s} < t \leq 20 \text{ s}$ and $30 \text{ s} < t \leq 40 \text{ s}$. Figure 12 indicates that the battery is charging/discharging according to the desired level to balance the power in the microgrid within the acceptable SoC limit. An increment in load voltage is experienced at $t = 9$ s due to off-grid operation of the HRES during power receiving mode and at $t = 20$ s due decrease in load demand. Similarly, the load voltage decreases at $t = 10$ s due to the increase in the load demand and at $t = 30$ s due to connection of the grid to the HRES during the power transferring mode (Figures 13 and 14).

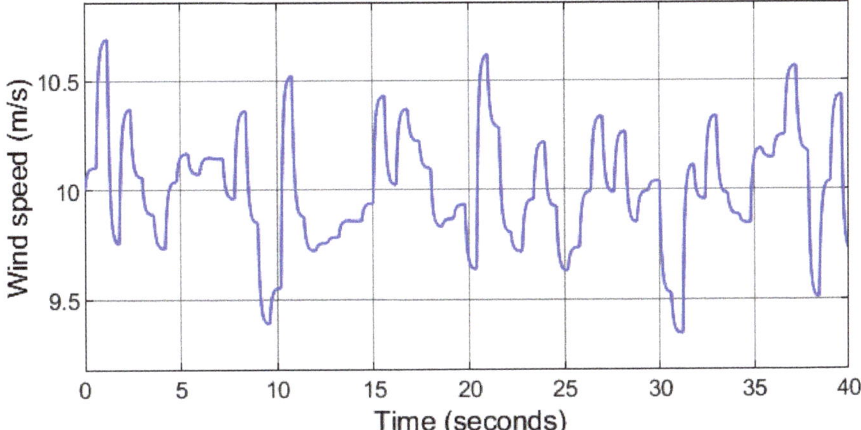

Figure 6. Case 1: Random wind speed profile.

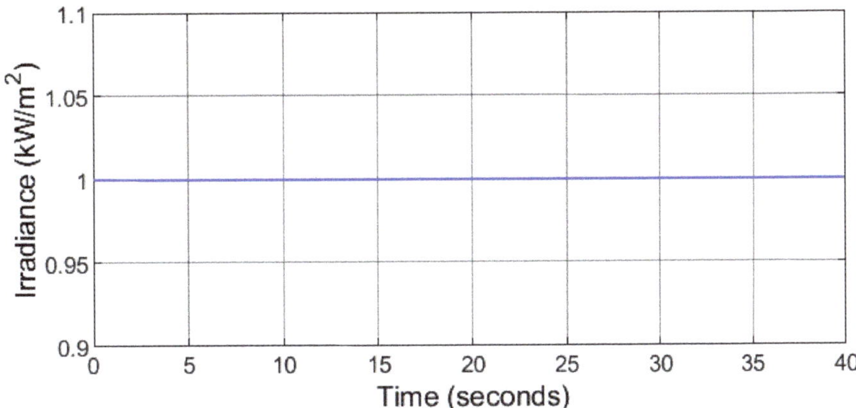

Figure 7. Case 1: Constant solar irradiation.

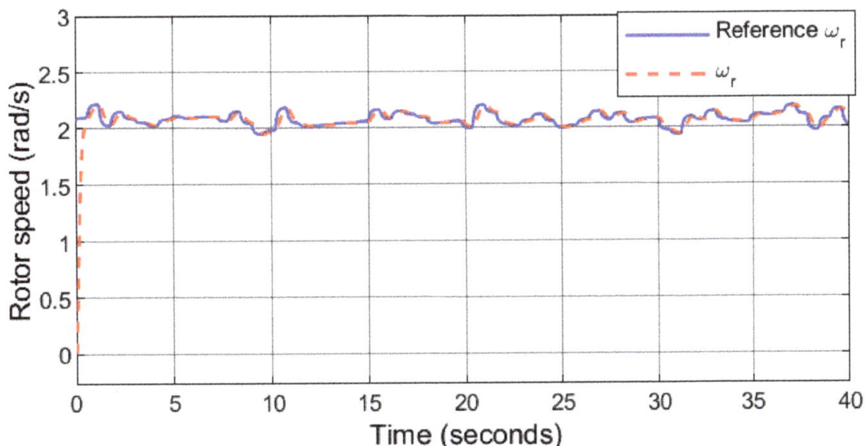

Figure 8. Case 1: Rotor speed.

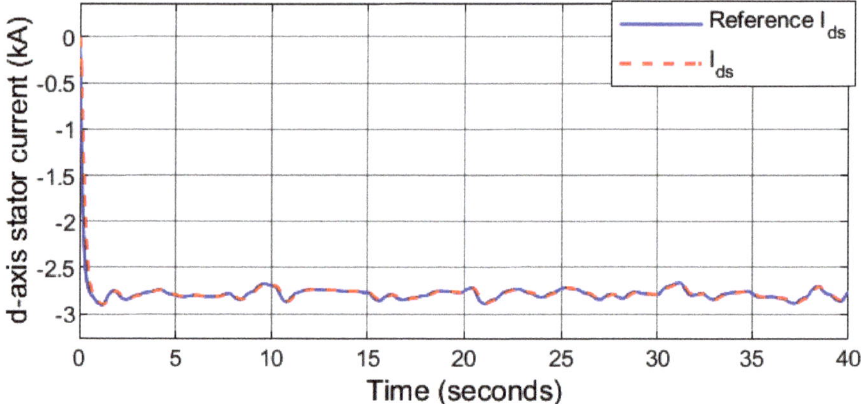

Figure 9. Case 1: *d*-axis stator current.

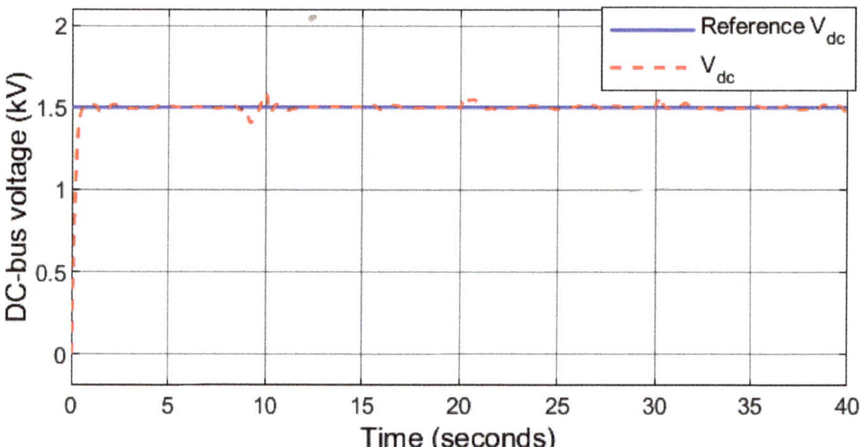

Figure 10. Case 1: DC-bus voltage.

It can be observed that the controller can restore the load voltage to the desired level regardless of the load demand variations and at the transition intervals of the HRES between the standalone and grid-connected modes. Moreover, the controller keeps U_{ql} close to zero to minimize the reactive power. The transition of the system from islanded to grid-connected mode and vice-versa are smooth due to the robustness and accuracy of the proposed controller.

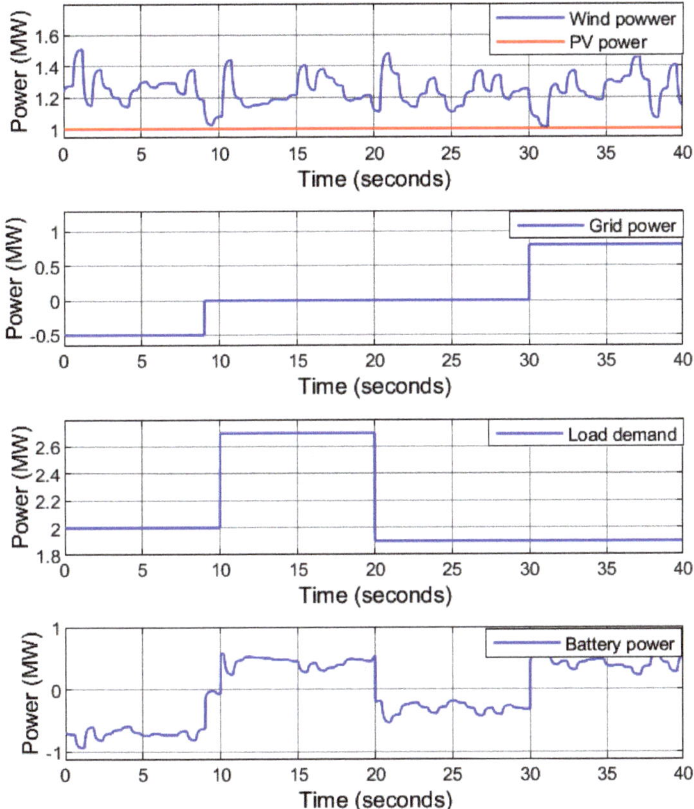

Figure 11. Case 1: Power flow.

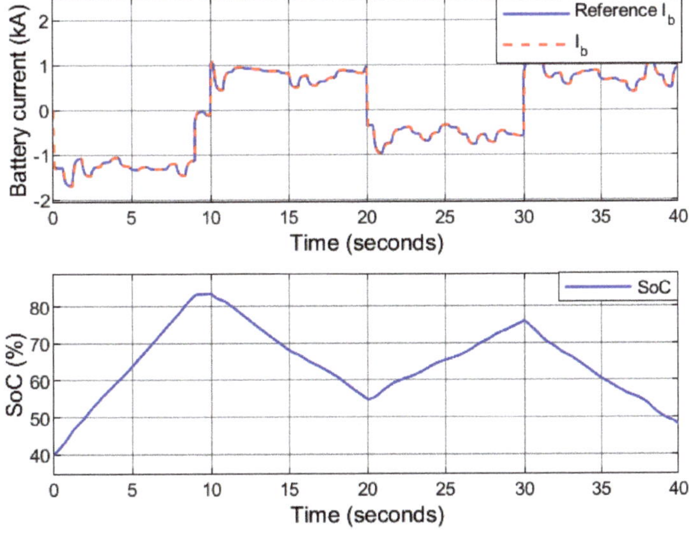

Figure 12. Case 1: The battery current (**above**) and the SoC (**below**).

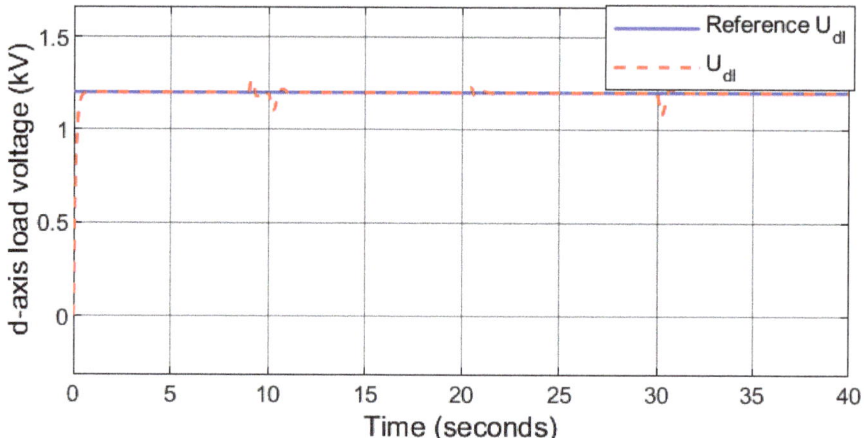

Figure 13. Case 1: *d*-axis load voltage.

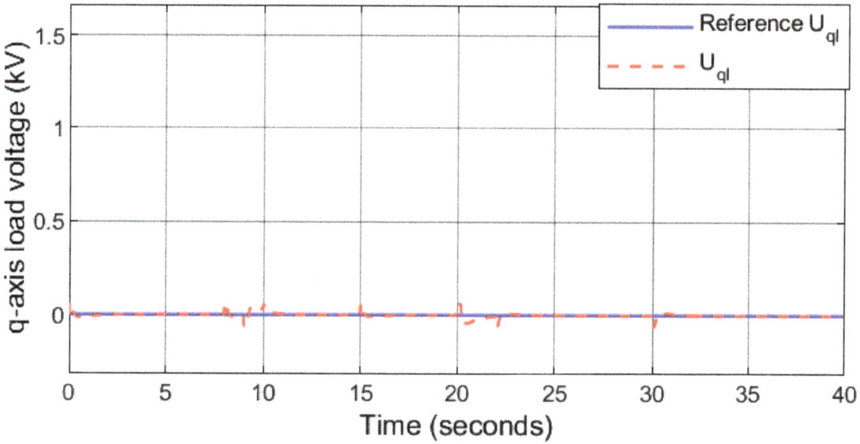

Figure 14. Case 1: *q*-axis load voltages.

6.2. Case 2: Step Change in Wind and Solar with Varying Load

In this case, a step changing wind speed profile and solar irradiance level are considered. As depicted in Figure 15, the wind speed rises from 7.7 m/s to 9.2 m/s, 11 m/s and finally decreases to 10.1 m/s at $t = 8, 15$ and 22 s, respectively. The solar irradiance level decreases from 1000 W/m^2 to 965 W/m^2, 933 W/m^2 and finally increases to 1000 W/m^2 at $t = 10$ s, 15 s and 20 s, respectively (Figure 16). The rotor speed follows the desired speed under varying wind conditions (Figure 17), which indicates that the PMSG is rotating at the optimal speed computed by the MPPT algorithm ensuring maximum power generation under variable wind speed. The *d*-axis stator current tracks the desired current accurately as depicted in Figure 18, that allows wind power transfer with a unity power factor. Furthermore, results obtained using PI controller is also presented in this section. The comparative analysis highlights the performance between the proposed controller and benchmark PI control technique [33] based on the calculated optimal values of ω_r, I_{ds}, V_{dc}, I_b, U_{dl}, and U_{ql}.

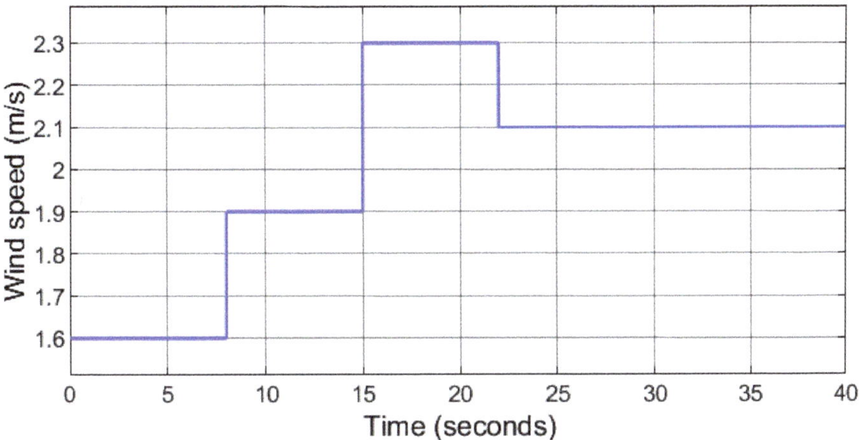

Figure 15. Case 2: Random wind speed profile.

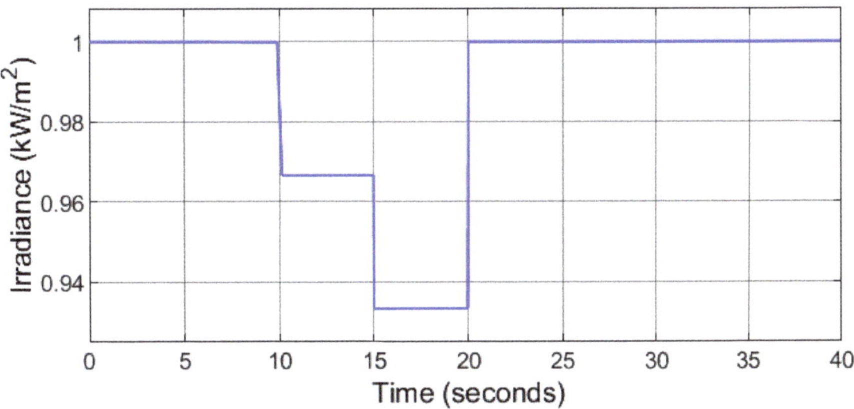

Figure 16. Case 2: Varying solar irradiation.

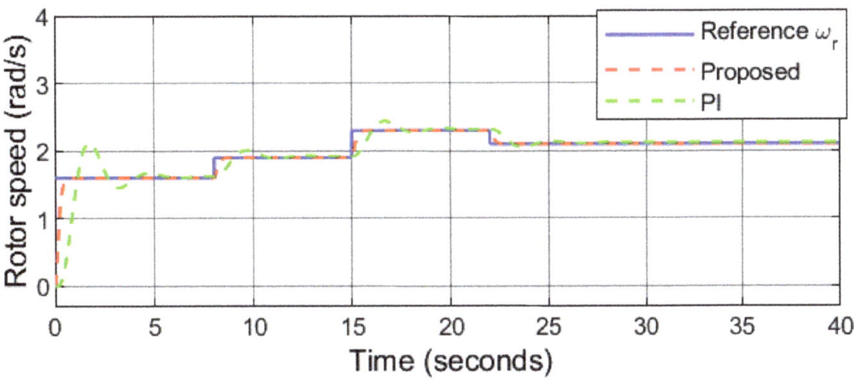

Figure 17. Case 2: Rotor speed.

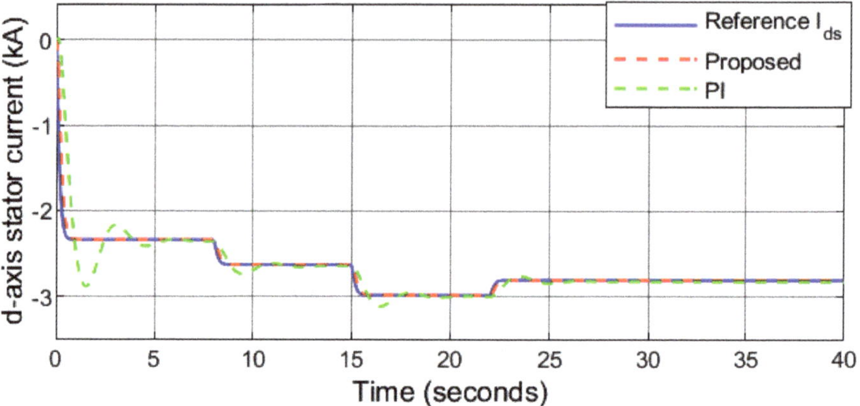

Figure 18. Case 2: *d*-axis stator current.

The irradiance level falls to 965 W/m^2 and 933 W/m^2 at 10 s $< t \leq$ 20 s. During this interval, the PV MPPT voltage also falls, and thus the reference DC-bus voltage is set as the nominal value instead of the PV MPPT voltage to maintain a constant DC-bus voltage. It can be observed that the proposed controller keeps the DC-bus voltage stable and constant value under varying PV power, wind power, and load demand as depicted in Figure 19. The load demand profile is similar to case 1 as shown in Figure 20. The utility grid is receiving power at 0 $< t \leq$ 9 s and sending power at 9 s $< t \leq$ 20 s and 30 s $< t \leq$ 40 s. The HRES is off-grid at 20 s $< t \leq$ 30 s. When 0 s $< t \leq$ 9 s and 20 s $< t \leq$ 30 s, the surplus power is transferred to the battery. When 9 s $< t \leq$ 20 s, the battery is unable to cover the power deficit as the maximum power it can safely deliver is 1 MW. As a result, 0.9 MW of the load demand is shedded for protecting the battery and maintain the power balance of the HRES.

Figure 21 describes that the ESS is charging/discharging in accordance with the demand to balance the microgrid. Accordingly, the power flow is balanced and no power is transferred into or out of the battery and its SoC remains constant (60%) at 30 s $< t \leq$ 40 s. The load voltage is kept at a constant level by the proposed controller as depicted in Figures 22 and 23. In addition, U_{ql} converges to zero, thereby minimizing the reactive power. It can be observed that the effects of varying load demand, grid transition from power receiving mode to power transferring mode, and grid transition from islanded mode to grid-connected mode and vice versa are mitigated by the proposed controller.

The performance of the benchmark PI controller proves to have an acceptable performance under external disturbances. The performance of the PI controller in this condition is compared with that of the proposed robust sliding mode controller. The analytical results are depicted in Figures 17–23. It can be observed that the responses of the proposed controller in following the reference signals are quite satisfactory as they reach them in 0.90 s, 0.63 s, 1.02 s, 0.04 s, 0.88 s, and 0.63 s for ω_r, I_{ds}, V_{dc}, I_b, U_{dl}, and U_{ql}, respectively. However, the responses under PI controller have comparatively higher overshoots, undershoots, and longer settling time highlighting the effectiveness of the proposed controller.

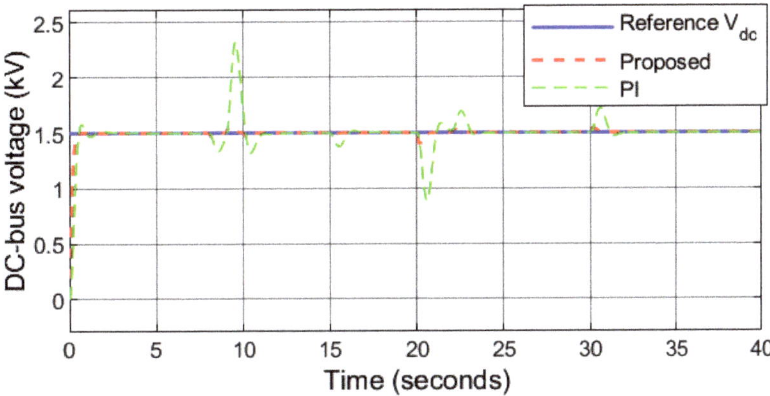

Figure 19. Case 2: DC-bus voltage.

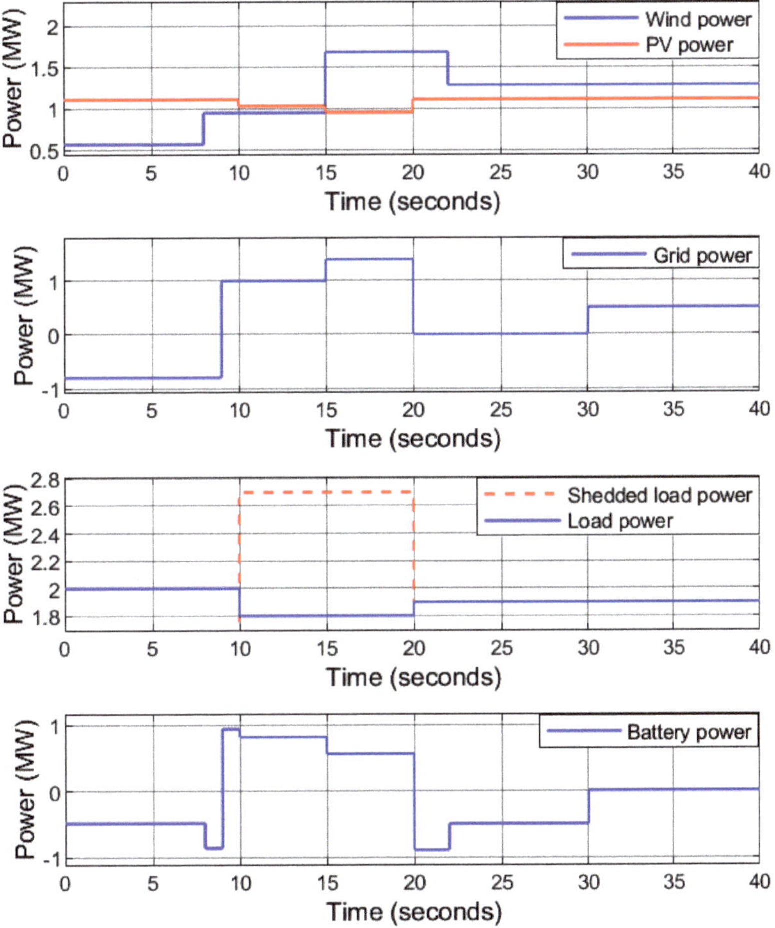

Figure 20. Case 2: Power flow.

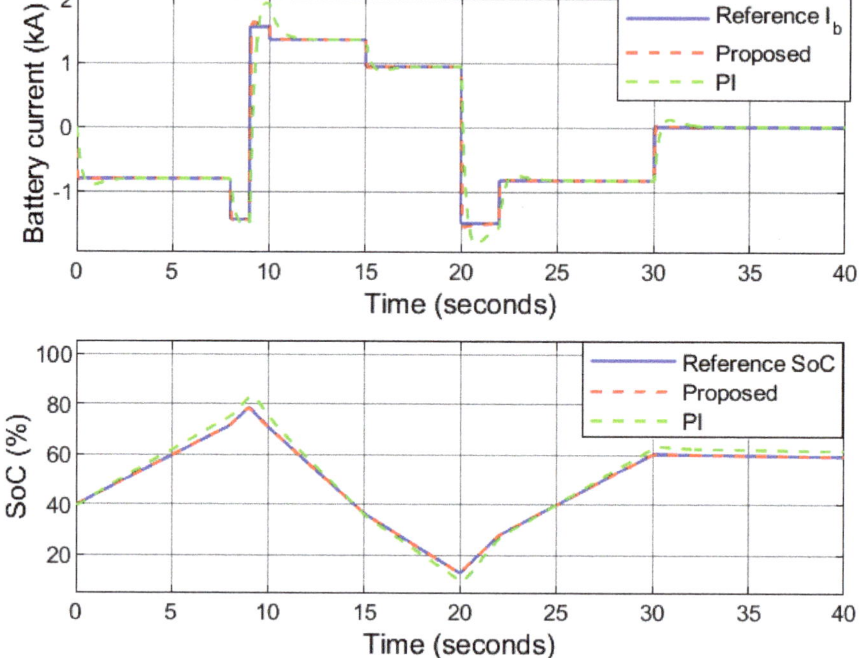

Figure 21. Case 2: Battery current.

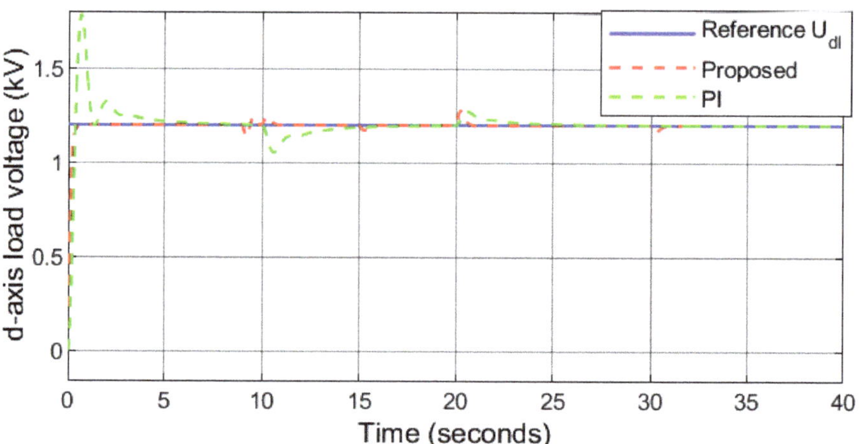

Figure 22. Case 2: *d*-axis load voltage.

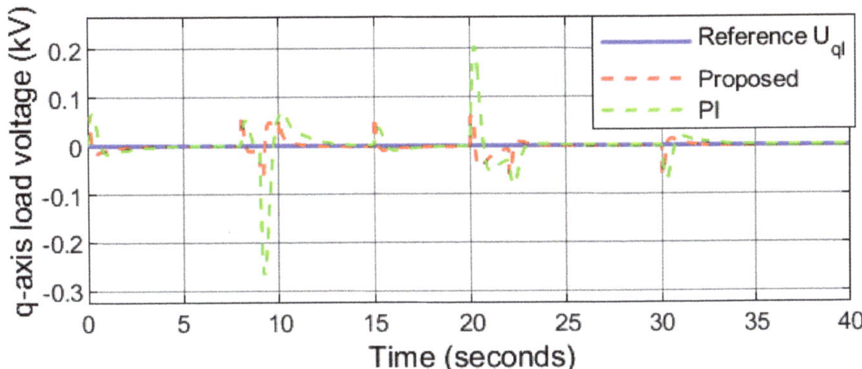

Figure 23. Case 2: *q*-axis load voltage.

7. Conclusions

This paper presented an integrated control and power management of a hybrid renewable energy system (HRES) under different external generation/load disturbances and internal parameter uncertainties for both standalone and grid-connected operational modes. The HRES consists of solar PV, wind energy source, and battery. The proposed robust sliding mode control successfully achieves maximum power point tracking (MPPT) for both the solar PV and wind energy sources while regulating the load voltages and maintaining the DC-bus voltage at 1.5 kV. The stability and convergence of the closed-loop system have been guaranteed using the Lyapunov candidate function. Furthermore, a comparative analysis is presented with a conventional PI-based controller. The results obtained highlight a significantly improved robustness and better power management in terms of overshoot and settling time with enhanced tracking capability towards the calculated optimal operation of the HRES. Furthermore, as the control theory has low complexity it can be extended to include different types of renewable energy sources and the power quality of the HRES can be further enhanced by including the hybrid energy storage systems as auxiliary support.

Author Contributions: Conceptualization, M.M., K.K. and M.K.; data curation, M.M.; formal analysis, M.M., K.K. and M.K.; funding acquisition, M.K.; investigation, M.M. and K.K.; methodology, M.M. and K.K.; project administration, M.K.; resources, M.K.; software, K.K.; supervision, M.K.; validation, M.M., K.K. and M.K.; visualization, M.K.; writing—original draft, M.M. and M.K.; writing—review and editing, K.K. and M.K. All authors have read and agreed to the published version of the manuscript.

Funding: This research was funded by the Deanship of Research Oversight and Coordination (DROC) at King Fahd University of Petroleum and Minerals (KFUPM) through project No. DF191011. The authors would also like to acknowledge the support provided by the King Abdullah City for Atomic and Renewable Energy (K.A. CARE).

Acknowledgments: The authors would like to acknowledge the support provided by the Deanship of Research Oversight and Coordination (DROC) at King Fahd University of Petroleum and Minerals (KFUPM) for funding this work through project No. DF191011. In addition, we would like to acknowledge the funding support provided by the King Abdullah City for Atomic and Renewable Energy (K.A.CARE).

Conflicts of Interest: The authors declare no conflict of interest.

Abbreviations

The following abbreviations are used in this manuscript:

A	P-N junction factor
B_{mode}	Operational mode of the bi-directional converter of battery
B2B	Back-to-back
β	Pitch angle
C_{dc}	DC-bus capacitance
C_p	Power coefficient
D_1	Duty cycle of the battery converter during charge mode
D_2	Duty cycle of the battery converter during discharge mode
DFIG	Doubly-fed induction generator
ESS	Energy storage system
E_g	Band gap energy of the semiconductor material
E_{th}	Thevenin's voltage
η_b	Efficiency of the battery
FLC	Fuzzy logic control
GSC	Grid-side converter
HRES	Hybrid renewable energy system
I_b	Battery current
I_{di}	GSC d-axis output AC current
I_{dl}	d-axis load current
I_{ds}	d-axis stator current
I_g	Grid current
I_{gdc}	GSC DC current
I_{pv}	PV output current
I_{qi}	GSC q-axis output AC current
I_{ql}	q-axis load current
I_{qs}	q-axis stator current
I_{sc}	PV short circuit current
J	Inertia of the mechanical shaft
K_B	Boltzman's constant
L_b	Battery inductance
L_d	d-axis self-inductance
L_f	Grid-side filter inductance
L_l	Load inductance
L_g	Line inductance
Li-ion	Lithium-ion
λ	Tip speed ratio
I_l	Load current
L_q	q-axis self-inductance
L_{th}	Thevenin's inductance
Λ_r	Rotor flux
MIMO	Multi-input-multi-output
MPPT	Maximum power point tracking
N_p	Number of parallel connected modules
N_s	Number of series connected modules
ω_g	Electrical angular speed
ω_r	Angular speed of wind turbine
P_b	Battery power
P_g	Grid power
P_g^-	Power received by the grid
P_g^+	Power transferred by the grid
P_l	Load demand active power
P_{net}	Net power

P_{pv}	Solar power
P_w	Wind power
P&O	Perturb and Observe
PMSG	Permanent magnet synchronous generator
PV	Photo-voltaic
ψ_i	Temperature coefficient
q	Electron charge
Q_b	Battery capacity
Q_l	Load demand reactive power
R	Radius of wind turbine
RES	Renewable energy source
RSC	Rotor-side converter
R_b	Battery internal resistance
R_l	Load resistance
R_s	Stator resistance
R_{se}	Equivalent series resistors
R_{sh}	Equivalent shunt resistors
R_{th}	Thevenin's resistance
ρ	Density of air
S	Solar irradiance level
SMC	Sliding mode control
SoC	State-of-charge of battery
SoC^{max}	Upper limit of SoC
SoC^{min}	Lower limit of SoC
T	Ambient temperature
T_e	Electrical torque
T_m	Mechanical torque
T_r	Operating temperature of the PV module
Θ_I	Stator current angle
Θ_s	Stator power factor angle
Θ_V	Stator voltage angle
U_{dl}	d-axis load voltage
U_g	Grid voltage
U_{ll}	Line-to-line RMS voltage
U_{ql}	q-axis load voltage
V_b	Battery voltage
V_D	Diode voltage in the PV circuit
V_{dc}	DC-bus voltage
V_{ds}	d-axis stator voltage
V_{pv}	PV output voltage
V_{qi}	GSC q-axis output voltage
V_{qs}	q-axis stator voltage
V_w	Wind speed
WECS	Wind energy conversion system
Z_l	Load impedance
Z_{th}	Thevenin's impedance

References

1. Zhou, P.; Jin, R.; Fan, L. Reliability and economic evaluation of power system with renewables: A review. *Renew. Sustain. Energy Rev.* **2016**, *58*, 537–547. [CrossRef]
2. Ghazal, H.M.; Khan, K.A.; Alismail, F.; Khalid, M. Maximizing capacity credit in generation expansion planning for wind power generation and compressed air energy storage system. In Proceedings of the 2021 IEEE PES Innovative Smart Grid Technologies Europe (ISGT Europe), Espoo, Finland, 18–21 October 2021; pp. 1–5.
3. Manwell, J.; McGowan, J.; Breger, D. A design and analysis tool for utility scale power systems incorporating large scale wind, solar photovoltaics and energy storage. *J. Energy Storage* **2018**, *19*, 103–112. [CrossRef]
4. Maaruf, M.; Khalid, M. Power quality control of hybrid wind/electrolyzer/fuel-cell/BESS microgrid. In Proceedings of the 2021 IEEE PES Innovative Smart Grid Technologies-Asia (ISGT Asia), Brisbane, Australia, 5–8 December 2021; pp. 1–5.

5. Wang, G.; Konstantinou, G.; Townsend, C.D.; Pou, J.; Vazquez, S.; Demetriades, G.D.; Agelidis, V.G. A review of power electronics for grid connection of utility-scale battery energy storage systems. *IEEE Trans. Sustain. Energy* **2016**, *7*, 1778–1790. [CrossRef]

6. Maaruf, M.; Khan, K.A.; Khalid, M. Integrated Power Management and Nonlinear-Control for Hybrid Renewable Microgrid. In Proceedings of the 2021 IEEE Green Technologies Conference (GreenTech 2021), Virtual Conference, 7–9 April 2021.

7. Al-Humaid, Y.M.; Khan, K.A.; Abdulgalil, M.A.; Khalid, M. Two-stage stochastic optimization of sodium-sulfur energy storage technology in hybrid renewable power systems. *IEEE Access* **2021**, *9*, 162962–162972. [CrossRef]

8. Liang, X. Emerging power quality challenges due to integration of renewable energy sources. *IEEE Trans. Ind. Appl.* **2016**, *53*, 855–866. [CrossRef]

9. Khan, K.A.; Shafiq, S.; Khalid, M. A strategy for utilization of reactive power capability of PV inverters. In Proceedings of the 2019 9th International Conference on Power and Energy Systems (ICPES), Perth, Australia, 10–12 December 2019; pp. 1–6.

10. Maaruf, M.; Khalid, M. Global sliding-mode control with fractional-order terms for the robust optimal operation of a hybrid renewable microgrid with battery energy storage. *Electronics* **2022**, *11*, 88. [CrossRef]

11. Ding, Z.; Hou, H.; Yu, G.; Hu, E.; Duan, L.; Zhao, J. Performance analysis of a wind-solar hybrid power generation system. *Energy Convers. Manag.* **2019**, *181*, 223–234. [CrossRef]

12. Alhumaid, Y.; Khan, K.; Alismail, F.; Khalid, M. Multi-input nonlinear programming based deterministic optimization framework for evaluating microgrids with optimal renewable-storage energy mix. *Sustainability* **2021**, *13*, 5878. [CrossRef]

13. Binbing, W.; Abuduwayiti, X.; Yuxi, C.; Yizhi, T. RoCoF droop control of PMSG-based wind turbines for system inertia response rapidly. *IEEE Access* **2020**, *8*, 181154–181162. [CrossRef]

14. Jurasz, J.; Canales, F.; Kies, A.; Guezgouz, M.; Beluco, A. A review on the complementarity of renewable energy sources: Concept, metrics, application and future research directions. *Sol. Energy* **2020**, *195*, 703–724. [CrossRef]

15. Bett, P.E.; Thornton, H.E. The climatological relationships between wind and solar energy supply in Britain. *Renew. Energy* **2016**, *87*, 96–110. [CrossRef]

16. Elobaid, L.M.; Abdelsalam, A.K.; Zakzouk, E.E. Artificial neural network-based photovoltaic maximum power point tracking techniques: A survey. *IET Renew. Power Gener.* **2015**, *9*, 1043–1063. [CrossRef]

17. Salman, U.; Khan, K.; Alismail, F.; Khalid, M. Techno-economic assessment and operational planning of wind-battery distributed renewable generation system. *Sustainability* **2021**, *13*, 6776. [CrossRef]

18. Nadeem, F.; Hussain, S.S.; Tiwari, P.K.; Goswami, A.K.; Ustun, T.S. Comparative review of energy storage systems, their roles, and impacts on future power systems. *IEEE Access* **2018**, *7*, 4555–4585. [CrossRef]

19. Khan, K.A.; Khalid, M. Improving the transient response of hybrid energy storage system for voltage stability in DC microgrids using an autonomous control strategy. *IEEE Access* **2021**, *9*, 10460–10472. [CrossRef]

20. Tian, Y.; Bera, A.; Benidris, M.; Mitra, J. Stacked revenue and technical benefits of a grid-connected energy storage system. *IEEE Trans. Ind. Appl.* **2018**, *54*, 3034–3043. [CrossRef]

21. Khan, K.A.; Khalid, M. Hybrid energy storage system for voltage stability in a DC microgrid using a modified control strategy. In Proceedings of the 2019 IEEE Innovative Smart Grid Technologies-Asia (ISGT Asia), Chengdu, China, 21–24 May 2019; pp. 2760–2765.

22. Majeed, M.A.; Asghar, F.; Hussain, M.I.; Amjad, W.; Munir, A.; Armghan, H.; Kim, J.T. Adaptive dynamic control based optimization of renewable energy resources for grid-tied microgrids. *Sustainability* **2022**, *14*, 1877. [CrossRef]

23. Alhammad, H.I.; Alsada, M.F.; Khan, K.A.; Khalid, M. A voltage sensitivity framework for optimal allocation of battery energy storage systems. In Proceedings of the 2021 IEEE Green Technologies Conference (GreenTech), Virtual Conference, 7–9 April 2021; pp. 558–562.

24. Morshed, M.J.; Fekih, A. A Novel Fault Ride Through Scheme for Hybrid Wind/PV Power Generation Systems. *IEEE Trans. Sustain. Energy* **2020**, *11*, 2427–2436. [CrossRef]

25. Fatu, M.; Blaabjerg, F.; Boldea, I. Grid to Standalone Transition Motion-Sensorless Dual-Inverter Control of PMSG With Asymmetrical Grid Voltage Sags and Harmonics Filtering. *IEEE Trans. Power Electron.* **2014**, *29*, 3463–3472. [CrossRef]

26. Ashabani, S.M.; Mohamed, Y.A.I. A Flexible Control Strategy for Grid-Connected and Islanded Microgrids With Enhanced Stability Using Nonlinear Microgrid Stabilizer. *IEEE Trans. Smart Grid* **2012**, *3*, 1291–1301. [CrossRef]

27. Li, X.; Zhang, H.; Shadmand, M.B.; Balog, R.S. Model Predictive Control of a Voltage-Source Inverter with Seamless Transition Between Islanded and Grid-Connected Operations. *IEEE Trans. Ind. Electron.* **2017**, *64*, 7906–7918. [CrossRef]

28. Sekhar, P.C.; Tupakula, R.R. Model predictive controller for single-phase distributed generator with seamless transition between grid and off-grid modes. *IET Gener. Transm. Distrib.* **2019**, *13*, 1829–1837. [CrossRef]

29. Chen, J.; Hou, S.; Chen, J. Seamless mode transfer control for master–slave microgrid. *IET Power Electron.* **2019**, *12*, 3158–3165. [CrossRef]

30. Blaabjerg, F.; Ma, K. Future on Power Electronics for Wind Turbine Systems. *IEEE J. Emerg. Sel. Top. Power Electron.* **2013**, *1*, 139–152. [CrossRef]

31. Akram, U.; Khalid, M. A coordinated frequency regulation framework based on hybrid battery-ultracapacitor energy storage technologies. *IEEE Access* **2017**, *6*, 7310–7320. [CrossRef]

32. Maaruf, M.; El Ferik, S.; Mahmoud, M. Integral Sliding Mode Control With Power Exponential Reaching Law for DFIG. 2020 17th International Multi-Conference on Systems, Signals & Devices (SSD), Monastir, Tunisia, 20–23 July 2020; pp. 1122–1127.

33. Radwan, A.A.A.; Mohamed, Y.A.I. Grid-Connected Wind-Solar Cogeneration Using Back-to-Back Voltage-Source Converters. *IEEE Trans. Sustain. Energy* **2020**, *11*, 315–325. [CrossRef]

34. Chishti, F.; Murshid, S.; Singh, B. PCC voltage quality restoration strategy of an isolated microgrid based on adjustable step adaptive control. *IEEE Trans. Ind. Appl.* **2020**, *56*, 6206–6215. [CrossRef]

35. Bae, S.; Kwasinski, A. Dynamic modeling and operation strategy for a microgrid with wind and photovoltaic resources. *IEEE Trans. Smart Grid* **2012**, *3*, 1867–1876. [CrossRef]

36. Puchalapalli, S.; Tiwari, S.K.; Singh, B.; Goel, P.K. A Microgrid Based on Wind-Driven DFIG, DG, and Solar PV Array for Optimal Fuel Consumption. *IEEE Trans. Ind. Appl.* **2020**, *56*, 4689–4699. [CrossRef]

37. Puchalapalli, S.; Singh, B. A single input variable FLC for DFIG-based WPGS in standalone mode. *IEEE Trans. Sustain. Energy* **2019**, *11*, 595–607. [CrossRef]

38. Nguyen, D.; Fujita, G. Dynamic response evaluation of sensorless MPPT method for hybrid PV-DFIG wind turbine system. *J. Int. Counc. Electr. Eng.* **2016**, *6*, 49–56. [CrossRef]

39. Mousa, H.H.; Youssef, A.R.; Mohamed, E.E. State of the art perturb and observe MPPT algorithms based wind energy conversion systems: A technology review. *Int. J. Electr. Power Energy Syst.* **2021**, *126*, 106598. [CrossRef]

40. Kavya, M.; Jayalalitha, S. Developments in perturb and observe algorithm for maximum power point tracking in photo voltaic panel: A review. *Arch. Comput. Methods Eng.* **2021**, *28*, 2447–2457. [CrossRef]

41. Motahhir, S.; El Hammoumi, A.; El Ghzizal, A. The most used MPPT algorithms: Review and the suitable low-cost embedded board for each algorithm. *J. Clean. Prod.* **2020**, *246*, 118983. [CrossRef]

42. Putri, R.I.; Pujiantara, M.; Priyadi, A.; Ise, T.; Purnomo, M.H. Maximum power extraction improvement using sensorless controller based on adaptive perturb and observe algorithm for PMSG wind turbine application. *IET Electr. Power Appl.* **2018**, *12*, 455–462. [CrossRef]

43. Ardjal, A.; Mansouri, R.; Bettayeb, M. Fractional sliding mode control of wind turbine for maximum power point tracking. *Trans. Inst. Meas. Control* **2019**, *41*, 447–457. [CrossRef]

44. Kihal, A.; Krim, F.; Laib, A.; Talbi, B.; Afghoul, H. An improved MPPT scheme employing adaptive integral derivative sliding mode control for photovoltaic systems under fast irradiation changes. *ISA Trans.* **2019**, *87*, 297–306. [CrossRef]

45. Twaha, S.; Ramli, M.A. A review of optimization approaches for hybrid distributed energy generation systems: Off-grid and grid-connected systems. *Sustain. Cities Soc.* **2018**, *41*, 320–331. [CrossRef]

46. Housseini, B.; Okou, A.F.; Beguenane, R. Robust Nonlinear Controller Design for On-Grid/Off-Grid Wind Energy Battery-Storage System. *IEEE Trans. Smart Grid* **2018**, *9*, 5588–5598. [CrossRef]

47. Basaran, K.; Cetin, N.S.; Borekci, S. Energy management for on-grid and off-grid wind/PV and battery hybrid systems. *IET Renew. Power Gener.* **2017**, *11*, 642–649. [CrossRef]

48. Wu, B.; Lang, Y.; Zargari, N.; Kouro, S. *Power Conversion and Control of Wind Energy Systems*; John Wiley & Sons: Hoboken, NJ, USA, 2011; Volume 76.

49. Maaruf, M.; Shafiullah, M.; Al-Awami, A.T.; Al-Ismail, F.S. Adaptive Nonsingular Fast Terminal Sliding Mode Control for Maximum Power Point Tracking of a WECS-PMSG. *Sustainability* **2021**, *13*, 13427. [CrossRef]

50. Chatrenour, N.; Razmi, H.; Doagou-Mojarrad, H. Improved double integral sliding mode MPPT controller based parameter estimation for a stand-alone photovoltaic system. *Energy Convers. Manag.* **2017**, *139*, 97–109. [CrossRef]

51. Azeem, M.K.; Armghan, H.; Ahmad, I.; Hassan, M. Multistage adaptive nonlinear control of battery-ultracapacitor based plugin hybrid electric vehicles. *J. Energy Storage* **2020**, *32*, 101813. [CrossRef]

52. Sofia, B. Online battery state-of-charge estimation methods in micro-grid systems. *J. Energy Storage* **2020**, *30*, 101518.

53. Alepuz, S.; Calle, A.; Busquets-Monge, S.; Kouro, S.; Wu, B. Use of stored energy in PMSG rotor inertia for low-voltage ride-through in back-to-back NPC converter-based wind power systems. *IEEE Trans. Ind. Electron.* **2012**, *60*, 1787–1796. [CrossRef]

54. Chen, C.T. *Linear System Theory and Design*; New York, NY, USA; Oxford, UK, 1999.

55. Youssef, A.R.; Ali, A.I.; Saeed, M.S.; Mohamed, E.E. Advanced multi-sector P & O maximum power point tracking technique for wind energy conversion system. *Int. J. Electr. Power Energy Syst.* **2019**, *107*, 89–97.

56. Chishti, F.; Murshid, S.; Singh, B. LMMN-Based Adaptive Control for Power Quality Improvement of Grid Intertie Wind–PV System. *IEEE Trans. Ind. Inform.* **2019**, *15*, 4900–4912. [CrossRef]

57. Dhar, R.K.; Merabet, A.; Al-Durra, A.; Ghias, A.M.Y.M. Power Balance Modes and Dynamic Grid Power Flow in Solar PV and Battery Storage Experimental DC-Link Microgrid. *IEEE Access* **2020**, *8*, 219847–219858. [CrossRef]

58. Stroe, D.I.; Knap, V.; Swierczynski, M.; Stroe, A.I.; Teodorescu, R. Operation of a grid-connected lithium-ion battery energy storage system for primary frequency regulation: A battery lifetime perspective. *IEEE Trans. Ind. Appl.* **2016**, *53*, 430–438. [CrossRef]

59. Li, J.; Danzer, M.A. Optimal charge control strategies for stationary photovoltaic battery systems. *J. Power Sources* **2014**, *258*, 365–373. [CrossRef]

60. Xu, D.; Dai, Y.; Yang, C.; Yan, X. Adaptive fuzzy sliding mode command-filtered backstepping control for islanded PV microgrid with energy storage system. *J. Frankl. Inst.* **2019**, *356*, 1880–1898. [CrossRef]

 *sustainability*

Review

Recent Advances in Energy Storage Systems for Renewable Source Grid Integration: A Comprehensive Review

Muhammed Y. Worku

Interdisciplinary Research Center for Renewable Energy and Power Systems (IRC-REPS), Research Institute, King Fahd University of Petroleum and Minerals, Dhahran 31261, Saudi Arabia; muhammedw@kfupm.edu.sa; Tel.: +96-65-5971-3973; Fax: +96-61-3860-3535

Abstract: The reduction of greenhouse gas emissions and strengthening the security of electric energy have gained enormous momentum recently. Integrating intermittent renewable energy sources (RESs) such as PV and wind into the existing grid has increased significantly in the last decade. However, this integration hampers the reliable and stable operation of the grid by posing many operational and control challenges. Generation uncertainty, voltage and angular stability, power quality issues, reactive power support and fault ride-through capability are some of the various challenges. The power generated from RESs fluctuates due to unpredictable weather conditions such as wind speed and sunshine. Energy storage systems (ESSs) play a vital role in mitigating the fluctuation by storing the excess generated power and then making it accessible on demand. This paper presents a review of energy storage systems covering several aspects including their main applications for grid integration, the type of storage technology and the power converters used to operate some of the energy storage technologies. This comprehensive review of energy storage systems will guide power utilities; the researchers select the best and the most recent energy storage device based on their effectiveness and economic feasibility.

Keywords: renewable energy sources; power fluctuation; energy storage systems; selection criteria

Citation: Worku, M.Y. Recent Advances in Energy Storage Systems for Renewable Source Grid Integration: A Comprehensive Review. *Sustainability* **2022**, *14*, 5985. https://doi.org/10.3390/su14105985

Academic Editor: Nicu Bizon

Received: 29 March 2022
Accepted: 12 May 2022
Published: 15 May 2022

Publisher's Note: MDPI stays neutral with regard to jurisdictional claims in published maps and institutional affiliations.

1. Introduction

Power generation using renewable energy sources has minimized the use of hydrocarbons for power generation and transportations. Power generated from renewable energy sources can be integrated to the grid in grid connected mode or can act as an independent power island (island mode) [1–3]. Renewable energy supplies 14.8% of the total industrial energy demand mainly for low temperature industries. Nevertheless, for heavy industries such as iron and steel, cement and chemicals, renewable energy accounts for just less than 1% of the combined energy demand. Currently, an energy mix of electricity, solar, wind, and nuclear is being used to supply the loads in various countries of the world and the other forms of energy contributed just less than 1% of the total energy demand [4,5].

The intermittent nature of renewable resources hinders the performance of the grid by introducing issues with system stability, reliability, and power quality. The variability and uncertainty of power output are the two fundamental issues that hinder the bulk integration of renewable energy sources with the existing grid. Introducing energy storage systems (ESSs) to the grid can address the variability issue by decoupling the power generation from demand. In addition, the ESSs improve the power quality of the grid by providing ancillary services [6–8]. The demand for energy storage will continue to grow as the penetration of renewable energy into the electric grid increases year by year.

ESSs are enabling technologies for well-established and new applications such as power peak shaving, electric vehicles, the integration of renewable energies, etc. [9]. ESSs make the grid more reliable by acting as a power source or providing different functions such as spinning reserve, load leveling, power quality improvement and power fluctuation minimization from renewable energy sources. Large ESSs are routinely used alongside renewable generation such as wind to stabilize the power output. The authors of [10–12] presented a comprehensive review of different energy storage systems that are used for grid integration of large-scale renewable energy sources. There is a big opportunity to transition to a carbon-free energy future by integrating ESS with renewable power. ESSs with high ratings and a long duration will play a great role in reducing the environmental impact of the conventional power source.

According to estimates, the worldwide revenue from energy storage for renewables integration will exceed $23 billion by 2026 and the requirements for storing energy will become triple the present values by 2030 [13]. Solar energy has reached grid parity in several locations around the globe and no longer requires policy incentives to incentivize deployment in many markets. However, energy storage mechanisms also face many challenges as well [14] as there being no one storage type that has the complete characteristics required by the modern grid. Limitations such as storage capacity, response time, efficiency, cost and implementation requirements are to name a few. Some ESSs such as batteries also have an environmental effect by releasing toxic gas [15].

This review paper provides a comprehensive review of electrical energy storage technologies used to integrate renewable energy sources to the grid. Recent advances and maturity level of the ESSs is also addressed. ESSs are compared based on efficiency, response time and storing capacity and will help researchers and power utilities identify the best storage technology for their system. The rest of the paper is organized as follows. Section 2 presents the global renewable installation while Section 3 describes the necessity of storing electrical energy. Section 4 presents Energy storage systems while Section 5 presents discussion and recommendation and Section 6 concludes the paper.

2. Global Renewable Installation

The total global installed renewable generation capacity at the end of 2020 reached 2799 GW. Hydropower takes the lion share of the global total with an installed capacity of 1211 GW. Wind and solar come second and third with a total installed capacity of 733 GW and 714 GW, respectively. Other renewables installed include bioenergy with an installed capacity of 127 GW, geothermal 14 GW and marine energy 0.5 GW [16].

There was a 10.3% increase in renewable generation capacity in 2020 with installed capacity of 261 GW. Solar energy leads the installed capacity with an increase of 127 GW (+22%) followed by wind with 111 GW (+18%). Hydropower capacity increased by 20 GW (+2%) and bioenergy by 2 GW (+2%). Geothermal energy increased by 164 MW. Along with the renewed growth of hydropower, this exceptional growth in wind and solar led to the highest annual increase in renewable generating capacity ever seen. Figure 1 depicts the share of the renewable generation capacity. Figure 2 shows the total wind installed capacity for the years 2010–2020. Wind power accounted for a substantial share of electricity generation in several countries in 2020. Global capital expenditures committed to offshore wind power in 2020 surpassed investments in offshore oil and gas. Figure 3 represents the total PV installed capacity for the years 2010–2020 and solar PV had another record-breaking year in 2020. Favorable economics have boosted interest in distributed rooftop systems. Competition and price pressures continued to motivate investment to improve efficiencies.

The energy consumption of different countries is variable and depends on economic development, lifestyle, and weather. The top ten highest consuming countries in descending order are China, USA, India, Russia, Japan, Canada, Germany, South Korea, and Brazil [17]. The per capita consumption of electricity is also highly variable in different countries. Table 1 presents region based renewable generation capacity.

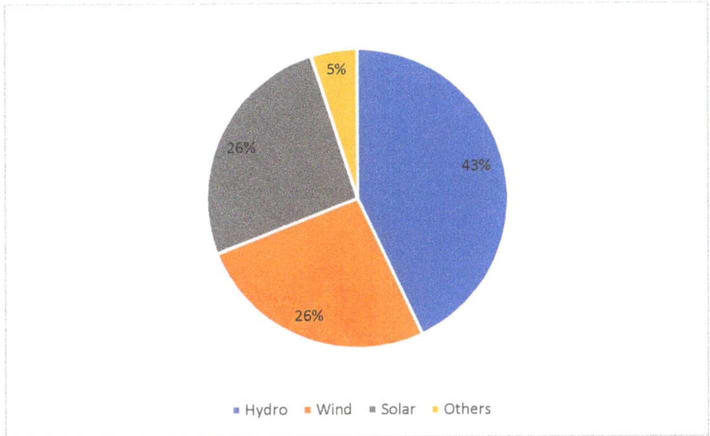

Figure 1. Energy source based renewable generation capacity.

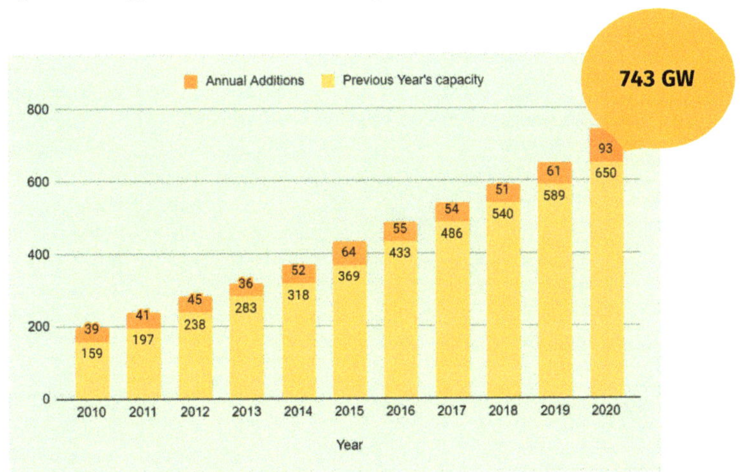

Figure 2. Wind Power Global Capacity and Annual Additions, 2010–2020. Source: [16].

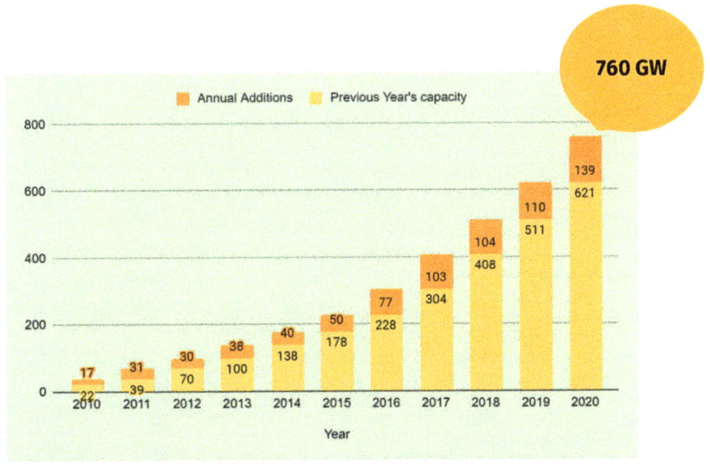

Figure 3. Solar PV Global Capacity and Annual Additions, 2010–2020 [16].

Table 1. Renewable generation capacity by region [17].

Region	Capacity	Global Share	Change	Growth
Asia	1286 GW	46%	+167.6 GW	+15%
Eurasia	116 GW	4%	+6.2 GW	+6%
Europe	609 GW	22%	+34.3 GW	+6%
North America	422 GW	15%	+32.1 GW	+8.2%
South America	233 GW	8%	+9.2 GW	+4.1%
Central America and the Caribbean	16 GW	1%	+0.3 GW	+2.1%
Middle East	24 GW	1%	+1.2 GW	+5.2%
Africa	54 GW	2%	+2.6 GW	+5%
Oceania	44 GW	2%	+6.9 GW	+18.5

Asia's installed capacity reached 1.29 TW in 2020 by increasing its capacity by 167.6 GW. Asia only accounts for 46% of the global total. A huge part of this increase occurred in China. Capacity in Europe and North America expanded by 34 GW (+6.0%) and 32 GW (+8.2%) respectively, with a notably large expansion in the USA. Africa continued to expand steadily with an increase of 2.6 GW (+5.0%), slightly more than in 2019. Although its share of global capacity is small, Oceania remained the fastest growing region (+18.4%).

3. Energy Storage Necessity

The demand for energy fluctuates from peak to off-peak due to individual needs and climatic effects. Storing the excess power during off-peak hours might be an urgent need as generation may surpass the total demand. The power mismatch challenge between generation and demand becomes more relevant because of the intermittency of the RES [18–21]. The conventional grid reliability is affected by the large scale integration of renewable energy sources. It is generally agreed that more than 20% penetration from intermittent renewables can greatly destabilize the grid system. Large scale ESSs can alleviate many of the inherent inefficiencies and deficiencies of the conventional grid and facilitate the full scale integration of renewable energy sources [22–27]. Generally, ESSs can balance supply and demand, reduce power fluctuations, decrease environmental pollution, and increase grid reliability and efficiency.

Recent studies have shown that energy storage facilities, when properly scheduled, are capable of assuring firm power (up to 90% on average of their nameplate capacity) during peak loading conditions. By charging during valleys of net demand and discharging during peak hours, ESSs can make a profit from the differences in energy prices while at the same time enhancing the overall load factor, thereby reducing the need for expensive peak generators, and preventing renewable energy from being spilled. This should be supported by enhanced forecasting and control techniques, and be fully coordinated with demand-side flexibility. Additional markets that could enhance the business case for storage might also emerge in the near future; for example, providing advanced grid functions such as synthetic/virtual inertia/frequency regulation to support system stability.

Small-scale ESS are finding their place in households or small businesses. There might be two main reasons. On the one hand, they can store self-generated energy, typically from PV systems, for later consumption. On the other hand, if connection tariffs are in place, they might be used in order to decrease the network connection sizing, to support consumption at peak times by storing network energy at valley times, regardless of a self-generation system being installed or not. The economics of both applications are dependent on the tariff structure. Electric vehicles (EVs), including transitional technologies such as plug-in hybrids, are expected to play a relevant role.

Large scale energy storage with a capacity of 100 MW is being installed frequently around the world from 2020. According to statistics from the CNESA, the total energy storage installed capacity globally reached 191.1 GW by the end of 2020; an increase of 3.4 % from the previous year [28]. The largest share (around 90%) of the energy storage capacity is covered by pumped hydro with 172.5 GW. The second largest energy storage installed is electrochemical energy storage with an installed capacity of 14.1 GW. Battery energy storage tops the electrochemical storage technologies with an installed capacity of 13.1 GW (Lithium-ion type). In 2020, the scale of electrochemical energy storage projects newly put into operation in the world reached 4.73 GW, and the scale of planned and under construction projects exceed 36 GW; most of them are applied in wind and solar power generation projects. Figure 4 presents the global energy storage installed capacity for the years 2000–2020. Figure 5 shows the electrochemical energy storage types whereas Figure 6 presents the installed electrochemical energy storage capacity for the years 2000–2020. Figure 7 depicts the regional electrochemical energy storage installed capacity for 2020.

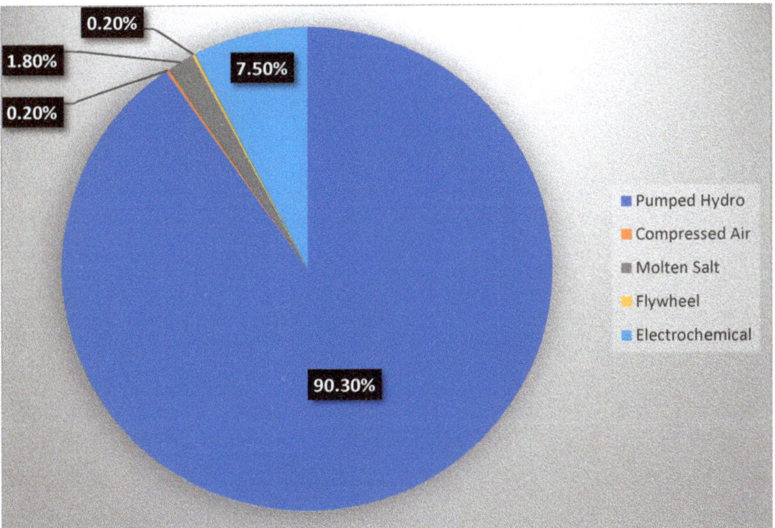

Figure 4. Global energy storage market by total installed capacity (2000–2020).

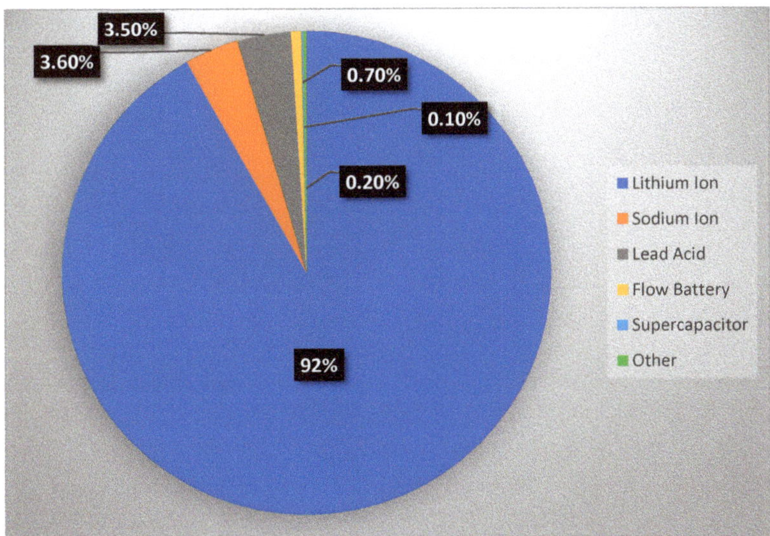

Figure 5. Electrochemical energy storage types.

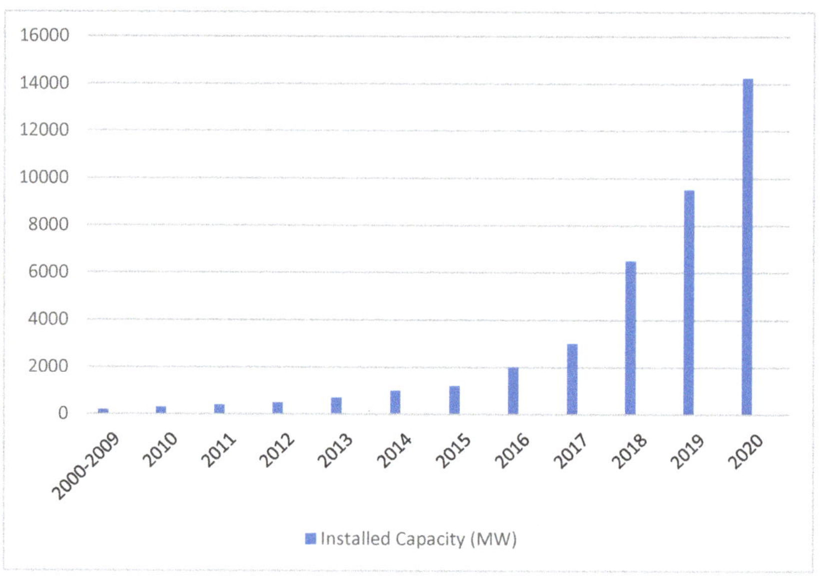

Figure 6. Global electrochemical energy storage market size by cumulative installed capacity (2000–2020) [28].

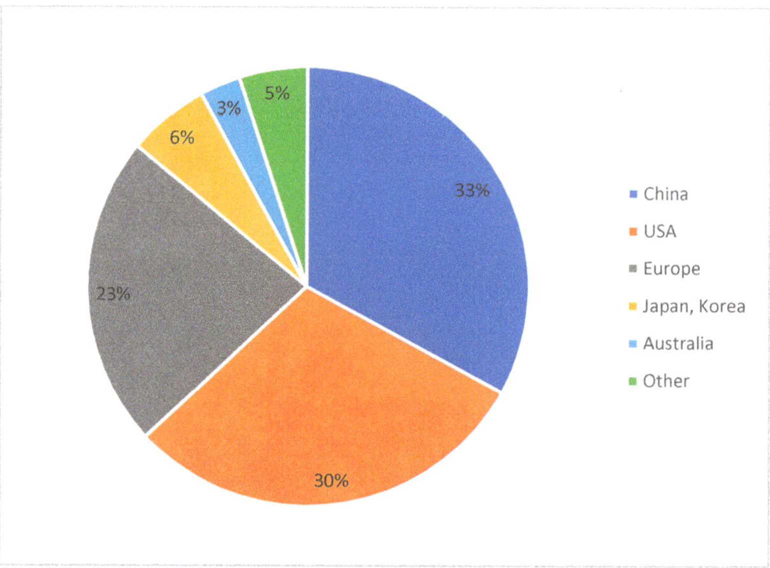

Figure 7. Regional distribution by new installed electrochemical energy storage capacity in 2020 (MW%).

4. Energy Storage Systems

Electrical energy in an AC system cannot be stored electrically. However, energy can be stored by converting the AC electricity and storing it electromagnetically, electrochemically, kinetically, or as potential energy. Each energy storage technology usually includes a power conversion unit to convert the energy from one form to another. Energy storage systems (ESSs) make the power system more reliable and efficient by providing a wide array of solutions including spinning reserves, frequency control, load leveling and shifting, voltage regulation and VAR support, power quality improvement and relief of overloaded transmission lines. The use of artificial intelligence to optimally integrate energy storage systems and renewable energy sources is presented in [29]. The authors of [30] presented a review of machine learning tools for the integration of energy storage systems with renewable sources.

Depending on the method of operation, there are a variety of ESSs such as flywheels, pumped hydro, batteries, supercapacitors, super magnetic energy storage, and compressed air energy storage. Thus, choosing a storage device that can perform the required function efficiently is a preliminary step, as the majority of storage devices are expensive.

Long-term storage may favor chemical fuels as the cost of renewable power generation is decreasing and the curtailment of excess generated power provides an opportunity to convert the renewable power to fuel or chemicals when combining hydrogen with sequestrated or recycled carbon dioxide. Pumped hydro is well established, efficient as well as versatile, and has been around for nearly one hundred years; however, its expansion is limited by geographical, as well as environmental, constraints. Many of the suitable locations for hydro dams are within protected areas, where constructing a dam wall will have an important impact on the eco-system. Underground pumped hydro seems to be a promising alternative in flat regions, but it is still at the design or prototype stage. Compressed air energy storage (CAES) combined with natural gas for incineration in gas turbines appears on all candidate lists, yet only a handful of industrial facilities exist worldwide. Research efforts that are currently underway on the much more efficient adiabatic CAES systems that store the heat generated during compression, to re-inject it during expansion still raise concerns about the technical and economic feasibility of such facilities. Electrochemical batteries are perhaps the most versatile technology (given their outstanding ramping and start-up/shut-down capabilities), but their costs need to be significantly reduced and their life cycle extended. Fast-response AC/DC power converters with sophisticated control strategies are used to integrate ESSs to the electric network. Figure 8 shows the different classification of energy storage systems used in power systems.

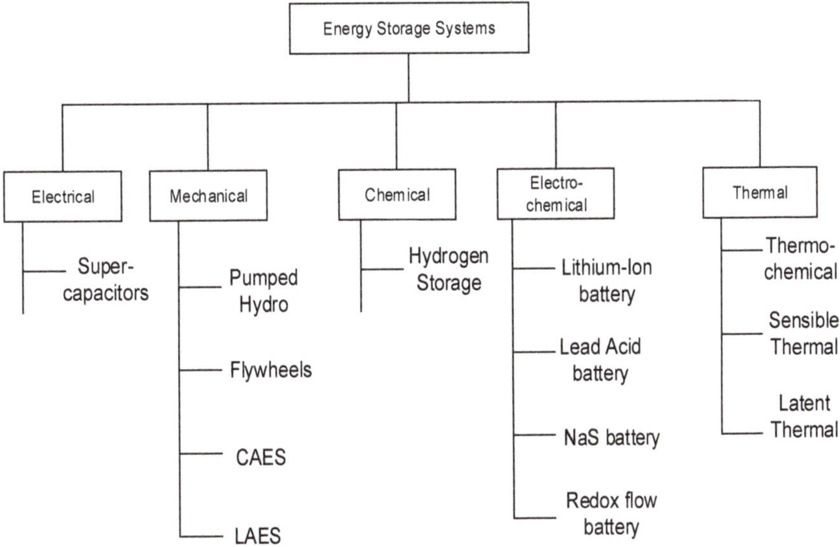

Figure 8. Classification of different energy storage systems.

The amount of energy they can store versus the response speed varies depending on the energy storage selected. A correlation between these two attributes does exist. For instance, supercapacitors are able to store up to about 1 kWh to release in about 1 s, whereas pumping stations can store 10 GWh or more on daily or weekly cycles. Some technologies, such as hydrogen electro-synthesis, would be able to store even greater amounts of energy for even longer periods. Some technologies, such as pumped storage, are quite mature whereas other ones, such as CAES, are still in the research and development (R&D) phase. A review of energy storage systems used in renewable energy resources is presented in [31–33]. Figure 9 shows the technological maturity of the different technologies.

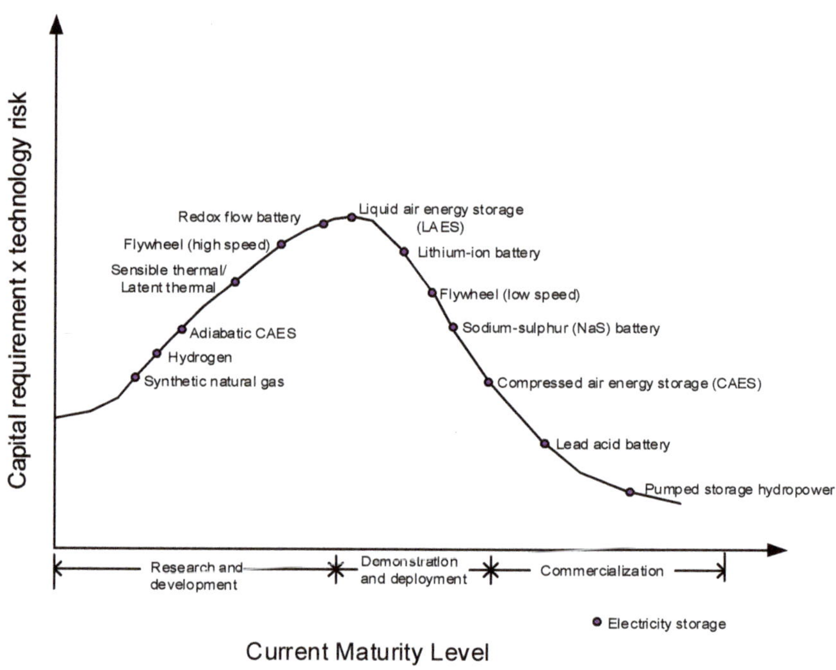

Figure 9. Technology Maturity level of different ESS.

Technological progress is the root to achieving a better energy storage system. In 2020, there were advances in battery technology because of the breakthrough of the cost inflection point of lithium-iron phosphate batteries. In addition, there has been good progress in the development of non-lithium storage systems such as liquid flow batteries, CAES, and sodium ion batteries. CAES is a potential competent of PHS with the advancement of speed reduction technology. Hydrogen storage systems are developing more rapidly and more advanced hydrogen systems will be available in the market. A review of hydrogen energy storage and the impact it will have on the future of renewable source integration is described in [34]. The authors of [35] presented a techno-economic assessment of hydrogen energy storage systems for renewable grid integration. They performed a mixed-integer linear programming formulation to identify key factors that affect cost-effectiveness. To reduce the fluctuation caused by renewable sources, the authors of [36] proposed a nuclear based energy storage system using data-driven stochastic emulators. The role of thermal energy storage integrated with concentrated solar power (CSP) is presented in [37]. The authors concluded that the combination of CSP with thermal energy storage has small role in adding flexibility to the grid. A fuel cell energy storage system integrated with renewable energy sources for reactive scheduling and control is discussed in [38]. A review of artificial intelligence and numerical models for a fuel cell energy storage system integrated with hybrid renewable energy systems are presented in [39]. The authors of [40] studied the economic analysis and optimization of different energy storage systems integrated with renewable energy sources in the island mode. They optimized and compared nine different off-grid renewable energy sources and studied the impact of self-discharge on the energy cost. A review of modeling variable renewable energy and storage in the long-term electric sector is discussed in [41]. A critical overview of energy storage systems, specifically thermal and electrochemical energy storage and their synergies with the development of renewable energy source technologies, is discussed in [42]. A review of hybrid electrochemical energy storage systems for electrified vehicle and smart grid applications is presented in [43]. An effective method for sizing electrical energy storage systems for standalone and grid-connected hybrid systems using energy balance is presented in [44,45]. Some of the energy storage systems used in power systems are explained in detail below.

4.1. Battery Energy Storage Systems (BESS)

Batteries store energy electrochemically and are made of several modules connected in parallel or series to achieve the desired rating. Power electronics converters are required to convert the DC stored energy in batteries to connect it to the AC grid. Batteries have several advantages including high energy density, high efficiency, high life span, and cycling capability [46,47]. Batteries can be designed for bulk energy storage or for rapid charge/discharge [48,49]. The disadvantage of batteries is that they cannot operate at high power levels for a long time due to chemical kinetics. Improving the energy and power density and charging characteristics are active research areas. The other disadvantage of battery energy storage systems is that batteries release toxic gas during battery charge/discharge. The disposal of hazardous materials presents some battery disposal problems [50,51].

Battery energy storage systems are playing a great role in integrating solar photovoltaic power generation to the grid and in reducing the fluctuations. Systems equipped with battery energy storage can deliver both active and reactive power and improve the system voltage and frequency. Beyond these applications focusing on system stability, energy storage control systems can also be integrated with energy markets to make the solar resource more economical [52]. A review of battery energy storage systems with its historical overview and analysis for renewable integration is discussed in [53]. Among the different battery storage systems, the most mature battery technology at this moment is the lead–acid battery [54,55]. A sustainability analysis of a battery energy storage system integrated with a hybrid renewable energy source in the island mode is presented in [56]. Recent advances in non-Vanadium redox chemistries for flow batteries for grid-scale energy storage are discussed in [57]. A case study of a microbrewery under demand response for optimal energy management of a grid-connected photovoltaic system with battery storage is discussed in [58]. A thorough assessment of battery energy storage systems, describing the features and capabilities of each type of battery storage technology including the benefits and drawbacks of each innovation is presented in [59]. A battery energy storage system for the supervisory energy management of a hybrid renewable energy source based on a combined fuzzy logic controller and high order sliding mode methods is discussed in [60]. A case study of the environmental benefit and emissions reductions thresholds of flow battery energy storage systems is presented in [61].

4.2. Flywheel Energy Storage (FES)

Flywheel energy storage stores energy as rotational energy and works by accelerating a cylindrical rotor called a flywheel at high speed. The energy is stored as kinetic energy with the rotating rotor and the storage capacity depends on the mass, shape and the maximum available angular velocity of the rotor. Mechanical inertia is the basis of this storage method and the energy is stored in the rotational mass as kinetic energy. The discharge process begins when an electric generator is connected to the flywheel. Conversely, when a torque is applied to the flywheel, the system is charged. The storage time can be prolonged by keeping the friction as minimum as possible by placing the flywheel in a vacuum [62]. Generally, depending on the speed of operation, FES are divided into two groups. The first group has a maximum speed of 10,000 rpm while the second group has a rotational speed of up to 36,000 rpm [63,64]. FES has a round-trip efficiency of 70–80% with equal discharge and recharge time. FES has approximately 100,000 full charge/discharge cycles and has a power density that is almost ten times greater than that of batteries. Currently, one of the most encountered flywheel applications is the microgrid [65]. The market value of FES is growing fast due to increasing industrial development and population growth causing an increase in power demand [66]. Figure 10 presents the operation principle of a flywheel energy storage system.

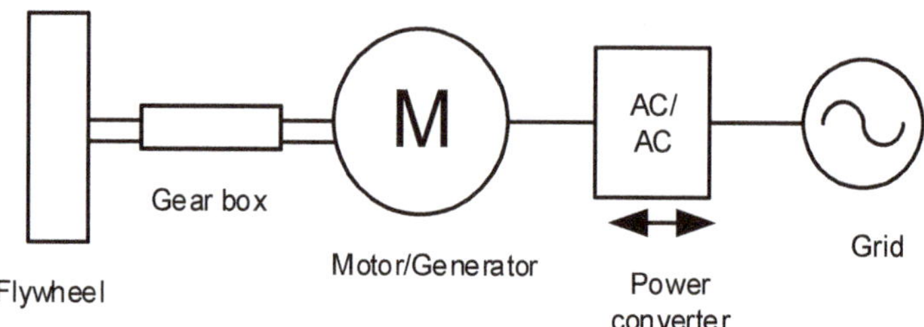

Figure 10. Flywheel Energy storage system.

4.3. Compressed Air Energy Storage (CAES)

The basic working principle of CAES is to drive compressors using motors to compress air and store it in suitable storage vessels. An expander is used to expand the compressed air and release the stored energy. The expander drives a generator to convert the stored energy to produce electricity [67]. A burning natural gas can be used to boost the output power but this will release CO2 emissions and affect the environment [68]. More advanced CAES can store heat during air compression and release it during the expansion phase. CAES are cost effective and promising for bulk grid services as they have a high power rating and storage capacity, a long life time and low self-discharge. However, the start-up time is usually high [69,70]. The economic and reliability impacts of grid-scale storage in a high penetration renewable energy system are presented in [71]. The authors concluded that energy storage systems, specifically CAES, will support the grid inertia if it is synchronously connected for a long duration.

CAES can be used together with renewable energy sources to compress the air using the power generated from renewable energy sources during off-peak hours. During peak-hours the air can be released and converted back to electrical power to make sure that there is no curtailment in the renewable source. Storing fresh air in salt caverns is a proven, reliable and safe method of ensuring that excess energy is not wasted [72–75]. The authors of [76] compared CAES and battery energy storage systems based on a levelized cost of storage. They concluded that the adoption of CAES systems can lead to a better economic performance with respect to battery technologies. The use of combined heat and CAES for wind power peak shaving is presented in [77]. There are only two commissioned CAES worldwide. The first one was commissioned in 1978 in Huntorf, Germany and is 290 MW. The second one is located in Alabama, USA, is 360 MW, and was commissioned in 1991.

4.4. Pumped Hydro Storage (PHS)

PHS is the most mature energy storage technology and has the highest installed generation and storage capacity in the world. It is a type of hydroelectric energy storage which has two water reservoirs (upper and lower) at different elevations that can generate power as water moves down from one to the other (discharge), passing through a turbine.

The lower reservoir is usually a river or lake while the upper reservoir can be an artificial lake [78,79]. The stored water is released during peak demand to hit a turbine and convert it to electrical power similar to a conventional hydropower station. During off peak demand, the upper reservoir is recharged using low cost power or a power generated from renewable energy sources. Similar to CAES, PHS is used for large scale renewable integration and helps the grid in many respects, such as reactive power support, frequency control, and synchronous or virtual inertia and black-start capabilities. The operating cost per energy unit has been reported as the cheapest in the PHS. However, the construction of reservoirs and other infrastructures needs very high investment [80,81]. A review of low-head pumped hydro storage and its application for renewable source integration is presented in [82]. A case study on the potential of a pumped hydropower storage (PHS) system and its contribution to hybrid renewable energy power fluctuation minimization is presented in [83]. The authors used an optimization technique to decrease the PHS sites required for renewable energy source grid integration. The use of PHS with renewable energy sources to fully supply the Barbados grid with a renewable source is discussed in [84]. The authors used open source modelling and concluded that an 80% share of renewable energy sources is cost optimal; however, 100% of renewable systems face flexibility. A comparison between PHS and a fuel cell on a hybrid renewable energy system based on diesel/PV is discussed in [85]. The authors concluded that the use of PHS is more cost effective than fuel cells. A case study to techno-economically compare battery and micro PHS for renewable energy sources is presented in [86]. It was concluded that the use of a hybrid PV-wind-battery storage system is the best option in terms of economic benefits and reliability. Figure 11 depicts the basic operation of a pumped hydro storage system.

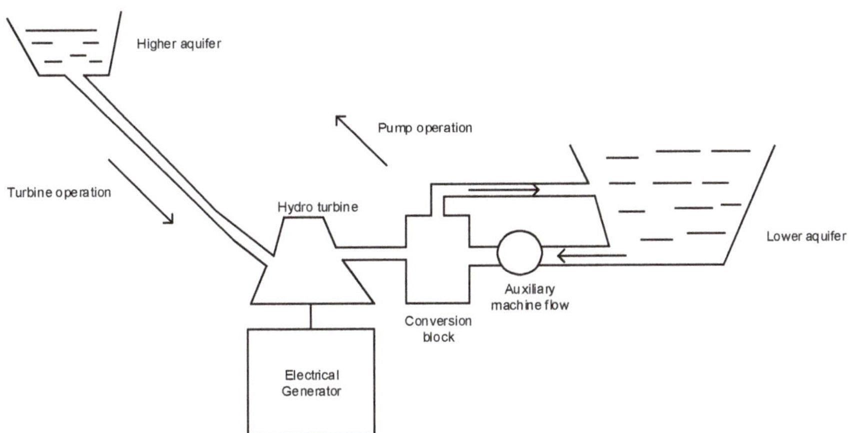

Figure 11. Working principle of Pump storage system.

4.5. Superconducting Magnetic Energy Storage (SMES)

SMES were proposed as an energy storage system because of their high response and efficiency (charge–discharge efficiency over 95%) [87]. The basic configuration of SMES consists of a refrigeration system, superconducting coils and a power conditioning unit. The energy is stored in the superconducting coil at a very low temperature. Figure 12 presents the operation of the superconducting energy storage system. The stored power in the coil can be absorbed or released depending on demand requirements. SMES have applications in load leveling, damping control and the load frequency control of power systems [88–92]. Generally, due to the high costs implied by the superconductive wire and refrigeration, SMES systems are used for military applications or energy storage over short periods of time [93].

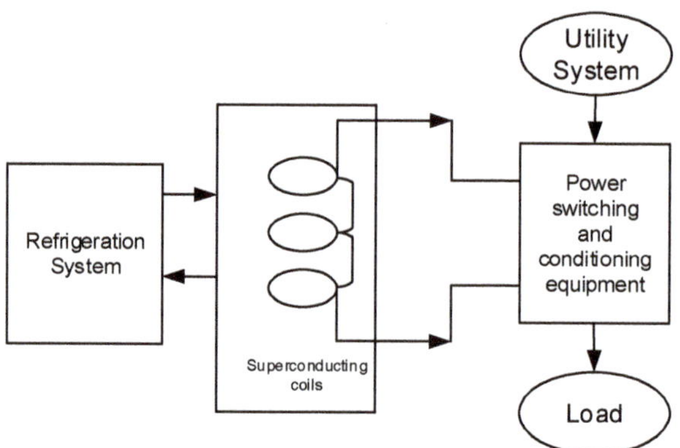

Figure 12. Superconducting Energy storage system.

4.6. Supercapacitor Energy Storage Systems (SCESS)

Among the different energy storage systems, SCESS have been a significant attraction for researchers due to their extraordinary characteristics such as fast charging–discharging, greater power density, lower maintenance cost and environmental-friendliness. Attributed to their outstanding performances, supercapacitors have found applications in diversified areas, e.g., uninterruptible power supplies (UPS), power electronics, renewables integration, and hybrid energy storage. However, the energy density is less than expected [94,95]. The most important advantage of supercapacitors as compared to rechargeable batteries is that supercapacitors in general possess a relatively low internal resistance and can store and deliver energy at a higher power rating.

SCESS help the grid ride through the fault, regulate the voltage and control the frequency and improve the power quality issues [96,97]. They have a high cycle life of around 12 years [98]. Currently, supercapacitors are used together with the batteries especially in smart grid applications due to their shorter discharge time. Figure 13 shows the operation principle of a supercapacitor.

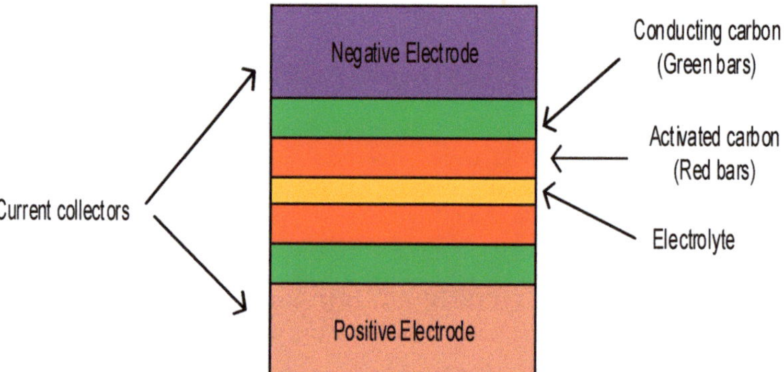

Figure 13. Supercapacitor energy storage system structure.

The use of supercapacitors to minimize the fluctuation of the power generated from PV and wind sources is reported in [99]. The authors connected the supercapacitor with a bi-directional buck-boost converter at the DC link to exchange power with the grid and renewable energy sources. Supercapacitors are also used to ride through a fault. During fault events, the power generated from the renewable energy sources will be stored in the supercapacitor and will be later used when the fault is cleared. Figures 14 and 15 show the topology of supercapacitors used in a PV source. The application of a battery supercapacitors hybrid energy storage system for microgrids is presented in [100]. An optimal design and energy management of an island mode fully renewable based microgrid integrated with battery and supercapacitors is described in [101].

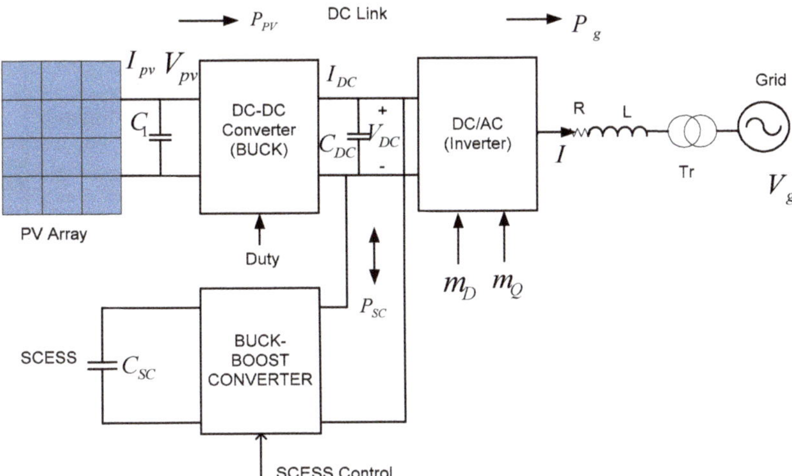

Figure 14. Supercapacitors connected to PV source to minimize the power fluctuation.

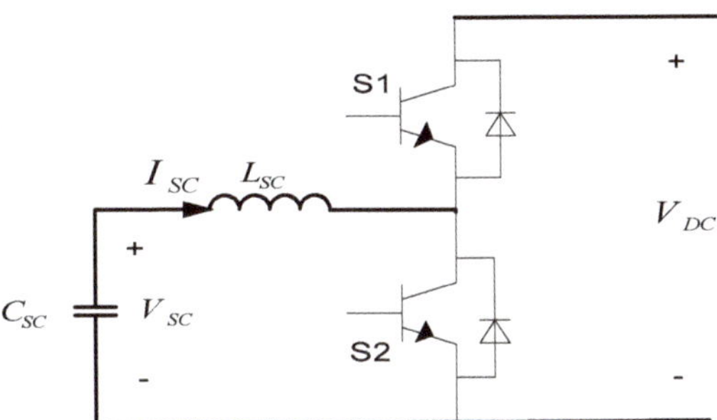

Figure 15. Supercapacitor connected with a bidirectional buck boost converter.

5. Discussion and Recommendation

Except for a few notable exceptions, such as pumped hydro, energy storage technologies are still in their infancy, and significant improvements and cost reductions are expected within a decade as they follow their anticipated learning curve. The life span and cycle life comparison of different energy storage systems is presented in Figure 16. A comparison of different energy storage systems in terms of power density, energy density, response time and efficiency is tabulated in Table 2.

29. Abdalla, A.N.; Nazir, M.S.; Tao, H.; Cao, S.; Ji, R.; Jiang, M.; Yao, L. Integration of energy storage system and renewable energy sources based on artificial intelligence: An overview. *J. Energy Storage* **2021**, *40*, 102811. [CrossRef]

30. Rangel-Martinez, D.; Nigam, K.; Ricardez-Sandoval, L.A. Machine learning on sustainable energy: A review and outlook on renewable energy systems, catalysis, smart grid and energy storage. *Chem. Eng. Res. Des.* **2021**, *174*, 414–441. [CrossRef]

31. Rahman, M.; Oni, A.O.; Gemechu, E.; Kumar, A. Assessment of energy storage technologies: A review. *Energy Convers. Manag.* **2020**, *223*, 113295. [CrossRef]

32. Aneke, M.; Wang, M. Energy storage technologies and real life applications—A state of the art review. *Appl. Energy* **2016**, *179*, 350–377. [CrossRef]

33. Salkuti, S.R.; Jung, C.M. Comparative analysis of storage techniques for a grid with renewable energy sources. *Int. J. Eng. Technol.* **2018**, *7*, 970–976. [CrossRef]

34. Arsad, A.Z.; Hannan, M.; Al-Shetwi, A.Q.; Mansur, M.; Muttaqi, K.; Dong, Z.; Blaabjerg, F. Hydrogen energy storage integrated hybrid renewable energy systems: A review analysis for future research directions. *Int. J. Hydrogen Energy* **2022**, *47*, 17285–17312. [CrossRef]

35. Ross, M.; Bindra, H. Estimating energy storage size for Nuclear-Renewable hybrid energy systems using data-driven stochastic emulators. *J. Energy Storage* **2021**, *40*, 102787. [CrossRef]

36. Wu, D.; Wang, D.; Ramachandran, T.; Holladay, J. A techno-economic assessment framework for hydrogen energy storage toward multiple energy delivery pathways and grid services. *Energy* **2022**, *249*, 123638. [CrossRef]

37. Kennedy, K.M.; Ruggles, T.H.; Rinaldi, K.; Dowling, J.A.; Duan, L.; Caldeira, K.; Lewis, N.S. The role of concentrated solar power with thermal energy storage in least-cost highly reliable electricity systems fully powered by variable renewable energy. *Adv. Appl. Energy* **2022**, *6*, 100091. [CrossRef]

38. Pravin, P.S.; Misra, S.; Bhartiya, S.; Gudi, R.D. A reactive scheduling and control framework for integration of renewable energy sources with a reformer-based fuel cell system and an energy storage device. *J. Process Control* **2020**, *87*, 147–165. [CrossRef]

39. Al-Othman, A.; Tawalbeh, M.; Martis, R.; Dhou, S.; Orhan, M.; Qasim, M.; Olabi, A.G. Artificial intelligence and numerical models in hybrid renewable energy systems with fuel cells: Advances and prospects. *Energy Convers. Manag.* **2022**, *253*, 115154. [CrossRef]

40. Javed, M.S.; Ma, T.; Jurasz, J.; Canales, F.A.; Lin, S.; Ahmed, S.; Zhang, Y. Economic analysis and optimization of a renewable energy based power supply system with different energy storages for a remote island. *Renew. Energy* **2021**, *164*, 1376–1394. [CrossRef]

41. Bistline, J.; Blanford, G.; Mai, T.; Merrick, J. Modeling variable renewable energy and storage in the power sector. *Energy Policy* **2021**, *156*, 112424. [CrossRef]

42. De Rosa, M.; Afanaseva, O.; Fedyukhin, A.V.; Bianco, V. Prospects and characteristics of thermal and electrochemical energy storage systems. *J. Energy Storage* **2021**, *44*, 103443. [CrossRef]

43. Zhang, L.; Hu, X.; Wang, Z.; Ruan, J.; Ma, C.; Song, Z.; Dorrell, D.G.; Pecht, M.G. Hybrid electrochemical energy storage systems: An overview for smart grid and electrified vehicle applications. *Renew. Sustain. Energy Rev.* **2020**, *139*, 110581. [CrossRef]

44. Kichou, S.; Markvart, T.; Wolf, P.; Silvestre, S.; Chouder, A. A simple and effective methodology for sizing electrical energy storage (EES) systems based on energy balance. *J. Energy Storage* **2022**, *49*, 104085. [CrossRef]

45. Jansen, G.; Dehouche, Z.; Corrigan, H. Cost-effective sizing of a hybrid Regenerative Hydrogen Fuel Cell energy storage system for remote & off-grid telecom towers. *Int. J. Hydrogen Energy* **2021**, *46*, 18153–18166. [CrossRef]

46. Hill, C.A.; Such, M.C.; Chen, D.; Gonzalez, J.; Grady, W.M. Battery Energy Storage for Enabling Integration of Distributed Solar Power Generation. *IEEE Trans. Smart Grid* **2012**, *3*, 850–857. [CrossRef]

47. Divya, K.C.; Østergaard, J. Battery energy storage technology for power systems—An overview. *Electr. Power Syst. Res.* **2009**, *79*, 511–520. [CrossRef]

48. Alharbi, H.; Bhattacharya, K. Optimal Sizing of Battery Energy Storage Systems for Microgrids. In Proceedings of the 2014 IEEE Electrical Power and Energy Conference, Calgary, AB, Canada, 12–14 November 2014; pp. 275–280. [CrossRef]

49. Murty, V.V.S.N.; Kumar, A. Retracted article: Multi-objective energy management in microgrids with hybrid energy sources and battery energy storage systems. *Prot. Control Mod. Power Syst.* **2020**, *5*, 1–20. [CrossRef]

50. Lee, Y.; Kim, S.; Hempelmann, R.; Jang, J.H.; Kim, H.J.; Han, J. Nafion membranes with a sulfonated organic additive for the use in vanadium redox flow batteries. *J. Appl. Polym. Sci.* **2019**, *136*, 47547. [CrossRef]

51. Zhang, Z.; Ding, T.; Zhou, Q.; Sun, Y.; Qu, M.; Zeng, Z.; Chi, F.; Ju, Y.; Li, L.; Wang, K. A review of technologies and applications on versatile energy storage systems. *Renew. Sustain. Energy Rev.* **2021**, *148*, 111263. [CrossRef]

52. Martinez, M.; Molina, M.G.; Frack, F.; Mercado, P.E. Dynamic Modeling, Simulation and Control of Hybrid Energy Storage System Based on Compressed Air and Supercapacitors. *IEEE Lat. Am. Trans.* **2013**, *11*, 466–472. [CrossRef]

53. Wali, S.B.; Hannan, M.A.; Reza, M.S.; Ker, P.J.; Begum, R.A.; Rahman, M.S.A.; Mansor, M. Battery storage systems integrated renewable energy sources: A biblio metric analysis towards future directions. *J. Energy Storage* **2021**, *35*, 102296. [CrossRef]

54. Gur, T.M. Review of electrical energy storage technologies, materials and systems: Challenges and prospects for large-scale grid storage. *Energy Environ. Sci.* **2018**, *11*, 2696. [CrossRef]

55. Krishan, O.; Suhag, S. An updated review of energy storage systems: Classification and applications in distributed generation power systems incorporating renewable energy resources. *Int. J. Energy Res.* **2019**, *43*, 6171–6210. [CrossRef]

56. Qi, X.; Wang, J.; Królczyk, G.; Gardoni, P.; Li, Z. Sustainability analysis of a hybrid renewable power system with battery storage for islands application. *J. Energy Storage* **2022**, *50*, 104682. [CrossRef]
57. Emmett, R.K.; Roberts, M.E. Recent developments in alternative aqueous redox flow batteries for grid-scale energy storage. *J. Power Sources* **2021**, *506*, 230087. [CrossRef]
58. Kusakana, K. Optimal energy management of a grid-connected dual-tracking photovoltaic system with battery storage: Case of a microbrewery under demand response. *Energy* **2020**, *212*, 118782. [CrossRef]
59. Olabi, A.; Wilberforce, T.; Sayed, E.T.; Abo-Khalil, A.G.; Maghrabie, H.M.; Elsaid, K.; Abdelkareem, M.A. Battery energy storage systems and SWOT (strengths, weakness, opportunities, and threats) analysis of batteries in power transmission. *Energy* **2022**, *2022*, 123987. [CrossRef]
60. Soliman, M.S.; Belkhier, Y.; Ullah, N.; Achour, A.; Alharbi, Y.M.; Al Alahmadi, A.A.; Abeida, H.; Khraisat, Y.S.H. Supervisory energy management of a hybrid battery/PV/tidal/wind sources integrated in DC-microgrid energy storage system. *Energy Rep.* **2021**, *7*, 7728–7740. [CrossRef]
61. Tian, S.; He, H.; Kendall, A.; Davis, S.J.; Ogunseitan, O.A.; Schoenung, J.M.; Samuelsen, S.; Tarroja, B. Environmental benefit-detriment thresholds for flow battery energy storage systems: A case study in California. *Appl. Energy* **2021**, *300*, 117354. [CrossRef]
62. Wang, L.; Yu, J.-Y.; Chen, Y.-T. Dynamic stability improvement of an integrated offshore wind and marine-current farm using a flywheel energy-storage system. *IET Renew. Power Gener.* **2011**, *5*, 387–396. [CrossRef]
63. Pena-Alzola, R.; Sebastian, R.; Quesada, J.; Colmenar, A. Review of flywheel-based energy storage systems. In Proceedings of the 2011 International Conference on Power Engineering, Energy and Electrical Drives, Malaga, Spain, 11–13 May 2011; pp. 1–6. [CrossRef]
64. Yulong, P.; Cavagnino, A.; Vaschetto, S.; Feng, C.; Tenconi, A. Flywheel energy storage systems for power systems application. In Proceedings of the 2017 6th International Conference on Clean Electrical Power (ICCEP), Santa Margherita Ligure, Italy, 27–29 June 2017; pp. 492–501. [CrossRef]
65. Choudhury, S. Flywheel energy storage systems: A critical review on technologies, applications, and future prospects. *Int. Trans. Electr. Energy Syst.* **2021**, *31*, e13024. [CrossRef]
66. Olabi, A.G.; Wilberforce, T.; Abdelkareem, M.A.; Ramadan, M. Critical Review of Flywheel Energy Storage System. *Energies* **2021**, *14*, 2159. [CrossRef]
67. Breeze, P. Compressed Air Energy Storage. In *Power System Energy Storage Technologies*; Breeze, P., Ed.; Academic Press: Cambridge, MA, USA, 2018; pp. 23–31, ISBN 978012812902. [CrossRef]
68. Dooner, M.; Wang, J. Compressed-Air Energy Storage. In *Future Energy*, 3rd ed.; Letcher, T.M., Ed.; Elsevier: Amsterdam, The Netherlands, 2020; pp. 279–312. [CrossRef]
69. Lemofouet, S.; Rufer, A. A Hybrid Energy Storage System Based on Compressed Air and Supercapacitors with Maximum Efficiency Point Tracking (MEPT). *IEEE Trans. Ind. Electron.* **2006**, *53*, 1105–1115. [CrossRef]
70. Swider, D.J. Compressed Air Energy Storage in an Electricity System With Significant Wind Power Generation. *IEEE Trans. Energy Convers.* **2007**, *22*, 95–102. [CrossRef]
71. Johnson, S.C.; Papageorgiou, D.J.; Harper, M.R.; Rhodes, J.D.; Hanson, K.; Webber, M.E. The economic and reliability impacts of grid-scale storage in a high penetration renewable energy system. *Adv. Appl. Energy* **2021**, *3*, 100052. [CrossRef]
72. Safaei, H.; Keith, D.W.; Hugo, R.J. Compressed air energy storage (CAES) with compressors distributed at heat loads to enable waste heat utilization. *Appl. Energy* **2013**, *103*, 165–179. [CrossRef]
73. Molina, M.G. Energy Storage and Power Electronics Technologies: A Strong Combination to Empower the Transformation to the Smart Grid. *Proc. IEEE* **2017**, *105*, 2191–2219. [CrossRef]
74. Tiano, F.A.; Rizzo, G. Use of an Under-Water Compressed Air Energy Storage (UWCAES) to Fully Power the Sicily Region (Italy) With Renewable Energy: A Case Study. *Front. Mech. Eng.* **2021**, *7*, 37. [CrossRef]
75. Mousavi, S.B.; Ahmadi, P.; Pourahmadiyan, A.; Hanafizadeh, P. A comprehensive techno-economic assessment of a novel compressed air energy storage (CAES) integrated with geothermal and solar energy. *Sustain. Energy Technol. Assess.* **2021**, *47*, 101418. [CrossRef]
76. Salvini, C.; Giovannelli, A. Techno-economic comparison of diabatic CAES with artificial air reservoir and battery energy storage systems. *Energy Rep.* **2022**, *8*, 601–607. [CrossRef]
77. Zhao, P.; Wang, P.; Xu, W.; Zhang, S.; Wang, J.; Dai, Y. The survey of the combined heat and compressed air energy storage (CH-CAES) system with dual power levels turbomachinery configuration for wind power peak shaving based spectral analysis. *Energy* **2021**, *215*, 119167. [CrossRef]
78. Rehman, S.; Al-Hadhrami, L.M.; Alam, M.M. Pumped hydro energy storage system: A technological review. *Renew. Sustain. Energy Rev.* **2015**, *44*, 586–598. [CrossRef]
79. IRENA. *Innovation Landscape Brief: Innovative Operation of Pumped Hydropower Storage*; International Renewable Energy Agency: Abu Dhabi, United Arab Emirates, 2020.
80. Blakers, A.; Stocks, M.; Lu, B.; Cheng, C. A review of pumped hydro energy storage. *Prog. Energy* **2021**, *3*, 022003. [CrossRef]
81. Nadeem, F.; Hussain, S.M.S.; Tiwari, P.K.; Goswami, A.K.; Ustun, T.S. Comparative review of energy storage systems, their roles, and impacts on future power systems. *IEEE Access* **2019**, *7*, 4555–4585. [CrossRef]

82. Hoffstaedt, J.; Truijen, D.; Fahlbeck, J.; Gans, L.; Qudaih, M.; Laguna, A.; De Kooning, J.; Stockman, K.; Nilsson, H.; Storli, P.-T.; et al. Low-head pumped hydro storage: A review of applicable technologies for design, grid integration, control and modelling. *Renew. Sustain. Energy Rev.* **2022**, *158*, 112119. [CrossRef]

83. Qiu, L.; He, L.; Lu, H.; Liang, D. Pumped hydropower storage potential and its contribution to hybrid renewable energy co-development: A case study in the Qinghai-Tibet Plateau. *J. Energy Storage* **2022**, *51*, 104447. [CrossRef]

84. Harewood, A.; Dettner, F.; Hilpert, S. Open-source modelling of scenarios for a 100% renewable energy system in Barbados incorporating shore-to-ship power and electric vehicles. *Energy Sustain. Dev.* **2022**, *68*, 120–130. [CrossRef]

85. Makhdoomi, S.; Askarzadeh, A. Impact of solar tracker and energy storage system on sizing of hybrid energy systems: A comparison between diesel/PV/PHS and diesel/PV/FC. *Energy* **2021**, *231*, 120920. [CrossRef]

86. Shabani, M.; Dahlquist, E.; Wallin, F.; Yan, J. Techno-economic comparison of optimal design of renewable-battery storage and renewable micro pumped hydro storage power supply systems: A case study in Sweden. *Appl. Energy* **2020**, *279*, 115830. [CrossRef]

87. Department of Trade and Industry Report. *Review of Electrical Energy Storage Technologies and Systems and Their Potential for the UK*; DG/DTI/00055/00/00, URN Number 04/1876; UK Department of Trade and Industry: London, UK, 2004; pp. 1–34.

88. Buckles, W.; Hassenzahl, W. Superconducting Magnetic Energy Storage. *IEEE Power Eng. Rev.* **2000**, *20*, 16–20. [CrossRef]

89. Chen, H.; Cong, T.N.; Yang, W.; Tan, C.; Li, Y.; Ding, Y. Progress in electrical energy storage system: A critical review. *Prog. Nat. Sci.* **2009**, *19*, 291–312. [CrossRef]

90. Amaro, N.; Pina, J.M.; Martins, J.; Ceballos, J.M. Superconducting Magnetic Energy Storage–A Technological Contribute to Smart Grid Concept Implementation. In Proceedings of the 1st International Conference on Smart Grids and Green IT Systems (SMARTGREENS), Porto, Portugal, 19–20 April 2012; pp. 113–120. [CrossRef]

91. Hassenzahl, W.V. Superconducting magnetic energy storage. *Proc. IEEE* **1983**, *71*, 1089–1098. [CrossRef]

92. Tixador, P. Superconducting Magnetic Energy Storage: Status and Perspective. *IEEE/CSC ESAS Eur. Supercond. News Forum* **2008**, *3*, 1–14.

93. Ali, M.H.; Wu, B.; Dougal, R.A. An Overview of SMES Applications in Power and Energy Systems. *IEEE Trans. Sustain. Energy* **2010**, *1*, 38–47. [CrossRef]

94. Wang, S.; Wei, T.; Qi, Z. Supercapacitor Energy Storage Technology and its Application in Renewable Energy Power Generation System. In *Proceedings of ISES World Congress 2007*; Goswami, D.Y., Zhao, Y., Eds.; Springer: Berlin/Heidelberg, Germany, 2008; Volumes I–V, pp. 2805–2809. [CrossRef]

95. Bostrom, A.; von Jouanne, A.; Brekken, T.K.A.; Yokochi, A. Supercapacitor energy storage systems for voltage and power flow stabilization. In Proceedings of the 2013 1st IEEE Conference on Technologies for Sustainability (SusTech), Portland, OR, USA, 1–2 August 2013; pp. 230–237. [CrossRef]

96. Chandra Sekhara, P. Design of Supercapacitor Energy Storage System. *Turk. J. Comput. Math. Educ.* **2021**, *12*, 6699–6706.

97. Fares, A.M.; Kippke, M.; Rashed, M.; Klumpner, C.; Bozhko, S. Development of a Smart Supercapacitor Energy Storage System for Aircraft Electric Power Systems. *Energies* **2021**, *14*, 8056. [CrossRef]

98. Navarro, G.; Blanco, M.; Torres, J.; Nájera, J.; Santiago, Á.; Santos-Herran, M.; Ramírez, D.; Lafoz, M. Dimensioning Methodology of an Energy Storage System Based on Supercapacitors for Grid Code Compliance of a Wave Power Plant. *Energies* **2021**, *14*, 985. [CrossRef]

99. Worku, M.Y.; Abido, M.A.; Iravani, R. Power fluctuation minimization in grid connected photovoltaic using supercapacitor energy storage system. *J. Renew. Sustain. Energy* **2016**, *8*, 013501. [CrossRef]

100. Khalid, M. A Review on the Selected Applications of Battery-Supercapacitor Hybrid Energy Storage Systems for Microgrids. *Energies* **2019**, *12*, 4559. [CrossRef]

101. Elmorshedy, M.F.; Elkadeem, M.; Kotb, K.M.; Taha, I.B.; Mazzeo, D. Optimal design and energy management of an isolated fully renewable energy system integrating batteries and supercapacitors. *Energy Convers. Manag.* **2021**, *245*, 114584. [CrossRef]

 *sustainability*

Article

Data-Driven Optimal Battery Storage Sizing for Grid-Connected Hybrid Distributed Generations Considering Solar and Wind Uncertainty

Abdul Rauf [1,2], **Mahmoud Kassas** [1,2] **and Muhammad Khalid** [1,2,3,*]

[1] Electrical Engineering Department, King Fahd University of Petroleum and Minerals, Dhahran 31261, Saudi Arabia
[2] Center for Renewable Energy and Power Systems, King Fahd University of Petroleum and Minerals, Dhahran 31261, Saudi Arabia
[3] SDAIA-KFUPM Joint Research Center for Artificial Intelligence, Dhahran 31261, Saudi Arabia
* Correspondence: mkhalid@kfupm.edu.sa

Abstract: A large-scale renewable-based sustainable power system requires multifaced techno-economic optimization and energy penetration. Due to the volatile and non-periodic nature of renewable energy, the uncertainty of renewables combined with load uncertainties significantly impacts the operational efficiency of renewable integration. The complexities in balancing demand, generation, and maintaining system reliability have introduced new challenges in the current distribution system. Most of the associated challenges can be effectively reduced by using a battery energy storage system (BESS) and the right techniques for handling uncertainties. In this paper, a distributionally robust optimization (DRO) technique with a linear decision rule is formulated for the unit commitment (UC) framework for optimal scheduling of a distribution network that consists of a wind farm, solar PV, a distributed generator (DG), and BESS. To cut the energy cost per unit, BESS plays an important role by storing energy at an off-peak time for on-peak-time use with relatively lower prices. For the all-time minimum overall systems cost, the distribution system requires an optimal size of the BESS to be connected to provide optimal scheduling of DGs. Three case studies are formulated using an IEEE 14 bus system (converted from MW to kW to match the BESS size available in the market) and solved with the proposed distributionally robust optimization technique to achieve the maximum operating point with an optimal capacity of BESS, i.e., wind, solar and hybrid. Each case study has its own optimal 30-min interval schedule for DGs along with the optimal capacity of BESS. The cost comparison with and without BESS and its impact on the start-up and shut down of DGs is reported with all the dynamic economic dispatch results, including the battery's state-of-charge profile. The proposed technique can handle the uncertainties in renewables and allows economical energy dispatch and optimal BESS sizing with comparatively lower computational processing and complexities.

Keywords: unit commitment; battery energy storage systems; wind-farm uncertainty; distributionally robust optimization; solar pv uncertainty; distributed generators

Citation: Rauf, A.; Kassas, M.; Khalid, M. Data-Driven Optimal Battery Storage Sizing for Grid-Connected Hybrid Distributed Generations Considering Solar and Wind Uncertainty. *Sustainability* **2022**, *14*, 11002. https://doi.org/10.3390/su141711002

Academic Editor: George Kyriakarakos

Received: 13 June 2022
Accepted: 1 August 2022
Published: 2 September 2022

Publisher's Note: MDPI stays neutral with regard to jurisdictional claims in published maps and institutional affiliations.

1. Introduction

1.1. Motivation

Due to depletion of fossil fuels decade after decade with the rise in energy demand, renewable energy sources are a better alternative for the energy demands of the emerging population of the earth. In addition, the climate of the earth is severely impacted due to carbon emissions coming out of the energy generation plants using these types of fuels. To cut the usage of fossil fuels, renewable energy sources play a major role, i.e., solar photo-voltaic (PV) and wind power. These sources are the most abundantly available energy sources around the world [1,2]. Due to the volatile and non-periodic nature of renewable energy, the complexities in balancing load with generation and maintaining system reliability have introduced new challenges in the current distribution system [3]. These uncertainties cost money in terms of making the whole system capable of dealing with unwanted circumstances, i.e., optimum economic dispatch and sensible unit commitment, while considering other sources of energy to minimize the total system cost.

1.2. Literature Review

Unit commitment determines when a particular unit should be started so that, at a given time when the power is needed, it can be provided because each unit needs time to startup and shutdown. These startups and shutdowns have an economic impact in terms of fuel, manpower, apparatus, and other related costs [4,5]. On the other hand, the economic dispatch of the generating units determine how much power each unit produces at a certain time to cope with the load and overcome losses due to the transmission network. Ultimately, to minimize the cost of the generating units, scheduling of the generating units is performed based on the load profile data, whether it is one day ahead or one hour [6]. The problem of economic dispatch and UC becomes more complex when there is an intermittent and uncertain renewable energy source, i.e., solar PV.

For better approximations and more reliable results, more historical data is needed to access the true distribution of the data, i.e., the normal distribution, mean, variance, standard deviation, and probabilities of specific events that behave periodically [7]. Similarly, high penetration of renewable sources means more and more historical data is needed to deal with the uncertain profile of those renewable sources [8–10]. Handling this data distribution is performed in [11] using a method proposed to adjust the parameter of the distribution in distributionally robust optimization. Unit commitment and energy management/economic dispatch in [12,13] use distributionally robust optimization with chance constrained. A method proposed in this system uses mean and variance for the distribution set to handle uncertainty in renewable energy sources for the micro-grid and performs the energy dispatch by not considering unit commitment as constrained to reduce the start-up/shut-down cost.

The uncertainty of wind and solar power needs to be modeled to use for the current power generation system [13,14]. There are many models proposed to deal with the uncertainty of wind power generation so that the UC of all the generating units becomes possible. From the numerous uncertainty models proposed for UC over the last decade, stochastic programming [15,16] is being extensively explored to improve the scheduling of UC decisions under wind farm uncertainty [17,18]. The method proposed in [19] uses an alternative scenario selection method to check consistency with the moments of a wind time series and to explicitly specify the modelers that are believed to significantly influence the performance of the unit commitment schedule. Nevertheless, stochastic programming performs a lot of computations pertaining to various scenarios to outline and formulate the expected results. This leads to undesirable computational time and space as well as requiring a bigger processing unit. To address a large number of scenarios for a given problem, some advanced scenario selection algorithms [18] and decomposition techniques have been developed, i.e., at the first stage, mixed-integer programming provides the results and modeling approaches for deterministic UC, i.e., priority listing, Lagrangian relaxation, and dynamic programming; later on, stochastic programming changes the conventional way of solving such UC problems [20,21]. Then, combined stochastic UC and robust UC formulations are introduced for unit commitment decisions to overcome problem solving time required due to the larger set of scenarios in stochastic optimization and the conservativeness of the robust optimization; the weights of the events for the robust and stochastic parts in the objective function are adjusted by the system operator to obtain optimal and time-saving results [22]. However, a feasible and effective realization requires an optimal compromise between the degree of accuracy and computational time.

Besides the aforementioned methods above, a comprehensive formulation of security-constrained unit commitment with compressed air energy storage and wind generation is developed in [23]. Simultaneous optimization of energy and ancillary services with storage is proposed and justified with case studies and results. With the help of case studies, the impact of compressed air energy storage on both economical and technical aspects is reported. Pumped hydro energy storage is also an option for storing the energy in a dynamic economic dispatch problem where unit commitment is a major part of reducing the cost of power produced by thermal generators. This type of energy storage is being used in [24], where deterministic unit commitment and interval unit commitment formulations are modeled that co-optimize the UC and pumped hydro energy storage decisions considering hydraulic limits constraining the pumped hydro energy storage. The study is performed on the system for day-ahead decisions and the results are reported. It employs the stochastic unit commitment model to solve the small-scale (available) wind power.

The optimal sizing of BESS in a system that includes renewable resources is discussed in many papers due to the uncertainties of the renewable resources that could be overcome with the integration of the BESS. Reference [25] suggested a method to find the optimal size of BESS integrated with wind turbines and considered wind and load uncertainties. Two parts of the operational strategy are proposed in [26] for distribution companies. Those two parts are day-ahead and real-time, where, in day-ahead, the load, wind, solar, and prices should be forecasted and in real-time, the gap between the forecasted and real value is considered. The objective here is to locate BESS in the network to minimize the cost and the loss by developing a highly nonlinear model for the network and BESS. Stochastic planning formulation for BESS in a micro-grid using Monte Carlo simulations (MCS) is proposed in [27]. Although uncertainties in wind generation are considered, the network constraints are neglected. Reference [28] used a similar Stochastic approach but with a Radial IEEE distribution system nonlinear model to simulate the line and voltage limits. A deterministic approach to optimize the battery sizing and location in the distribution network is proposed in [29]. The effects of uncertainty in wind power resources using the point estimate method (PEM) are enlisted in [30], where the naturally inspired PSO is combined with the tabu search (TS) to solve the problem. For the optimum size and location of the storage system, the stochastic mixed integer linear programming (MILP) technique is used in the distribution system to minimize the costs of investment and operation [31]. In [32], two levels of the profit-maximizing strategy were introduced, including planning and control. For optimal BESS planning and control in the primary control market, a framework has been developed to achieve the goal. This developed framework provides the balance between the capital cost and the operating cost considering the energy capacity factor. BESS degradation was considered here as a weighting factor that depends on the BESS lifetime.

The charging and discharging rate (C), or the rate at which the power is provided by the battery at any instant, is directly related to battery size. For the better operation of energy storage, an optimum value of the charging and discharging rates of the battery play a major role [33,34]. By using a smaller battery size but larger power requirement, the battery will discharge more quickly than its rated power discharge, which will not only reduce the battery life but also damage the battery cell by increasing the temperature of the battery due to higher power losses inside the battery [35,36]. With a small storage capacity and lesser charging and discharging rates, the battery will not be fully charged/discharged as compared to the bigger storage size for the same power requirement.

For a high penetration of renewable energy sources, more historical data will be required for distribution to be more contracted and to achieve true distribution (i.e., normal distribution) for any uncertain parameter, which makes the optimization model more complex in terms of solving techniques, i.e., the quantization of the normal distribution curve for each event generated by the CPLEX compiler [8]. To deal with the distribution, ref. [11] proposed a method to address economic dispatch and energy management, using distributionally robust optimization with chance constrained, which uses only mean and variance for the distribution set to forecast wind for the micro-grid, and performs the energy dispatch, but unit commitment constraints are not considered for units scheduling. In [12], UC is performed with DRO using chance constrained for wind-farm uncertainty, but the impact of BESS on unit scheduling is not studied. According to the model solved through chance constrained, the total cost of the system increases with a higher confidence bound due to an uncertainty error. Due to rapid change in the total system cost using DRO and chance constrained, this model creates the research gap for improvements in terms of handling the distribution curve. In addition, the chance-constrained transformation increases the optimization complexity if the distribution is precisely modeled over a set of linear values for random numbers [37]. In addition, with higher uncertainties, chance constraint becomes more computationally complex for random numbers [38]. An ambiguity set method was developed in [39] which was based on historical data. In fact, more historical data gives more ambiguous sets and, consequently, a less conservative solution. The case study conducted in this paper shows that the more information and maximum confidence bound in uncertainty an error is, the lower will be the system total cost. A model discussed in [40] describes the estimated and real energy consumption by an energy storage system. The proposed model estimates the energy consumption by electric vehicles with traffic flow theory and mechanics of locomotion. The author focuses on the percentage usage of ESS in terms of estimated and actual usage of ESS by using floating cara data (FCD) with available data provided by information and communications technology (ICT) devices.

As discussed above, the impact of ESS on UC with uncertainties of RES is not discussed with the help of distributionally robust optimization. Some topics in the literature are near to the topic discussed in this research work. A robust optimization approach for designing an off-grid solar-powered charging station is proposed to provide electric vehicles (EVs) with electricity and hydrogen vehicles (HV) with hydrogen using deterministic mixed-integer linear programming (MILP) and robust optimization in [41]. Therein, a robust optimization approach was employed to design the charging station based on the different levels of system robustness against the uncertainties. This stochastic approach provides worst-case solutions for the uncertainties, i.e., either the smallest or the largest outcome for the uncertainty to minimize or maximize the objective function. Due to worst-case outcomes, this model gives a robust solution. The sizing and siting of the RES for the local distribution system having solar PV uncertainty are designed using DRO and cone relaxation techniques. By using the cone relaxation technique, the DRO becomes mixed-integer second-order cone programming (MISOCP) [42]. The second-order cone programming (SOCP) makes the system more complex, and the solution time also increases as compared to the linear decision rule where nonlinear functions (equations) are linearized by finding their respective linear coefficients for their cumulative linear function (equation).

1.3. Contribution

In this paper, a method is developed to demonstrate the impact of BESS on the DG schedule and total cost of the system under hybrid uncertainty using a distributionally robust optimization approach. In the proposed model using distributionally robust optimization with a linear decision rule, the ambiguity set gives the freedom to not only control the distribution but also the computational burden by providing the dual gap between the random variables and the auxiliary variables [43]. A linear decision rule has an advantage over the chanced constrained as this method provides the vicinity to the solver to adopt the complexity of choosing the value from the random distribution by using the duality gap given by the solver. These gaps are the stages for the linear piece-wise function over the quadratic function of the probability distribution which grows moderately with the stages that are involved by the solver in terms of the dual gap. In light of the above literature and propositions, BESS is not being considered with UC for solar PV and wind-farm uncertainty using distributionally robust optimization. However, BESS, as shown in Figure 1, not only reduces the startup cost of DGs under given conditions but also mitigates the problem/challenges associated with the randomness of the wind farm and solar PV by providing power to the system with high ramp up as well as acting as a storage tank [44–46]. The behavior of the BESS as a storage tank also provides the energy at peak load time by storing the energy at off-peak load time. In this way, at an off-peak time, the surplus energy produced by the generators having the least cost is stored to BESS to be used at peak load time, which reduces the running cost of the generators by generating power with a significant figure.

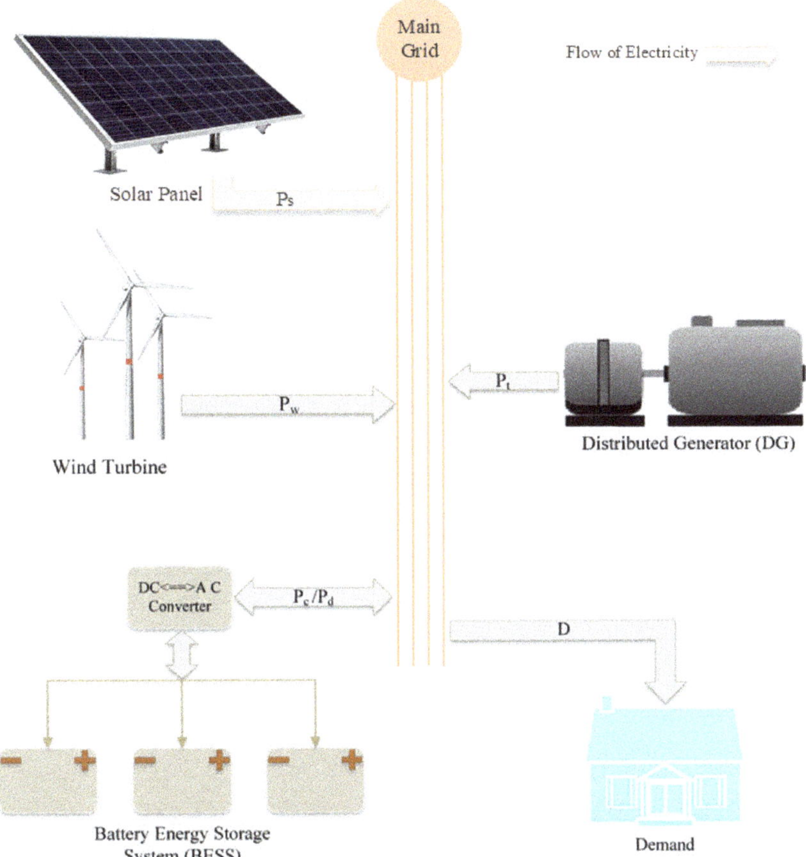

Figure 1. The schematic diagram of the proposed system having distributed generator, solar PV, wind-farm, and battery energy storage system.

The significance of this model includes its flexibility in adding more DGs, solar PVs, wind farms, and ESSs. The contribution of this paper includes:

1. Optimal battery energy storage system sizing with the unit commitment of DG's/thermal units on an IEEE 14 bus system, considering day-ahead solar PV and wind-farm uncertainties by using a distributionally robust optimization technique with a linear decision rule and distribution of the uncertain solar PV and wind output data.
2. Cost comparison with different sizes of battery energy storage system on the unit commitment of DG's/thermal units. Where the day-ahead 30-min duration of unit commitments with battery energy storage systems are discussed.

The inclusion of BESS has an important impact on the system, as it reduces the system startup cost as well as adhering to the uncertain behavior of solar PV energy.

1.4. Paper Organization

The remainder of the paper is organized as follows: Section 2 describes the problem at hand. Mathematical modeling of UC and ambiguity matrix construction with DRO using hybrid generation is discussed in Section 3. Four case studies are conducted in Section 4: 1—sample distributionally robust optimization, 2—an IEEE 14 bus system with optimal BESS (30-min of solar PV data), 3—an optimal BESS for an IEEE 14 bus system (30-min wind data), 4—the optimal BESS for hybrid RESs under given loads. The results obtained through these case studies illustrate the significance of the model proposed. Finally, the conclusion followed by future work, and acknowledgments, are listed in Section 5.

2. Problem Description

Unit commitment provides the necessary scheduling data required for optimal operation between generation and demand because each unit needs time to startup and shut down. These startups and shutdowns cost money in terms of fuel and some other costs related to manpower and apparatus. Therefore, to minimize that cost, scheduling of the generating units is performed based on the load profile data, whether it is one day ahead or one hour. The problem of economic dispatch and UC becomes more complex when there is an intermittent and uncertain renewable energy source, i.e., solar PV and wind farms. The uncertainty of solar PV and wind farms needs to be modeled for use in the current power generation system. There are many models proposed to deal with the uncertainty of solar PV and wind-farm generation so that the UC of all the generating units becomes possible. The complexity of the problem may be reduced if an energy storage system were to be introduced to overcome the sudden changes so, that the load duration curve and the uncertain output of solar PV and wind farm will become smooth and output power requirements from all the thermal generators will be met at all times at the minimum cost.

3. Problem Methodology

3.1. Mathematical Formulation of Unit Commitment

The unit commitment model listed below provides here and now solution while considering the deterministic wind power output in (1). Likewise, UC model with deterministic solar PV is listed in (2).

$$\Xi(\boldsymbol{x}, \boldsymbol{o}) = \min \sum_{t \in \tau} \left\{ \sum_{k \in \kappa} (c_{k,t} + z_{k,t}) + P^{sur} \sum_{b \in \beta} p_{b,t}^{cur} \right\} \tag{1}$$

$$\Xi(\boldsymbol{x}, \boldsymbol{\psi}) = \min \sum_{t \in \tau} \left\{ \sum_{k \in \kappa} (c_{k,t} + z_{k,t}) + P^{sur} \sum_{b \in \beta} p_{b,t}^{cur} \right\} \tag{2}$$

The deterministic mathematical model for the UC with hybrid RES is described in (3). This model is derived linearly from the above two models for wind farms and solar PV in (1) and (2), respectively.

$$\Xi(\boldsymbol{x}, \boldsymbol{o}, \boldsymbol{\psi}) = \min \sum_{t \in \tau} \left\{ \sum_{k \in \kappa} (c_{k,t} + z_{k,t}) + P^{sur} \sum_{b \in \beta} p_{b,t}^{cur} \right\} \tag{3}$$

s.t.

$$z_{k,t} \geq C_k^s(x_{k,t} - x_{k,(t-1)}) \quad \forall k \in \kappa, \quad t \in \tau \tag{4}$$

$$c_{k,t} \geq a_k p_{k,t}^t + b_k x_{k,t} \quad k \in \kappa, \quad t \in \tau \tag{5}$$

$$\sum_{k \in \kappa} p_{k,t}^t + \sum_{l \in \Lambda} p_{l,t}^s + \sum_{j \in \mathcal{J}} p_{j,t}^w + \sum_{b \in \beta} p_{b,t}^d = \sum_{b \in \beta} D_{b,t} + \sum_{b \in \beta} p_{b,t}^c + \sum_{b \in \beta} p_{b,t}^{cur} \quad \forall t \in \tau \tag{6}$$

$$\sum_{b \in \mathcal{B}} LCF_{g,b} \left(\sum_{k \in \kappa} p_{k,t}^t + \sum_{l \in \Lambda} p_{l,t}^s + \sum_{j \in \mathcal{J}} p_{j,t}^w + p_{b,t}^{cur} - D_{b,t} \right) \leq LC_g \quad \forall g \in \mathcal{L}, \quad \forall t \in \tau \tag{7}$$

$$-\sum_{b\in\mathcal{B}} LCF_{g,b}\left(\sum_{k\in\kappa} p^t_{k,t} + \sum_{l\in\Lambda} p^s_{l,t} + \sum_{l\in\mathcal{J}} p^w_{j,t} + p^{cur}_{b,t} - D_{b,t}\right) \le LC_g \quad \forall g\in\mathcal{L}, \quad \forall t\in\tau \qquad (8)$$

$$\underline{P}_k x_{k,t} \le p^t_{k,t} \le \overline{P}_k x_{k,t} \quad \forall k\in\kappa, \quad \forall t\in\tau \qquad (9)$$

$$0 \le p^s_{l,t} \le \overline{s}_{l,t} + \psi_{l,t} \quad \forall l\in\Lambda, \quad \forall t\in\tau \qquad (10)$$

$$0 \le p^w_{j,t} \le \overline{w}_{j,t} + o_{j,t} \quad \forall l\in\mathcal{J}, \quad \forall t\in\tau \qquad (11)$$

$$p^t_{k(t-1)} - p^t_{k,t} \le RD_k \cdot x_{k,t} + \overline{P}_k(1 - x_{k,t}) \quad \forall k\in\kappa, \quad \forall t\in\tau \qquad (12)$$

$$p^t_{k,t} - p^t_{k,(t-1)} \le RU_k \cdot x_{k,(t-1)} + \overline{P}_k\left(1 - x_{k,(t-1)}\right) \quad \forall k\in\kappa, \quad \forall t\in\tau \qquad (13)$$

In the above formulation, the objective function in (3) includes startup cost $z_{k,t}$, generation cost $c_{k,t}$ and power loss $p^{cur}_{k,t}$. Startup cost and generation cost is calculated through (4) and (5), respectively. In (4), C^s_k is the fixed startup cost for unit k, $x_{k,t}$ is the binary variable for generator status at time t, $x_{k(t-1)}$ is the binary variable for generator status at time $t-1$ and $x_{k,t} - x_{k(t-1)}$ gives the binary difference "1" if a particular DG was turned off at time $t-1$ and it is turned on at time t and "0", if a particular DG was turned on at time $t-1$ and it is still turned on at time t. In (5), a_k is the generator k^{th} first-order cost parameter, and b_k is the generator k^{th} second-order cost parameter. The Equation (6) represents the power balance between the generation and demand, as $p^t_{k,t}$ is the power generated by the kth DG unit at time t, $p^s_{l,t}$ is the power generated by the lth solar PV unit at time t, $p^w_{j,t}$ is the power generated by the jth wind-farm unit at time t, $p^d_{b,t}$ is battery discharge by the battery storage connected to b^{th} bus at time t, $p^c_{b,t}$ is battery charge (acting as a load) by the battery storage connected to b^{th} bus at time t and $D_{b,t}$ is the load demand at b^{th} bus in time t, $p^{cur}_{b,t}$ is the power surplus/loss at bus b in time t, if it is negative then there is power surplus at bus b in time t and if positive then it is load shedding at bus b in time t. The dc power flow model confines the power transmission below the line capacity LC_g in inequalities (7) and (8) [47]. The maximum and minimum capacity of the DG units is defined in (9). In (10), solar PV generation is less than the maximum capacity of the solar PV unit l at time t and (11) defines that the wind-farm generation is less than the maximum capacity of the wind-farm unit j at time t. The ramp-up RU_k and ramp-down RD_k limits of DG units are enforced in (12) and (13).

3.2. Power Flow Model through Transmission Lines

The distributionally robust optimization with the linear decision rule model used in this paper solves the UC problem by using mixed integer linear programming (MILP). The DRO with linear decision rule is a two-stage optimization model, where unit commitments are determined at the first stage with the help of MILP and economic dispatch and cost calculations in the second stage with the standard normal distribution of the random variables used in the model to obtain the expectation of the random variable presenting the uncertainty of wind and solar. The second stage of the optimization makes the system nonlinear, but DRO solves the model with MILP by using the linear decision rule. The linear decision rule transforms the nonlinear model to its equivalent linear model. Similar is the case for power flow through the transmission lines by using Newton–Raphson and Gauss–Seidel; the DRO model solves the equivalent linear model of the nonlinear equations related to voltages, phase angles, and impedances in a network containing nodes/buses. Instead of finding the linear model for these methods, dc power flow is also one of the linear models being used to give an estimation of power flow and overcome the challenge being faced by the DRO method. The Q-power flow and Q-losses are not considered at this stage to make the problem simple, but the AC power flow can also be conducted using the distflow method at a later stage to examine the exact Q-power flow and Q-losses and their impacts on ESS.

The power flows through the transmission lines in the proposed model are modeled using the DC power flow method as listed in (14), the net active power injection P_b at bus b is the difference between power generation and the power demand at that bus, i.e., $P_b = P_{Gb} - P_{Db}$.

$$P_b = \sum_{b'=1}^{N} B_{bb'}(\theta_b - \theta_{b'})$$ (14)

where $B_{bb'}$ is the reciprocal of the reactance between bus b and bus b', and θ_b and $\theta_{b'}$ are the voltage angles of bus b and bus b', respectively. The active power flow $P_{Lbb'}$ between buses b and b' can be calculated from (15). These line flows are confined in the form of (7) and (8), as discussed earlier.

$$P_{Lbb'} = \frac{1}{X_{Lbb'}}(\theta_b - \theta_{b'})$$ (15)

where $X_{Lbb'}$ is the reactance of line between bus b and b'.

3.3. Battery-Energy-Storage-System Modeling

The objective function for the battery energy storage system modeled is described in (16), where the cost and charge cycles of BESS are minimized so that the overall cost of the system is minimized, as discussed in the here and now objective functions in (1), (2) and (3)

$$C_B^T(o, \psi) = \min \sum_{b \in \beta} \left\{ \sum_{t \in \tau} C^{fix} p_{b,t}^d + C^{fix} SOC_b^{max} + B_{count,b}^T \right\}$$ (16)

s.t.

$$SOC_{b,t} = SOC_{b,t-1} + (p_{b,t}^c \eta_c - p_{b,t}^d / \eta_d)\Delta t \quad \forall b \in \beta, \quad \forall t \in \tau$$ (17)

$$p_{b,t}^{c,min} \leq p_{b,t}^c \leq p_{b,t}^{c,max} \quad \forall b \in \beta, \quad \forall t \in \tau$$ (18)

$$p_{b,t}^{d,min} \leq p_{b,t}^d \leq p_{b,t}^{d,max} \quad \forall b \in \beta, \quad \forall t \in \tau$$ (19)

$$SOC_{b,t}^{min} \leq SOC_{b,t} \leq SOC_{b,t}^{max} \quad \forall b \in \beta, \quad \forall t \in \tau$$ (20)

$$SOC_{b,t}^{max} \geq 0 \quad \forall b \in \beta, \quad \forall t \in \tau$$ (21)

The state of charge of the battery is also linked to its efficiency of charging η_c and discharging η_d in (17). Maximum/minimum charging and discharging of the battery are constrained in (18) and (19), respectively. In (20) battery's state of charge is also restricted with minimum and maximum limits. Most importantly, the constraint for battery sizing is listed in (21), where the maximum size of BESS with the minimum objective cost is determined through (23).

To minimize the BESS charge and discharge cycles for optimum battery life, an algorithm is listed in Algorithm 1 to count for charge ($B_{count,b}^c$) and discharge ($B_{count,b}^d$) cycles at bus b for a complete time period during which the optimization and UC is performed.

$$B_{count,b}^T \geq max(B_{count,b}^c, B_{count,b}^d) \quad \forall b \in \beta$$ (22)

$$BC_b = max(SOC_{b,t}^{max}) \quad \forall b \in \beta, \forall t \in \tau$$ (23)

The total charge/discharge cycles count ($B_{count,b}^T$) is calculated using (22). This constraint in the objective function forces the battery to optimally charge and discharge, which helps the battery to have a longer life. Long-lifespan BESS has the least O&M cost with respect to those having a shorter lifespan.

Algorithm 1 Battery charge/discharge cycle counter

$t \leftarrow 1$
$B^c_{count,b} \leftarrow 0$
$B^d_{count,b} \leftarrow 0$
if $P_d(t-1) > 0$ & $P_c(t) > 0$ & $SOC(t) \leq 20\%$ **then**
    $B^d_{count,b} \leftarrow B^d_{count,b} + 1$
end if
if $P_c(t-1) > 0$ & $P_d(t) > 0$ & $SOC(t) \geq 80\%$ **then**
    $B^c_{count,b} \leftarrow B^c_{count,b} + 1$
end if

3.4. Distributionally Robust Optimization Model

The models demonstrated in (3) and (16) are used to formulate the solar PV uncertainty based on (25) and wind-farm uncertainty based on (24) that have an average value ($\bar{s}_{l,t}$) of the solar PV output and average value ($\bar{w}_{j,t}$) of the wind-farm output produced with difference between expected and actual power ($\tilde{\psi}_{l,t}$) and ($\tilde{o}_{j,t}$), respectively.

$$\tilde{w}_{j,t} = \overline{w}_{j,t} + \tilde{o}_{l,t} \quad \forall j \in \mathcal{J}, \quad \forall t \in \tau \tag{24}$$

$$\tilde{s}_{l,t} = \overline{s}_{l,t} + \tilde{\psi}_{l,t} \quad \forall l \in \Lambda, \quad \forall t \in \tau \tag{25}$$

Both models share the same steps to derive the model for distributionally robust optimization. Therefore, in this paper, a distributionally robust optimization model with solar-PV uncertainty is derived to obtain the ambiguity matrix. Based on past data, the mean solar-PV output for the upcoming year can be determined. The difference between expected and actual power ($\tilde{\psi}_{l,t}$) is acting as an uncertain variable in the proposed model. The ambiguity set for uncertain variables in distributionally robust optimization is constructed using the information based on past data, i.e., mean, standard deviation, variance, and confidence bound. In the proposed two-stage problem, UC problem in (26)–(28) gives the here-and-now solution but the model (26)–(28) is intractable in its current form.

$$\min \sup_{\mathbb{M} \in \mathbb{I}} \mathbb{E}_{\mathbb{M}}\{\Xi(x, \tilde{o}, \tilde{\psi})\} \tag{26}$$

s.t.

$$x \in \zeta \tag{27}$$

$$x \in \{0,1\}^{|\kappa| \times |\tau|} \tag{28}$$

where κ is the set of all generators. The set ζ in (27) represents the feasibility region of x defined by minimum up/down time constraints. The objective function minimizes the expectation of function $\Xi(x, \tilde{\psi})$ under a distribution $\mathbb{M}$, which is the worst-case distribution over an ambiguity set $\mathbb{I}$. The function $\Xi(x, \tilde{\psi})$ indicates the economic dispatch cost associated with UC decision x, under solar PV output outcome ψ, which can be calculated by solving the second-stage problem. Simple deterministic UC problem considering mean solar-PV output ($\bar{s}_{l,t}$) can be modeled in a matrix form as in (29) and (30).

$$\min \Xi(x) \tag{29}$$

s.t.

$$Ax \leq b \tag{30}$$

However, considering uncertainty in the output as having the difference between actual and mean value of the solar-PV output ($\tilde{\psi}$) under worst-case distribution, the expected cost $\Xi(x)$ can be formulated using (31).

$$\Xi(x) = \sup_{\mathbb{M} \in \mathbb{I}} \mathbb{E}_{\mathbb{M}}\{\Xi(x, \tilde{o}, \tilde{\psi})\} \tag{31}$$

where

$$\Xi(\boldsymbol{x}, \tilde{\boldsymbol{o}}, \boldsymbol{\psi}) = \min \, \boldsymbol{q}^T \boldsymbol{y} \tag{32}$$

s.t.

$$\boldsymbol{A}\boldsymbol{x} + \boldsymbol{b}\boldsymbol{y} \leq h(\tilde{\boldsymbol{o}}, \boldsymbol{\psi}) \tag{33}$$

The facilitating vector $h(\tilde{\boldsymbol{o}}, \boldsymbol{\psi})$ is defined as

$$h(\tilde{\boldsymbol{o}}, \boldsymbol{\psi}) = h^0 + \sum_{e \in \mathcal{S}} h_r^o o_r + \sum_{r,e \in \mathcal{S}} h_e^\psi \psi_e \tag{34}$$

The second-stage recourse decisions are represented by a vector $(\boldsymbol{y})$ in (32). All the constraints for second stage problems are presented by (33), where the recourse decisions represented by $(\boldsymbol{y})$ are directly related to the first stage decisions $(\boldsymbol{x})$. All the constants in (33), i.e., h^0, h_r^o and h_e^ψ, define the function at right side of the constraint in (33) and these constants are directly affected by the uncertain variables o_r and $\boldsymbol{\psi}_e$.

3.4.1. Modeling Ambiguity Parameters for Solar-PV Uncertainty

To facilitate above model an ambiguity set construction is needed as in (35).

$$\mathbb{I}_\psi = \left\{ \mathrm{M} \in \mathcal{P}_\psi(\mathbb{R}^{|\mathcal{S}|}) : \begin{array}{l} \tilde{\boldsymbol{\psi}} \in \mathbb{R}^{|\mathcal{S}|} \\ \mathbb{E}_\mathrm{M}\{\tilde{\boldsymbol{\psi}}\} = 0 \\ \mathrm{M}\{\tilde{\boldsymbol{\psi}} \in \mathcal{V}\} = 1 \\ \mathbb{E}_\mathrm{M}\{v_i(\tilde{\boldsymbol{\psi}})\} \leq \Gamma_i \quad \forall i \in \mathcal{I} \end{array} \right\} \tag{35}$$

where $\mathcal{P}_\psi(\mathbb{R}^{|\mathcal{S}|})$ denotes the set of all probability distributions on $R^{|\mathcal{S}|}$. The expected value of the random variable $\tilde{\boldsymbol{\psi}}$ is 0, which enforces the mean error signal to be "0" and sum of the probability for all outcomes for $\tilde{\boldsymbol{\psi}}$ is 1. Function v_i in (35) incorporates distribution information into the ambiguity model. The uncertainty set in (36) specifies the lower bound $V_{l,t}^-$ and the upper bound $V_{l,t}^+$ of each random variable $\psi_{l,t}$.

$$\mathcal{V} = \left\{ \boldsymbol{\psi} \in \mathbb{R}^S : V_{l,t}^- \leq \psi_{l,t} \leq V_{l,t}^+ \quad \forall l \in \Lambda, \quad \forall t \in \tau \right\} \tag{36}$$

To obtain the expectation of each function v_i under ambiguous distributions easily, the ambiguity set in (35) needs to be further extended in (37). Here, an auxiliary variable is introduced to make the ambiguity set in (37) less conservative.

$$\mathbb{H}_\psi = \left\{ \mathbb{O}_\psi \in \mathcal{P}_\psi(\mathbb{R}^{|\mathcal{S}|} \times \mathbb{R}^{|\mathcal{I}|}) : \begin{array}{l} (\tilde{\boldsymbol{\psi}}, \tilde{\boldsymbol{\alpha}}) \in \mathbb{R}^{|\mathcal{S}|} \times \mathbb{R}^{|\mathcal{I}|} \\ \mathbb{E}_{\mathbb{O}_\psi}\{\tilde{\boldsymbol{\psi}}\} = 0 \\ \mathbb{O}_\psi\{(\tilde{\boldsymbol{\psi}}, \tilde{\boldsymbol{\alpha}}) \in \bar{\mathcal{V}}\} = 1 \\ \mathbb{E}_{\mathbb{O}_\psi}\{\tilde{\alpha}_i\} \leq \gamma_i \quad \forall i \in \mathcal{I} \end{array} \right\} \tag{37}$$

The ambiguity set in (37) needs to be extended further to incorporate joint distribution information for the auxiliary variable as well as the influence of the uncertain variable. The joint distribution for the auxiliary and uncertain variables is presented in (38).

$$\bar{\mathcal{V}} = \left\{ (\boldsymbol{\psi}, \boldsymbol{\alpha}) \in \mathbb{R}^{|\mathcal{S}|} \times \mathbb{R}^{|\mathcal{I}|} : \begin{array}{l} \boldsymbol{\psi} \in \mathcal{V} \\ v_i(\boldsymbol{\psi}) \leq \alpha_i \quad \forall i \in \mathcal{I} \\ \alpha_i \leq \max_{\boldsymbol{\psi} \in \mathcal{V}} v_i(\boldsymbol{\psi}) \quad \forall i \in \mathcal{I} \end{array} \right\} \tag{38}$$

In expression (38), the function v_i is restricted by the auxiliary variable a_i so that the ambiguity parameter $\mathbb{I}$ having all possible outcomes in (35) holds if set $\mathbb{H}$ is valid in (37). To solve the objective function with given constraints, the above ambiguity set needs to be linearized with a linear decision rule, as depicted in (39) having auxiliary random variables.

$$y_n(\boldsymbol{\psi}, \boldsymbol{\alpha}) = y_n^0 + \sum_{e \in \mathcal{S}} y_{ie}^\psi \psi_e + \sum_{r \in \mathcal{S}} y_{ir}^\alpha \alpha_r \quad \forall i \in \mathcal{I} \tag{39}$$

The Equations (31) and (33) are re-arranged in the form of Equations (40) and (41) so that the linear decision rule can be applied.

$$\Xi(x) = \sup_{M\in\mathbb{I}} \mathbb{E}_M\{\Xi(x,\tilde{\psi})\} = \sup_{M\in\mathbb{I}} \mathbb{E}_M\{q^T y(\tilde{\psi})\} \tag{40}$$

where y is the recourse decision which depends on $\tilde{\psi}$ and is expressed as

$$y(\psi) \in \arg\min\{q^T y : Tx + Wy \le h(\psi)\} \quad \forall \psi \in \mathcal{V} \tag{41}$$

By applying the principle of linear decision rule on Equations (40) and (41), the proposed model can be converted to (42)–(43).

$$\bar{\Xi}(x) = \min \sup_{\mathbb{O}\in\mathbb{H}} \mathbb{E}_{\mathbb{O}}\{q^T y(\psi,\alpha)\} \tag{42}$$

s.t.

$$Tx + Wy(\psi,\alpha) \le h(\psi) \quad \forall(\psi,\alpha) \in \hat{\mathcal{V}} \tag{43}$$

These results are taken from [47]. Now, the model discussed in (2) and (3) with the uncertainty model of solar PV can be solved by using an ambiguity set for the distribution of RES's data for uncertain variables, i.e., characteristics of the distribution with the help of (42)–(43), which makes the model tractable.

3.4.2. Modeling Ambiguity Parameters for Wind Uncertainty

The ambiguity-parameters modeling for wind uncertainty follows the same steps as discussed in the previous section. The ambiguity matrix for wind is derived from the same method as the ambiguity matrix for solar but with different variables and parameters. The ambiguity set and extended ambiguity set for wind farms are listed in (44) and (48).

$$\mathbb{I}_o = \left\{ M \in \mathcal{P}_o(\mathbb{R}^{|\mathcal{S}|}) : \begin{array}{l} \tilde{o} \in \mathbb{R}^{|\mathcal{S}|} \\ \mathbb{E}_M\{\tilde{o}\} = 0 \\ M\{\tilde{o} \in \mathcal{V}\} = 1 \\ \mathbb{E}_M\{v_i(\tilde{o})\} \le \Gamma_i \quad \forall i \in \mathcal{I} \end{array} \right\} \tag{44}$$

where $\mathcal{P}_o(\mathbb{R}^{|\mathcal{S}|})$ denotes the set of all probability distributions on $R^{|\mathcal{S}|}$. The expected value of the random variable $\tilde{o}$ is 0, which enforces the mean error signal to be "0" and sum of the probability for all outcomes for $\tilde{o}$ is 1. Function v_i in (44) incorporates distribution information into the ambiguity model. To obtain the expectation of each function v_i under ambiguous distributions easily, the ambiguity set in (44) needs to be further extended in (45). Here, an auxiliary variable is introduced to make the ambiguity set in (45) less conservative.

$$\mathbb{H}_o = \left\{ \mathbb{O}_o \in \mathcal{P}_o(\mathbb{R}^{|\mathcal{S}|} \times \mathbb{R}^{|\mathcal{I}|}) : \begin{array}{l} (\tilde{o},\tilde{s}) \in \mathbb{R}^{|\mathcal{S}|} \times \mathbb{R}^{|\mathcal{I}|} \\ \mathbb{E}_{\mathbb{O}_o}\{\tilde{o}\} = 0 \\ \mathbb{O}_o\{(\tilde{o},\tilde{s}) \in \hat{\mathcal{V}}\} = 1 \\ \mathbb{E}_{\mathbb{O}_o}\{\tilde{s}_i\} \le \gamma_i \quad \forall i \in \mathcal{I} \end{array} \right\} \tag{45}$$

The model discussed in (1) and (3) with the uncertainty model of wind farms is solved by using an ambiguity set (45) for the distribution of RES's data for uncertain variables, i.e., characteristics of the distribution.

4. Results and Discussions

Case studies were performed in this paper with the help of the ambiguity sets developed in the previous section and the objective functions discussed above are now tractable models due to the linear decision rule implemented in (42)–(43).

4.1. Example Case Study for DRO Illustration with One DG for an Instance

In this section, a sample case study was conducted with a distributed generator and a solar PV for an instance. This sample case study elaborates the recourse action for all solar PV outcomes in an instance to achieve the worst-case expected cost objective. Here, demand is $D = 280$ kW, mean solar PV output is $\bar{\psi} = 40$ kW, and penalty cost for load loss is $P^{sur} = 150\ USD/\text{kW}$. To simplify the problem and to elaborate on the initial method for solving with BESS, one can take $SOC^{max} = 0$ kWh, which means there is no storage in this example, and eventually power charge or discharge for BESS $p^c/p^d = 0$ kW. The generation cost coefficients a and b are 1.7×10^{-7} USD/kW and 16.57×10^{-3} USD/kW, respectively, where the startup cost for the generator is USD 890. The upper and lower limit for the generator is $\overline{P} = 240$ kW and $\underline{P} = 20$ kW, respectively. The ambiguity matrix for the solar PV uncertainty is listed in (46). The duality gap for this example is 0.01.

$$\mathbb{H} = \left\{ \mathbb{O} \in \mathcal{P}_0(\mathbb{R}^2): \ \mathbb{O} \left\{ \begin{array}{c} \mathbb{E}_\mathbb{O}\{\tilde{\psi}\} = 0 \\ -10 \leq \tilde{\psi} \leq 10 \\ \max\{\tilde{\psi}, 0\} \leq \alpha \\ \alpha \leq 10 \\ \mathbb{E}_\mathbb{O}\{\alpha\} \leq 5 \end{array} \right\} = 1 \right\} \tag{46}$$

In the ambiguity matrix construction in Equation (46), in line one, the mean value of error in solar PV output is 0; in lines, two to four, the upper and lower bound of the confidence for the mean value of the solar PV output is from -10 to 10 which means if the mean solar PV output at a particular instance is 40 kW, then the uncertainty in the output power ranges from 30 to 50 kW; in line five, the expected positive error should be less than 5 kW. The worst-case expected cost for the sample case is found to be USD 753.9106 using CPLEX 12.9.0.

4.2. Optimum Battery Sizing and Its Impact on Unit Commitment in an IEEE 14 Bus System

After promising results obtained from the above sample case study using the DRO technique, the following case studies include solar PV, wind farm, and a BESS using an IEEE 14 bus system having four generators each connected to bus 1 to 4, respectively, [48] for a 24 h time period with a 30-min interval. These case studies elaborate the recourse action for all solar-PV and wind-power outcomes in a day to achieve the minimum expected cost objective. Data for battery storage is shown in Table 1; similarly, each distributed generator is shown in Table 2. The load profile for this case study is shown in Table 3. Solar-PV-output power data for a 200 kW solar PV is taken from [49] where the ambiguity matrix for the solar-PV uncertainty is listed in (47). The wind-power output data for a 700 kW wind farm is taken from [50], where the ambiguity matrix for the wind uncertainty is listed in (48). To make this case study realistic in terms of BESS, wind farm, and solar PV, all the generator's data is converted from MW to kW, and their respective prices are also converted from USD/MW to USD/kW. The mean output power of the solar PV for 30-min intervals provided by the source, as shown in Figure 2, is a bit tailored and converted from MW to kW to show the UC schedule. Similar is the case with wind power, as shown in Figure 3. Load flow limits are considered as per IEEE 14 bus system data in kW having a load capacity factor LCF_{gb} for each transmission line in all the case studies conducted here. In all the below case studies, the duality gap of 0.001 and the penalty cost for load loss of $P^{sur} = 125\ USD/\text{kW}$ were used.

Table 1. The data for battery energy storage system (BESS) integrated with unit commitments and dynamic economic dispatch [49].

Parameter	SOC_0	SOC_{max}	P^d_{max}	P^d_{min}		P^c_{max}
Value	200 kWh	1 MWh	0.25C	0		0.25C

Parameter	P^c_{min}	η_c	η_d	C^{var}	C^{fix}
				O&M cost	
Value	0	95%	90%	USD 0.31/kWh	USD 10/kW-year

$$
\mathbb{H}_\psi = \left\{ \mathbb{O}_\psi \in \mathcal{P}_\psi(\mathbb{R}^2): \quad \mathbb{O}_\psi \left\{ \begin{array}{l} \mathbb{E}_{\mathbb{O}_\psi}\{\tilde{\psi}\} = 0 \\ \left\{ \begin{array}{l} -5\% \le \tilde{\psi} \le 5\% \\ \max\{\tilde{\psi}, 0\} \le a \\ a \le 5\% \end{array} \right\} = 1 \\ \mathbb{E}_{\mathbb{O}_\psi}\{a\} \le 5\% \end{array} \right. \right\}
\tag{47}
$$

$$
\mathbb{H}_o = \left\{ \mathbb{O}_o \in \mathcal{P}_o(\mathbb{R}^2): \quad \mathbb{O}_o \left\{ \begin{array}{l} \mathbb{E}_{\mathbb{O}}\{\tilde{o}\} = 0 \\ \left\{ \begin{array}{l} -5\% \le \tilde{o} \le 5\% \\ \max\{\tilde{o}, 0\} \le s \\ s \le 5\% \end{array} \right\} = 1 \\ \mathbb{E}_{\mathbb{O}_o}\{s\} \le 5\% \end{array} \right. \right\}
\tag{48}
$$

Table 2. Thermal unit's data for dynamic economic dispatch and unit commitments [51].

a_g (USD/kW)	b_g (USD/kW)	c_g (USD/kW)	P_g^{min} (kW)	P_g^{max} (kW)	RU_g^0 (kW)	RD_g^0 (kW)
1.2×10^{-7}	14.80×10^{-3}	89	28	200	40	40
1.7×10^{-7}	16.57×10^{-3}	83	20	290	30	30
1.9×10^{-7}	16.21×10^{-3}	70	20	260	50	50
1.5×10^{-7}	15.55×10^{-3}	100	30	190	30	30

The 30-min solar-PV output data from solar plates and wind-power output data from wind farms are shown in Figures 2 and 3, respectively. In ambiguity matrix (47) for solar PV, the upper and the lower confidence bound is ±5% of the mean value of the solar-PV output and the expected value of error is 0. The expected positive error should be less than 5% of the mean value of the solar-PV output at a particular instance. Likewise, in ambiguity matrix (48) for wind power, the upper and the lower confidence bound is ±5% of the mean value of the wind power output and the expected value of error is 0. The expected positive error should be less than 5% of the mean value of the wind power output.

Table 3. Load/demand profile used with IEEE 14 bus system having 30-min interval.

Time (hr)	Load (kW)	Time (hr)	Load (kW)	Time (h)	Load (kW)	Time (h)	Load (kW)
0:00	525.30	6:00	525.30	12:00	665.38	18:00	756.02
0:30	530.45	6:30	526.59	12:30	675.68	18:30	759.63
1:00	535.60	7:00	527.88	13:00	685.98	19:00	763.23
1:30	540.75	7:30	529.16	13:30	696.28	19:30	766.84
2:00	545.90	8:00	530.45	14:00	706.58	20:00	770.44
2:30	542.30	8:30	537.92	14:30	720.74	20:30	773.53
3:00	538.69	9:00	545.39	15:00	734.91	21:00	776.62
3:30	535.09	9:30	552.85	15:30	749.07	21:30	779.71
4:00	531.48	10:00	560.32	16:00	763.23	22:00	782.80
4:30	529.94	10:30	586.59	16:30	761.43	22:30	729.76
5:00	528.39	11:00	612.85	17:00	759.63	23:00	676.71
5:30	526.85	11:30	639.12	17:30	757.82	23:30	623.67

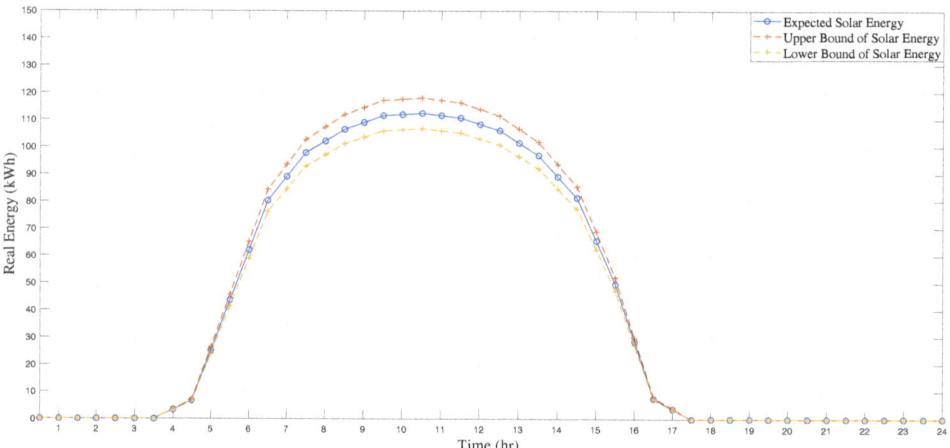

Figure 2. Expected solar-PV output and its upper/lower bounds.

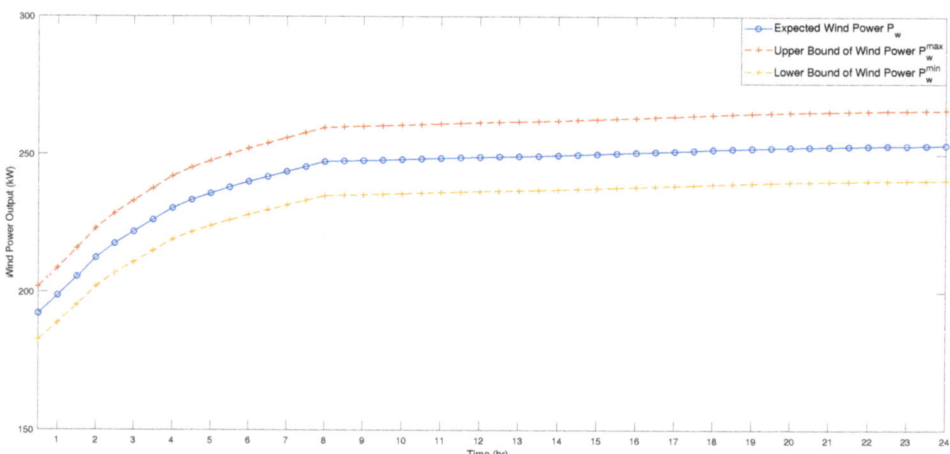

Figure 3. Expected wind output and its upper/lower bounds.

4.2.1. Solar-PV Uncertainty with 30-Minute Interval Unit Commitments

The "Here and Now" decisions of the unit scheduled for a 24-hour time period without BESS are shown in Figure 4; on the top of the figure, a colored square shows the status of a particular generator at a certain time instance as "turned on" while no square shows the status of a particular generator at a certain hour as "turned off". The "Wait and See" decisions in terms of economic dispatch for each generator are also shown in Figure 4 on the bottom, where all "turned on" generators are providing power with optimal dispatch and their individual contribution is shown to meet the load demand in the distributed generation system.

The UC schedule in Figure 4 shows that the generator P_{gen}^4 becomes active from 4:30 to 11:00 p.m. in a 24-hour time period. The load demand from 4:30 to 11:00 p.m. can also be achieved by using generator P_{gen}^2 power at an off-peak time using energy storage from 5:00 a.m. to 12:00 p.m. Utilizing power from other active generators at peak time, the generator P_{gen}^4 is avoided, which reduces the start-up cost as shown in Figure 5. By turning on/off the generator more than once a day, the total system's cost also increases. The frequency of turning on/off the generators is multiplied by the generator's start-up and shut-down cost, which increases the total system cost. The BESS can reduce more costs if a generator is turned on more than twice, because BESS can provide power at that time when the power is needed and store energy at off-peak times when there is a surplus to save costs from both sides. Therefore, the BESS also saves energy from being lost and provides the necessary backup when it is needed.

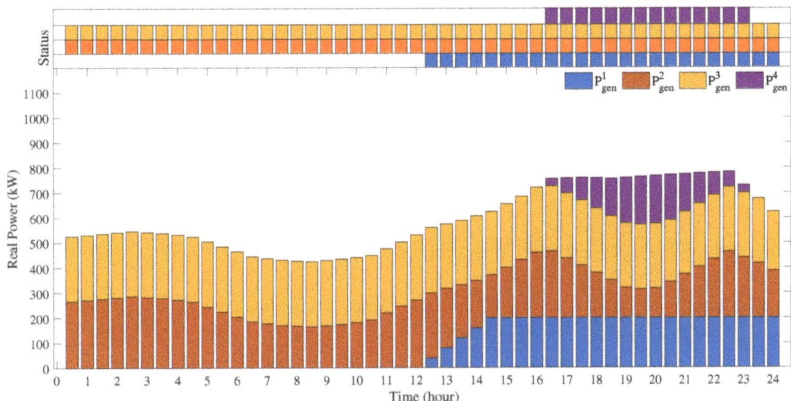

Figure 4. Unit commitment schedule with power contribution from each distributed generator after incorporating solar-PV output power to the system with 30-min duration without using BESS and $LCF_{gb} = 1.0$ for each bus.

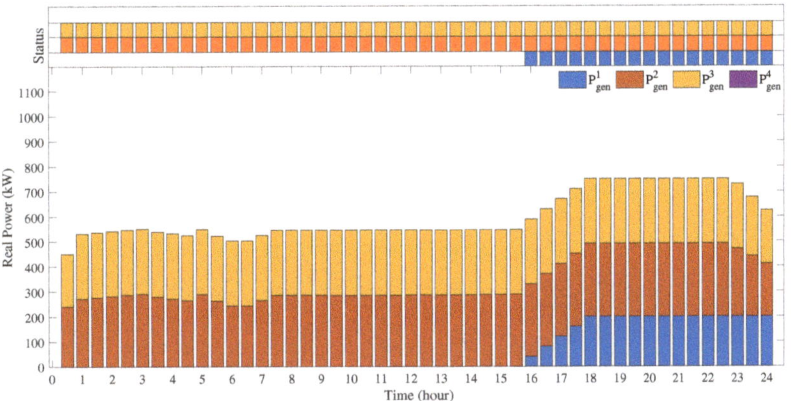

Figure 5. The power contribution and the status of each distributed generator after incorporating solar-PV output power to the system with optimal capacity of BESS and $LCF_{gb} = 1.0$ for each bus.

The total system cost observed without using BESS is USD 11,112.1 and the total system cost observed using optimal BESS of 1520 kWh having 0.25C of charge and discharge rating is USD 9334.1. The cost with optimal BESS has a significant improvement (+16%). Battery profile is also shown in Figure 6, where the BESS mostly stores the energy when it is available through the solar PV and generator P_{gen}^2. Later on, this energy is given to the system to relax not only generator P_{gen}^4 but generator P_{gen}^1, as shown in Figure 5.

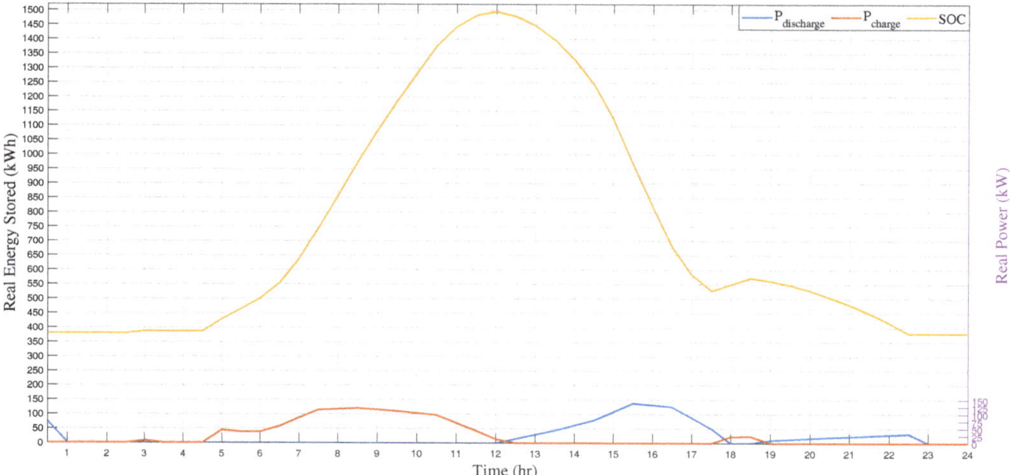

Figure 6. Charging and discharging power and hourly energy stored in the battery having BESS capacity optimized to 1520 kWh.

The empirical cumulative distribution function with the size of BESS is shown in Figure 7, where the reader has the freedom to choose the battery size based on the compromise with the total system's cost. The cost effectiveness observed with the optimal use of BESS is +16%, as listed above, so the cost effectiveness will be multiplied with normalized battery usage to obtain the significance of BESS over the current IEEE 14 bus system.

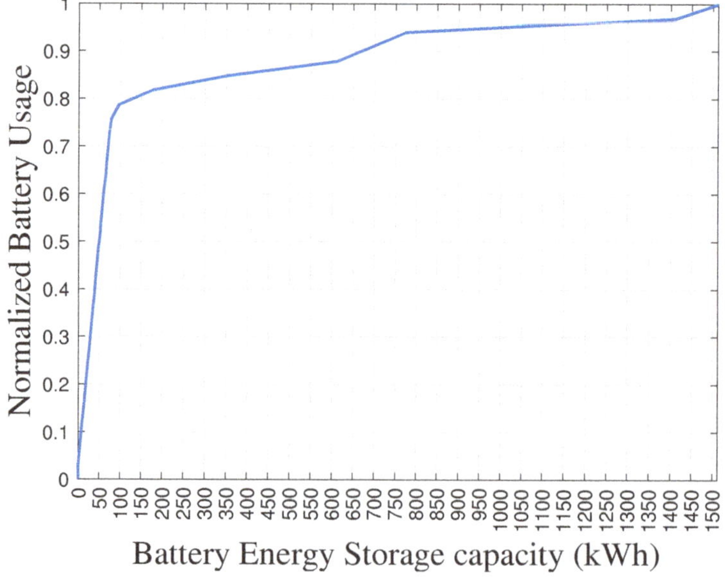

Figure 7. Empirical cumulative distribution graph considering solar PV having optimal BESS capacity of 1520 kWh.

The whole system took 139.04 s to solve the problem with optimum objective value for 24 h operation of all scheduled distributed generators.

4.2.2. Wind-Farm Uncertainty with 30-Minute Interval Unit Commitments

The UC schedule without BESS in Figure 8 shows that, from 3:30 a.m. to 10:00 a.m., the generator P_{gen}^2 and P_{gen}^3 are active but P_{gen}^2 is providing a small amount of power as compared to P_{gen}^3 due to the reason of being active during that period to avoid shut-down and startup costs, as after that period of time the system needs P_{gen}^2 badly. The start-up cost for generator P_{gen}^2 is higher than P_{gen}^3, but the running cost of generator P_{gen}^2 is lower than that of generator P_{gen}^3. Now comes the role of BESS, as shown in Figure 9, when the optimal size of the battery having a capacity of 558.13 kWh is placed inside the system. The generator P_{gen}^2 is producing power at its peak by providing the power to the system with a relatively low running cost per unit and the rest of the power is being provided by the BESS to meet the load for the time duration specified above.

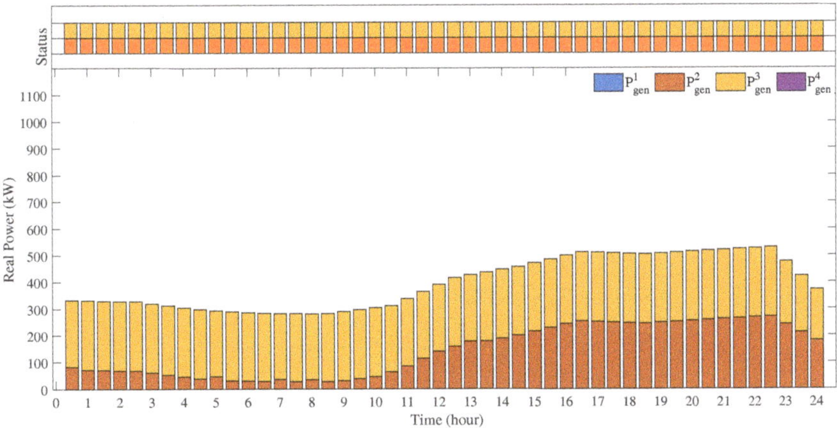

Figure 8. Unit commitment schedule with power contribution from each distributed generator after incorporating wind power to the system with 30-min duration without using BESS and $LCF_{gb} = 1.0$ for each bus.

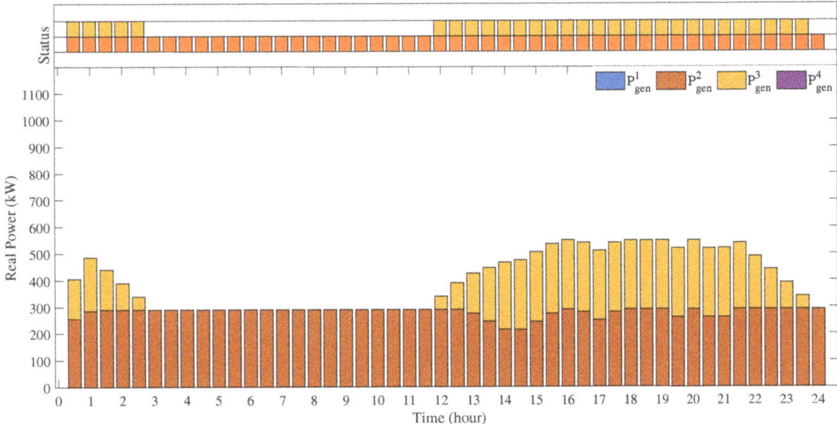

Figure 9. The power contribution and the status of each distributed generator after incorporating wind power to the system with optimal capacity of BESS and $LCF_{gb} = 1.0$ for each bus.

The total system cost observed without using BESS is USD 7499.8 and the total system cost observed using optimal BESS of 558.13 kWh having 0.35C of charge and discharge rating is USD 6336.3. The cost with optimal BESS has significant improvement (+15.51%). The battery profile is also shown in Figure 10, where the BESS initially stores the energy from 1:30 a.m. to 2:30 a.m. and then delivers this power to the system to turn off generator P_{gen}^3 for the above mentioned time. Later on, BESS stores energy to be used at the end of the day to turn off generator P_{gen}^3 again for reducing the cost incurred due to the generator's P_{gen}^3 higher running cost.

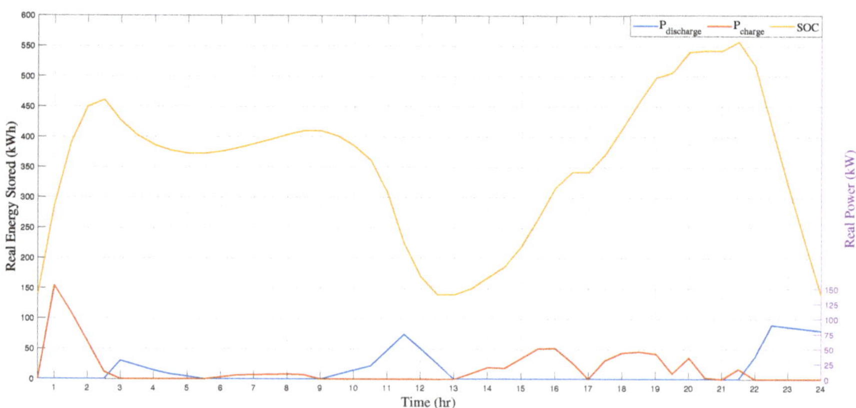

Figure 10. Charging and discharging power and hourly energy stored in the battery having battery capacity optimized to 558.13 kWh.

The empirical cumulative distribution function with the size of BESS is shown in Figure 11, where the reader has the freedom to choose the battery size based on the compromise with the total system's cost. The cost effectiveness observed with optimal use of BESS is +14.56%, as listed above, so the cost effectiveness will be multiplied with normalized battery usage to obtain the significance of BESS over the current IEEE 14 bus system.

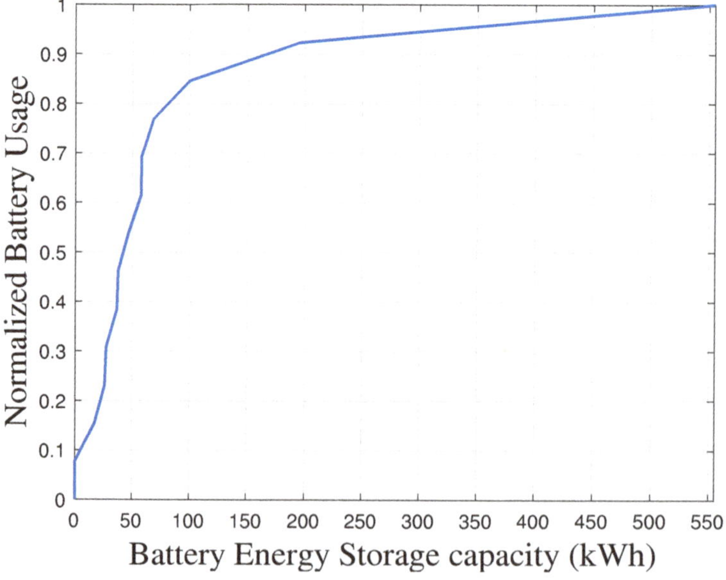

Figure 11. Empirical cumulative distribution graph considering wind farms having optimal BESS capacity of 558.13 kWh.

4.2.3. Hybrid Uncertainty with 30-Minute Interval Unit Commitments

The unit commitment schedule without BESS in Figure 12 shows that, from 0:00 a.m. to 5:00 a.m., the generator P_{gen}^2 and P_{gen}^3 are active but P_{gen}^2 is providing a small amount of power as compared to P_{gen}^3 and then is turned off from 5:30 a.m. to 12:00 p.m. due to the reason of being active during that period of time to avoid shut down and startup costs, as after that period of time the system needs P_{gen}^2 badly. The start-up cost for generator P_{gen}^2 is higher than P_{gen}^3 but the running cost of generator P_{gen}^2 is lower than that of generator P_{gen}^3. Now comes the role of BESS as shown in Figure 13, when the optimal size of the battery having a capacity of 712.99 kWh is placed inside the system. The generator P_{gen}^2 is providing the power to the system with a relatively low running cost per unit and the rest of the power is being provided by the BESS to meet the load for the time duration of 5:30 a.m. to 12:00 p.m. The most important thing can be observed by looking to Figures 12 and 13, the total system cost for a day without using BESS is USD 6404.1 while the system cost with BESS has reduced to USD 5874.9, which is 8.26% less than the total system cost without BESS.

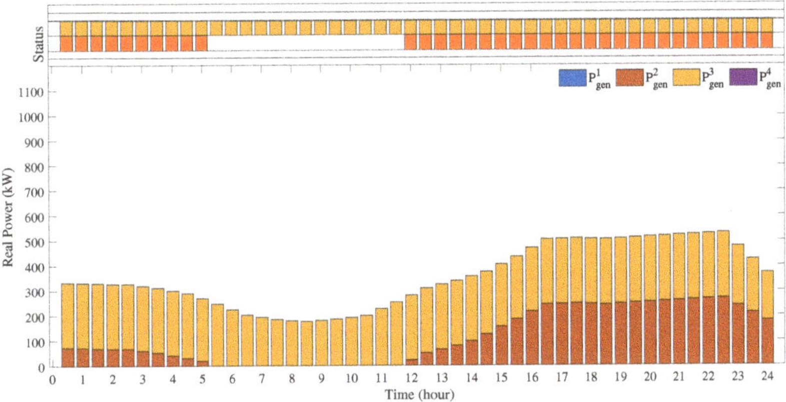

Figure 12. Unit commitment schedule with power contribution from each distributed generator after incorporating hybrid power to the system with 30-min duration without using BESS and $LCF_{gb} = 1.0$ for each bus.

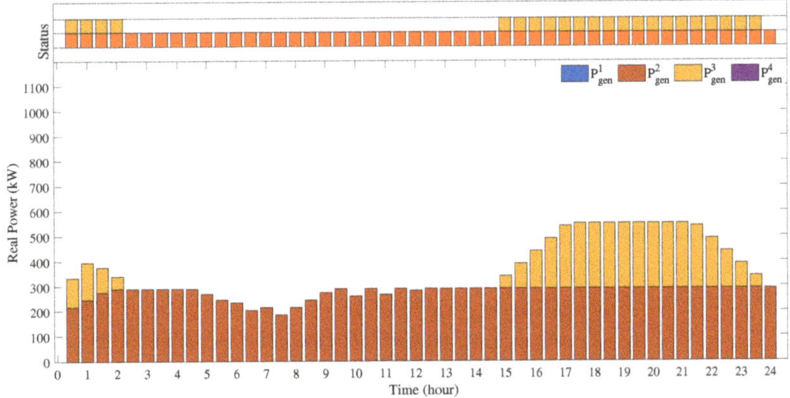

Figure 13. The power contribution and the status of each distributed generator after incorporating hybrid power to the system with optimal capacity of BESS and $LCF_{gb} = 1.0$ for each bus.

Total system costs also increase by turning on/off the generator more than once a day. The frequency of turning on/off the generators is multiplied by the generator's start-up and shut-down cost, which increases the total system cost. The BESS can reduce more costs if a generator is being turned on more than twice because BESS can provide power at that time when the power is needed and store energy at off-peak times when there is a surplus to save cost from both sides. Therefore, the BESS also saves energy from being lost and provides the necessary backup when it is needed.

There is a total of three cycles of charging and discharging being utilized for 24 h in the current IEEE 14 bus system. The battery behavior and its contribution to the system can be seen in Figure 14, where the battery's charge and discharge cycles and the state charge in the battery at any time for a 24-hour time period are shown. With a 0.35C of charging/discharging rate, the total cost of the system including BESS is changing with the storage size of BESS. Initially, without the battery energy storage system, the total system cost is USD 6404.14. The total system cost is at its peak, and this cost gradually decreases to USD 5874.9 and then increases due to the battery cost. The increasing behavior is due to the size of BESS as the size of BESS increases, but the battery utilization remains ineffective after a certain limit.

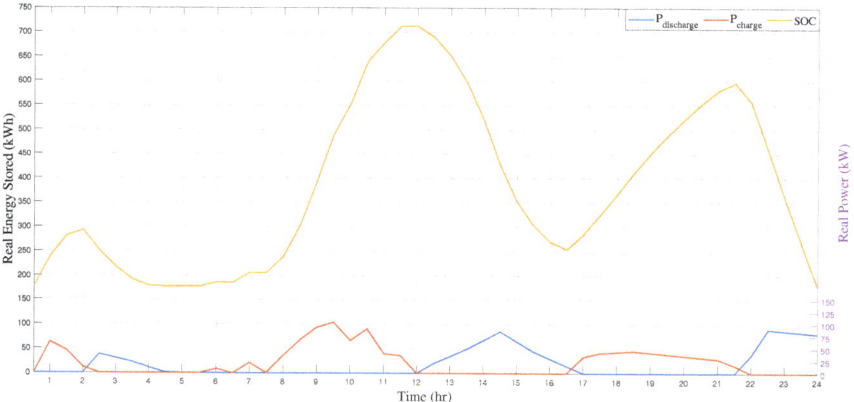

Figure 14. Charging and discharging power and hourly energy stored in the battery.

The BESS available in the market normally have a continuous charging and discharging rate ranging from (0.1C to 0.4C or 10% to 40% of C) of the battery energy storage system, as according to the General Electric with their energy storage unit "RSU-4000" [52]. This is due to several reasons, i.e., battery temperature, battery life, and its continuous charging and discharging capabilities. Due to high charging and discharging rates, the inner temperature of the battery increases, which needs to be cooled enough to make the optimum temperature for the battery, otherwise it may burn the battery or at least create power losses and reduce the battery life. For high charging and discharging rate batteries, cooling systems are needed which cost money. In addition, due to high temperature, power losses inside the battery increase, due to which the battery will become unable to supply the rated power for which it is designed. Therefore, the optimum range of the charging and discharging rate (C) for heavy-duty BESS in the industries is (0.25C to 0.35C).

The BESS power and capacity optimization can be observed in Figure 15, where the mesh plot shows the system objective cost for battery capacity and power (C-rate). In Figure 16, the empirical cumulative distribution function with the size of BESS is shown, which gives an overall perspective on selecting the battery size based on the compromise with the total system's cost. The cost-effectiveness observed with the optimal use of BESS is 8.26%, as listed above, so the cost effectiveness will be multiplied with normalized battery usage to obtain the significance of BESS over the current IEEE 14 bus system.

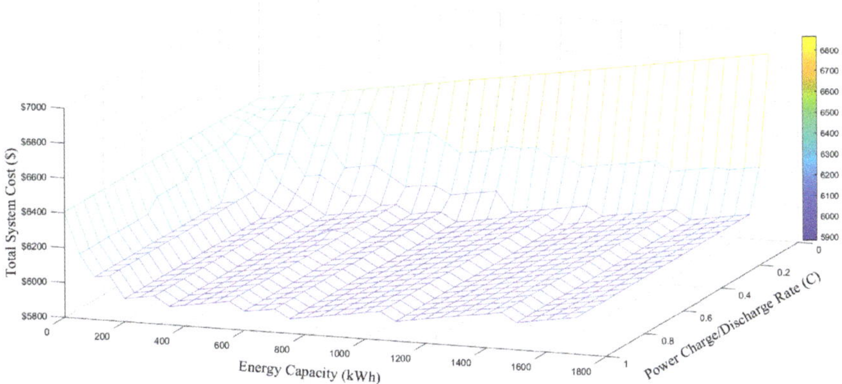

Figure 15. Total system cost vs battery capacity with charging/discharging rates ranging from 0.05C to 1C.

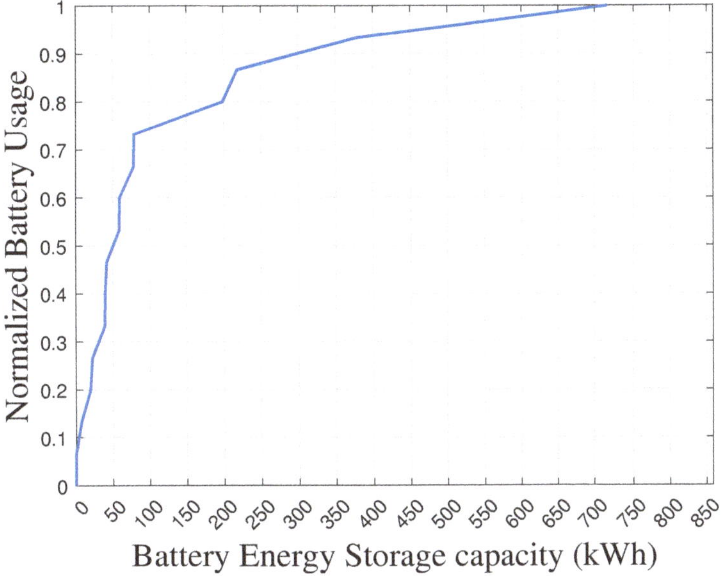

Figure 16. Empirical cumulative distribution graph considering hybrid system having optimal BESS capacity of 712.99 kWh.

Figure 17 shows the whole picture of the economic dispatch by distributed generators, wind farm, solar PV, and battery energy storage systems. The system is balanced, which means the load demand is fulfilled by the power supplied through sources discussed in this paper.

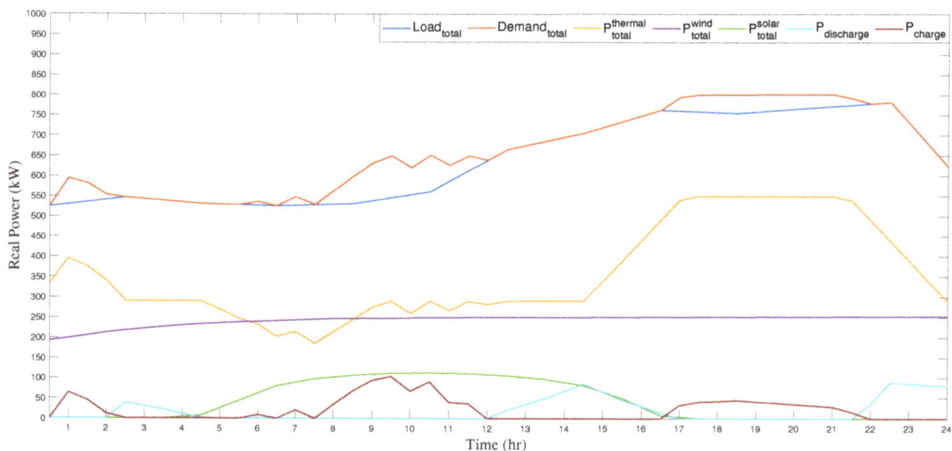

Figure 17. Total system's demand vs generation (including distributed generation, wind-farm, solar PV, and optimal capacity of BESS).

4.3. Comparison with Similar Studies

The IEEE RTS 24 bus system is investigated here with DRO as well as the stochastic through scenario approach in current research work. The results obtained through the DRO technique are compared with the stochastic approach presented in [53] with the IEEE 24-Bus system. The system used in [53] is an IEEE 24-Bus system with a time span divided into 24 1-hour time slots having two wind farms connected at buses 10 and 17, and three ESS installed on buses 10, 17, and 20. The system has ten thermal units for scheduling to obtain the overall minimum generating cost while undermining the line limits. The cost for energy not served used in that case study is 200 USD/MWh. The data for thermal units, ESS, hourly load, and mean wind power are provided in [53].

Total system cost objective comparison between distributionally robust optimizations and stochastic optimization with penetration level 100% over various forecasting accuracy variances for an IEEE RTS 24-Bus is presented in Table 4. The total system's cost obtained through the scenario approach is less than the deterministic cost against 100% penetration of RES and a variance of 0.1. This cost is the worst-case solution and, hence, conservative for the IEEE 24-Bus system. However, the total cost of the system with the same parameters obtained with DRO is slightly higher as compared to stochastic, which means the proposed method in the current research has a less conservative solution while handling uncertainties with more than one uncertain parameter. In addition, the value of the stochastic solution (VSS) with stochastic optimization in Table 4 has higher variations as compared to the VSS with DRO, which shows the robustness as well as the accuracy of the method.

Table 4. Total system cost objective comparison between distributionally robust optimizations and stochastic optimization with forecasting accuracy variance 0.1 over various penetration level (IEEE RTS 24-Bus).

Wind Forecast	Stochastic Optimization				Distributionally Robust Optimization			
Accuracy	Total Cost (USD)		VSS		Total Cost (USD)		VSS	
Percent	Deterministic	Stochastic	USD	%	Deterministic	DRO	USD	%
100	1,482,836	1,404,468	78,367	5.285	1,482,836	1,411,223	71,613	4.829
80	1,543,488	1,506,339	37,148	2.407	1,543,488	1,512,588	30,899	2.002
60	1,643,850	1,618,083	25,766	1.567	1,643,850	1,627,354	16,496	1.003
40	1,753,050	1,740,145	12,905	0.736	1,753,050	1,747,797	5252	0.300
20	1,884,530	1,880,827	3702	0.196	1,884,530	1,881,773	2756	0.146

Similarly, total system cost objective comparison between distributionally robust optimizations and stochastic optimization with penetration level 100% over various forecasting accuracy variances for an IEEE RTS 24-Bus is presented in Table 5. With different wind-forecast-accuracy variances, both models are tested to claim the same behavior of using the DRO model as discussed above. The results obtained through DRO with various wind-forecast-accuracy variances with 100% RES penetration also prove that the DRO model gives a less conservative and more robust solution as compared to stochastic optimization through scenarios.

Table 5. Total system cost objective comparison between distributionally robust optimizations and stochastic optimization with penetration level 100% over various forecasting-accuracies variances (IEEE RTS 24-Bus).

| Wind Forecast | Stochastic Optimization | | | | Distributionally Robust Optimization | | | |
| Accuracy | Total Cost (USD) | | VSS | | Total Cost (USD) | | VSS | |
Variance	Deterministic	Stochastic	USD	%	Deterministic	DRO	USD	%
0.1	1,482,836	1,404,468	78,367	5.285	1,482,836	1,411,223	71,613	4.829
0.075	1,465,605	1,403,389	62,216	4.245	1,465,605	1,408,875	56,730	3.871
0.05	1,447,095	1,399,777	47,318	3.27	1,447,095	1,419,318	27,777	1.919
0.025	1,427,307	1,396,437	30,869	2.163	1,427,307	1,410,691	16,616	1.164

Therefore, from the above three cases, it is concluded that BESS has a major impact on unit scheduling. It reduces the startup cost of the generators and extra power generated due to the ramp up and ram down limits is stored to BESS in order to minimize the generation cost for peak times. Similarly, the cold cost of all the generators participating even in a single hour of that particular day is saved. All of the above case studies were solved with Xprog V1.0, a MATLAB-based platform by using IBM ILOG CPLEX Optimization Studio V12.6.3; the system requirements were a 64-bit Windows 11-based operating system with 16GB RAM.

5. Conclusions

Optimal sizing of a battery energy storage system using the ambiguity-based parametric model of distributionally robust optimization with a linear decision rule for grid-connected distributed generators along with uncertain wind farm and solar PV was developed to achieve the optimal scheduling of the distributed generation with optimal battery energy storage system capacity and power rating. The uncertainty of wind farms and solar PV is modeled through DRO by developing their respective ambiguity sets with mean, minimum, and maximum output for each time step (30-min) and each RES. The DRO model employed in this research provides less conservative solutions under uncertainty as compared to other models, i.e., the robust optimization model and adopted robust optimization model. Accordingly, the effect of BESS on unit commitment is examined with three different case studies conducted with solar-PV, wind-farm, and hybrid uncertainties on distributed generators over a 24-hour time period. These case studies show that the schedule of UC observably changes with BESS placement in the system. Further, the optimal capacity of BESS also reduces the maximum operating point of distributed generators by providing power to the system at peak times and storing energy in BESS at off-peak times. The optimal size of BESS is established, which is inclusive of the practical startup and shut down of distributed generators. This study also discussed the effect of the economical aspect of BESS size on unit scheduling, although with an increment in BESS capacity reducing the start-up cost, nevertheless, the initial investment of BESS economically limits this function; this trade-off defines the efficacy of scheduling-based optimization algorithms. The case studies discussed in this paper have the beauty of not having a complex system to make it more complex for the reader to understand the model proposed here; for future work, these case studies can be extended to multiple energy storage systems to evaluate the cost-effectiveness.

Author Contributions: Authors' contribution are as follows: Conceptualization, A.R. and M.K. (Mahmoud Kassas); Data curation, A.R., M.K. (Mahmoud Kassas) and M.K. (Muhammad Khalid); Formal analysis, A.R. and M.K. (Mahmoud Kassas); Funding acquisition, M.K. (Muhammad Khalid); Investigation, A.R. and M.K. (Muhammad Khalid); Methodology, A.R., M.K. (Mahmoud Kassas) and M.K. (Muhammad Khalid); Project administration, M.K. (Mahmoud Kassas) and M.K. (Muhammad Khalid); Resources, M.K. (Mahmoud Kassas) and M.K. (Muhammad Khalid); Software, A.R. and M.K. (Mahmoud Kassas); Supervision, M.K. (Mahmoud Kassas) and M.K. (Muhammad Khalid); Validation, A.R. and M.K. (Mahmoud Kassas); Visualization, A.R. and M.K. (Muhammad Khalid); Writing—original draft, A.R., M.K. (Mahmoud Kassas) and M.K. (Muhammad Khalid); Writing—review and editing, M.K. (Muhammad Khalid). All authors have read and agreed to the published version of the manuscript.

Funding: The research is funded in part by the Interdisciplinary Research Center for Renewable Energy and Power Systems (IRC-REPS) at KFUPM under Project No. INRE2106, and SDAIA-KFUPM Joint Research Center for Artificial Intelligence (JRC-AI).

Acknowledgments: The authors would like to acknowledge the support provided by the Deanship of Research Oversight and Coordination (DROC) at the King Fahd University of Petroleum and Minerals (KFUPM), Dhahran 31261, Kingdom of Saudi Arabia.

Conflicts of Interest: The authors declare no conflict of interest and the research was conducted in the absence of any commercial or financial relationships that could be construed as a potential conflict of interest.

Nomenclature

Sets and Indices for UC Model

τ/t	Time interval set
ζ	Feasible time constraints for minimum up/down time
κ/k	DG unit set
β/b	Buses in the system
Λ/l	Solar PV units
$\mathcal{L}/g$	Transmission line
$\mathcal{J}/j$	Wind-farm units

Constants for UC Model

$\overline{P}_k$	Upper limit of generator k
$\underline{P}_k$	Lower limit of generator k
RD_k	RD limit of generator k
RU_k	RU limit of generator k
$p_{b,t}^{d,min}$	Minimum power that storage can provide while discharging at bus b at time t
$p_{b,t}^{d,max}$	Maximum power that storage can provide while discharging at bus b at time t
$p_{b,t}^{c,min}$	Maximum power that storage needs while charging at bus b at time t
$p_{b,t}^{c,max}$	Maximum power that storage needs while charging at bus b at time t
$D_{b,t}$	Load demand at bus b during time t
a_k	Generator k cost parameter
b_k	Generator k cost parameter
C_k^s	Fixed start up cost for unit k
$SOC_{b,t}^{min}$	Minimum state of charge at time t and bus b
$SOC_{b,t}^{max}$	Maximum state of charge at time t and bus b
Δ_t	Time step of storage for time t
η_c	Efficiency of charging the battery
η_d	Efficiency of discharging the battery
LC_g	Line capacity of line g
$LCF_{g,b}$	Load capacity factor of line g connected to bus b
p^{sur}	Penalty surcharge for load loss
C	Charging and discharging rates of the battery

Decision Variables for UC Model

$c_{k,t}$	Unit k generation cost at time t
$B^c_{count,b}$	Battery's charge cycle count at bus b
$B^T_{count,b}$	Total charge and discharge cycle count for battery at bus b
$p^w_{j,t}$	Wind-farm unit j generation at time t
$p^{cur}_{b,t}$	Power surplus/generation loss at bus b on time step t
$p^t_{k,t}$	DG unit k generation at time t
$p^s_{l,t}$	solar PV unit l generation at time t
$x_{k,t}$	Binary variable for generator status i.e.,UC
$z_{k,t}$	Unit k start up cost at time t
$p^d_{b,t}$	Power that battery can provide (as a source) at bus b in time t
$p^c_{b,t}$	Power that battery(as a load) needs at bus b in time t
$SOC_{b,t}$	State of the charge at time t and bus b
$B^d_{count,b}$	Battery's discharge cycle count at bus b

Parameter Set for Uncertainty Model

$\tilde{\psi}_{l,t}$	Random variable for solar PV uncertainty error for unit l at time t
$V^+_{l,t}$	Maximum error in solar PV uncertainty for unit l at time t
$V^-_{l,t}$	Minimum error in solar PV uncertainty for unit l at time t
$\mathcal{V}$	Set of all uncertainties for random variables under linear constraints
$\overline{\mathcal{V}}$	Extended set of uncertainties for random variables defined under linear constraints
$\mathbb{O}_\psi$	Distribution for occurrence of random variables $\tilde{\psi}$ and $\tilde{\alpha}$ together
$\tilde{s}$	Auxiliary random variable for wind
$\tilde{o}_{j,t}$	Random variable for wind-farm uncertainty error for unit j at time t
$\mathbb{E}_\mathbb{M}$	Expectation within distribution $\mathbb{M}$
$\mathbb{I}$	Ambiguity matrix with given distribution of random variable $\tilde{o}$
$\mathbb{H}$	Extended form of ambiguity set $\mathbb{I}$
$\mathcal{I}/i$	Index for distribution of random variable
$\mathbb{J}$	Events describing the distribution of each random variable $v_{l,t}$
$\mathbb{M}$	Random variables$\tilde{\psi}$ distribution
$\mathbb{O}_o$	Distribution for occurrence of random variables $\tilde{o}$ and $\tilde{s}$ together
$\mathcal{S}/e,r$	Set of random variables
$\tilde{\alpha}$	Auxiliary random variable for solar PV

Others

$\Xi(x)$	Expected worst-case distribution recourse cost
$\Xi(x,o,\psi)$	UC cost based on decision of x under economic dispatch,with wind-farm o and solar PV ψ
$C^T_B(o,\psi)$	The running cost of the battery under economic dispatch, considering the recourse action of wind-farm o solar PV ψ uncertainty

Abbreviations

The following abbreviations are used in this manuscript:

UC	Unit commitment
DRO	Distributionally robust optimization
DG	Distribution generation
RES	Renewable energy sources
RO	Robust optimization
BESS	Battery energy storage system
ESS	Energy storage system
PV	Photo voltaic
MILP	Mixed integer linear programming
SOCP	Second-order cone programing
C	Charge and discharge rate
VSS	Value of stochastic solution

References

1. De Vos, K.; Morbee, J.; Driesen, J.; Belmans, R. Impact of wind power on sizing and allocation of reserve requirements. *IET Renew. Power Gener.* **2013**, *7*, 1–9. [CrossRef]
2. Rauf, A.; Al-Awami, A.; Kassas, M.; Khalid, M. Optimizing a residential solar PV system based on net-metering approaches. *IET* **2021**, 3300–3304.
3. Karimi, A.; Heydari, S.L.; Kouchakmohseni, F.; Naghiloo, M. Scheduling and value of pumped storage hydropower plant in Iran power grid based on fuel-saving in thermal units. *J. Energy Storage* **2019**, *24*, 100753. [CrossRef]
4. Roukerd, S.P.; Abdollahi, A.; Rashidinejad, M. Uncertainty-based unit commitment and construction in the presence of fast ramp units and energy storages as flexible resources considering enigmatic demand elasticity. *J. Energy Storage* **2020**, *29*, 101290. [CrossRef]
5. Liu, Y.; Wu, L.; Li, J. Towards accurate modeling of dynamic startup/shutdown and ramping processes of thermal units in unit commitment problems. *Energy* **2019**, *187*, 115891. [CrossRef]
6. Lebotsa, M.E.; Sigauke, C.; Bere, A.; Fildes, R.; Boylan, J.E. Short term electricity demand forecasting using partially linear additive quantile regression with an application to the unit commitment problem. *Appl. Energy* **2018**, *222*, 104–118. [CrossRef]
7. Pei, Z.; Lu, H.; Jin, Q.; Zhang, L. Target-based distributionally robust optimization for single machine scheduling. *Eur. J. Oper. Res.* **2022**, *299*, 420–431. [CrossRef]
8. Salimi, A.A.; Karimi, A.; Noorizadeh, Y. Simultaneous operation of wind and pumped storage hydropower plants in a linearized security-constrained unit commitment model for high wind energy penetration. *J. Energy Storage* **2019**, *22*, 318–330. [CrossRef]
9. Maaruf, M.; Khan, K.; Khalid, M. Robust Control for Optimized Islanded and Grid-Connected Operation of Solar/Wind/Battery Hybrid Energy. *Sustainability* **2022**, *14*, 5673. [CrossRef]
10. Rauf, A.; Al-Awami, A.; Kassas, M.; Khalid, M. Optimal Sizing and Cost Minimization of Solar Photovoltaic Power System Considering Economical Perspectives and Net Metering Schemes. *Electronics* **2021**, *10*, 2713. [CrossRef]
11. Hou, W.; Zhu, R.; Wei, H.; TranHoang, H. Data-driven affinely adjustable distributionally robust framework for unit commitment based on Wasserstein metric. *IET Gener. Transm. Distrib.* **2019**, *13*, 890–895. [CrossRef]
12. Shi, Z.; Liang, H.; Huang, S.; Dinavahi, V. Distributionally robust chance-constrained energy management for islanded microgrids. *IEEE Trans. Smart Grid* **2019**, *10*, 2234–2244. [CrossRef]
13. Maaruf, M.; Khalid, M. Global sliding-mode control with fractional-order terms for the robust optimal operation of a hybrid renewable microgrid with battery energy storage. *Electronics* **2021**, *11*, 88. [CrossRef]
14. Al-Muhaini, M.; Bizrah, A.; Heydt, G.; Khalid, M. Impact of wind speed modelling on the predictive reliability assessment of wind-based microgrids. *IET Renew. Power Gener.* **2019**, *13*, 2947–2956. [CrossRef]
15. Zhang, Y.; Wang, J.; Ding, T.; Wang, X. Conditional value at risk-based stochastic unit commitment considering the uncertainty of wind power generation. *IET Gener. Transm. Distrib.* **2017**, *12*, 482–489. [CrossRef]
16. Akbari-Dibavar, A.; Mohammadi-Ivatloo, B.; Zare, K. Optimal stochastic bilevel scheduling of pumped hydro storage systems in a pay-as-bid energy market environment. *J. Energy Storage* **2020**, *31*, 101608. [CrossRef]
17. Tuohy, A.; Meibom, P.; Denny, E.; O'Malley, M. Unit commitment for systems with significant wind penetration. *IEEE Trans. Power Syst.* **2009**, *24*, 592–601. [CrossRef]
18. Papavasiliou, A.; Oren, S.S. Multiarea stochastic unit commitment for high wind penetration in a transmission constrained network. *Oper. Res.* **2013**, *61*, 578–592. [CrossRef]
19. Papavasiliou, A.; Oren, S.S.; O'Neill, R.P. Reserve requirements for wind power integration: A scenario-based stochastic programming framework. *IEEE Trans. Power Syst.* **2011**, *26*, 2197–2206. [CrossRef]
20. Zheng, Q.P.; Wang, J.; Liu, A.L. Stochastic Optimization for Unit Commitment—A Review. *IEEE Trans. Power Syst.* **2015**, *30*, 1913–1924. [CrossRef]
21. Niknam, T.; Khodaei, A.; Fallahi, F. A new decomposition approach for the thermal unit commitment problem. *Appl. Energy* **2009**, *86*, 1667–1674. [CrossRef]
22. Zhao, C.; Guan, Y. Unified stochastic and robust unit commitment. *IEEE Trans. Power Syst.* **2013**, *28*, 3353–3361. [CrossRef]
23. Yang, N.; Dong, Z.; Wu, L.; Zhang, L.; Shen, X.; Chen, D.; Zhu, B.; Liu, Y. A Comprehensive Review of Security-Constrained Unit Commitment. *J. Mod. Power Syst. Clean Energy* **2021**, *10*, 562–576. [CrossRef]
24. Wei, L.; Liu, G.; Yan, S.; Dai, R.; Tang, Y. Graph computing based security constrained unit commitment in hydro-thermal power systems incorporating pumped hydro storage. *CSEE J. Power Energy Syst.* **2021**, *7*, 485–496.
25. Sharma, R.; Dutta, S. Optimal Storage Sizing for Integrating Wind and Load Forecast Uncertainties. U.S. Patent 8,996,187, 31 March 2015.
26. Zheng, Y.; Dong, Z.Y.; Luo, F.J.; Meng, K.; Qiu, J.; Wong, K.P. Optimal allocation of energy storage system for risk mitigation of DISCOs with high renewable penetrations. *IEEE Trans. Power Syst.* **2013**, *29*, 212–220. [CrossRef]
27. Alharbi, H.; Bhattacharya, K. Stochastic optimal planning of battery energy storage systems for isolated microgrids. *IEEE Trans. Sustain. Energy* **2017**, *9*, 211–227. [CrossRef]
28. Zhang, Y.; Ren, S.; Dong, Z.Y.; Xu, Y.; Meng, K.; Zheng, Y. Optimal placement of battery energy storage in distribution networks considering conservation voltage reduction and stochastic load composition. *IET Gener. Transm. Distrib.* **2017**, *11*, 3862–3870. [CrossRef]

29. Sedghi, M.; Aliakbar-Golkar, M.; Haghifam, M.R. Distribution network expansion considering distributed generation and storage units using modified PSO algorithm. *Int. J. Electr. Power Energy Syst.* **2013**, *52*, 221–230. [CrossRef]

30. Pon Ragothama Priya, P.; Baskar, S.; Mala, K.; Tamilselvi, S.; Mohamed Umar Raja, M. Optimal Storage Planning in Active Distribution Network Considering Uncertainty of Wind Energy System. In Proceedings of the International Conference on Data Science and Applications, 7–11 September 2022, Fethiye, TURKEY; Springer: Singapore, 2022; pp. 285–296.

31. Blanco, R.F.; Dvorkin, Y.; Xu, B.; Wang, Y.; Kirschen, D. Optimal energy storage siting and sizing: A WECC case study. *IEEE Power Energy Soc. Gen. Meet.* **2017**, *8*, 733–743.

32. Zhang, Y.J.A.; Zhao, C.; Tang, W.; Low, S.H. Profit-maximizing planning and control of battery energy storage systems for primary frequency control. *IEEE Trans. Smart Grid* **2016**, *9*, 712–723. [CrossRef]

33. Li, N.; Uckun, C.; Constantinescu, E.M.; Birge, J.R.; Hedman, K.W.; Botterud, A. Flexible Operation of Batteries in Power System Scheduling With Renewable Energy. *IEEE Trans. Sustain. Energy* **2016**, *7*, 685–696. [CrossRef]

34. Alismail, F.; Abdulgalil, M.; Khalid, M. Optimal Coordinated Planning of Energy Storage and Tie-Lines to Boost Flexibility with High Wind Power Integration. *Sustainability* **2021**, *13*, 2526. [CrossRef]

35. Nemati, M.; Braun, M.; Tenbohlen, S. Optimization of unit commitment and economic dispatch in microgrids based on genetic algorithm and mixed integer linear programming. *Appl. Energy* **2018**, *210*, 944–963. [CrossRef]

36. Abdulgalil, M.A.; Khalid, M.; Alismail, F. Optimal sizing of battery energy storage for a grid-connected microgrid subjected to wind uncertainties. *Energies* **2019**, *12*, 2412. [CrossRef]

37. Wang, Y.; Yang, Y.; Fei, H.; Song, M.; Jia, M. Wasserstein and multivariate linear affine based distributionally robust optimization for CCHP-P2G scheduling considering multiple uncertainties. *Appl. Energy* **2022**, *306*, 118034. [CrossRef]

38. Georghiou, A.; Kuhn, D.; Wiesemann, W. The decision rule approach to optimization under uncertainty: Methodology and applications. *Comput. Manag. Sci.* **2019**, *16*, 545–576. [CrossRef]

39. Duan, C.; Jiang, L.; Fang, W.; Liu, J. Data-driven affinely adjustable distributionally robust unit commitment. *IEEE Trans. Power Syst.* **2018**, *33*, 1385–1398. [CrossRef]

40. Croce, A.I.; Musolino, G.; Rindone, C.; Vitetta, A. Energy consumption of electric vehicles: Models' estimation using big data (FCD). *Transp. Res. Procedia* **2020**, *47*, 211–218. [CrossRef]

41. Wang, Y.; Kazemi, M.; Nojavan, S.; Jermsittiparsert, K. Robust design of off-grid solar-powered charging station for hydrogen and electric vehicles via robust optimization approach. *Int. J. Hydrogen Energy* **2020**, *45*, 18995–19006. [CrossRef]

42. Lu, Z.; Xu, X.; Yan, Z. Data-driven stochastic programming for energy storage system planning in high PV-penetrated distribution network. *Int. J. Electr. Power Energy Syst.* **2020**, *123*, 106326. [CrossRef]

43. Zhao, B.; Qian, T.; Tang, W.; Liang, Q. A data-enhanced distributionally robust optimization method for economic dispatch of integrated electricity and natural gas systems with wind uncertainty. *Energy* **2022**, *243* , 123113. [CrossRef]

44. Khalid, M.; Aguilera, R.P.; Savkin, A.V.; Agelidis, V.G. On maximizing profit of wind-battery supported power station based on wind power and energy price forecasting. *Appl. Energy* **2018**, *211*, 764–773. [CrossRef]

45. Ghahramani, M.; Nazari-Heris, M.; Zare, K.; Mohammadi-Ivatloo, B. Energy and reserve management of a smart distribution system by incorporating responsive-loads/battery/wind turbines considering uncertain parameters. *Energy* **2019**, *183*, 205–219. [CrossRef]

46. Khalid, M.; AlMuhaini, M.; Aguilera, R.P.; Savkin, A.V. Method for planning a wind–solar–battery hybrid power plant with optimal generation-demand matching. *IET Renew. Power Gener.* **2018**, *12*, 1800–1806. [CrossRef]

47. Xiong, P.; Jirutitijaroen, P.; Singh, C. A distributionally robust optimization model for unit commitment considering uncertain wind power generation. *IEEE Trans. Power Syst.* **2017**, *32*, 39–49. [CrossRef]

48. Christie, R.; Iraj, D. IEEE 14 Bus System Data. In *IEEE 14 Bus System, Illinois Center for a Smarter Electric Grid*; 1962. Available online: https://icseg.iti.illinois.edu/ieee-14-bus-system (accessed on 1 June 2022).

49. Cole, W.J.; Frazier, A. *Cost Projections for Utility-Scale Battery Storage*; Technical Report; National Renewable Energy Lab. (NREL): Golden, CO, USA, 2019.

50. Data for Hourly Aggregated Wind Output on the Ercot Website. Available online: http://www.ercot.com/gridinfo/generation/ (accessed on 13 June 2022).

51. Soroudi, A. Energy Storage Systems. In *Power System Optimization Modeling in GAMS*; Springer International Publishing: Cham, Switzerland, 2017; Chapter 7, pp. 175–201. [CrossRef]

52. General Electric. Energy Storage: GE Energy Storage Unit RSU-4000. Available online: https://www.ge.com/renewableenergy/ (accessed on 13 June 2022).

53. Hemmati, R.; Saboori, H.; Saboori, S. Assessing wind uncertainty impact on short term operation scheduling of coordinated energy storage systems and thermal units. *Renew. Energy* **2016**, *95*, 74–84. [CrossRef]

 *sustainability*

Article

Coordinated Frequency Control of an Energy Storage System with a Generator for Frequency Regulation in a Power Plant

Lateef Onaadepo Ibrahim [1], In-Young Chung [2], Juyoung Youn [3], Jae Woong Shim [4], Youl-Moon Sung [1], Minhan Yoon [5],* and Jaewan Suh [6],*

[1] Department of Electrical Engineering, Kyungsung University, Busan 48434, Republic of Korea
[2] Western Power Research Institute, Korea Western Power Corporation, Daejeon 34056, Republic of Korea
[3] Doosan Heavy Industries & Construction, Yongin 16858, Republic of Korea
[4] Department of Electrical Engineering, Sangmyung University, Seoul 03016, Republic of Korea
[5] Department of Electrical Engineering, Kwangwoon University, Seoul 01897, Republic of Korea
[6] Department of Electrical Engineering, Dongyang Mirae University, Seoul 08221, Republic of Korea
* Correspondence: minhan.yoon@gmail.com (M.Y.); jwsuh@dongyang.ac.kr (J.S.); Tel.: +82-2-940-5108 (M.Y.)

Citation: Ibrahim, L.O.; Chung, I.-Y.; Youn, J.; Shim, J.W.; Sung, Y.-M.; Yoon, M.; Suh, J. Coordinated Frequency Control of an Energy Storage System with a Generator for Frequency Regulation in a Power Plant. *Sustainability* **2022**, *14*, 16933. https://doi.org/10.3390/su142416933

Academic Editor: Muhammad Khalid

Received: 11 November 2022
Accepted: 13 December 2022
Published: 16 December 2022

Publisher's Note: MDPI stays neutral with regard to jurisdictional claims in published maps and institutional affiliations.

Abstract: Considering the controllability and high responsiveness of an energy storage system (ESS) to changes in frequency, the inertial response (IR) and primary frequency response (PFR) enable its application in frequency regulation (FR) when system contingency occurs. This paper presents a coordinated control of an ESS with a generator for analyzing and stabilizing a power plant by controlling the grid frequency deviation, ESS output power response, equipment active power, and state of charge (SoC) limitation of the ESS in a power plant. The conventional generator and FR-ESS controllers were investigated and compared. To obtain the optimal frequency and power response, an ESS-based adaptive droop control method was proposed. The proposed control strategy was developed and implemented considering the changes and limitations of the dynamic characteristics of the system, FR requirements, and an ESS using the PSCAD/EMTDC software. The simulation results showed that the proposed method was more effective than the conventional droop-control-based FR-ESS, and the effectiveness of this method was validated.

Keywords: energy storage system; droop control; frequency regulation; inertia constant; state of charge; PSCAD/EMTDC

1. Introduction

Currently, ESSs, which are required to achieve stability and grid safety owing to the high penetration of renewable energy resources, have received wide attention from researchers [1]. The integration of ESSs and power reserve synchronization is an effective solution for overcoming renewable energy source (RES) intermittency and fluctuating effects. This is supported by the IEC T120 work program objectives, which identify ESSs as a solution that can efficiently deliver sustainable, economic, and secure electricity supplies [2].

The importance of frequency regulation (FR) in power systems cannot be overemphasized. FR can be achieved via three distinct control stages: primary (inertial response), secondary (governor response), and tertiary (automatic generation control (AGC) [3]. An imbalance in the supply and generation at the power-grid level causes frequency deviation. An increase in the utility grid frequency can be caused by excessive power generation, which in turn increases the speeds of rotating machines, whereas a lack of supply leads to a frequency decrease. When there is a significant deviation from statutory limits, generation plants and loads are disconnected from the network, which can lead to blackouts [4]. Therefore, to maintain the desired frequency (either 60 or 50 Hz) by the grid, the total generation should be equal to the system loads and electrical losses [5]. Although low-frequency fluctuations can be handled by generator participation in secondary frequency control, the capability of these generators for high-frequency load fluctuations may not be adequate owing to the requirements of these fluctuations [6]. Therefore, ESS application for FR has a faster (quick) response, is less expensive, has a lower capacity in power plants, and offers a precise control capability over many conventional generators [7,8]. The flexibility and rapid control of the charge and discharge capabilities of ESSs to regulate the system frequency not only improves the FR performance, but also reduces the reserve of traditional units [9]. However, the use of only ESS for FR would require large storage capacity and energy, which is economically expensive [6,10]. The use of ESSs in traditional power plants initially designated for FR can therefore increase the overall efficiency of the power system [11].

Traditionally, grid operators engage with the governor-free operation of thermal power systems or generators for FR in large-scale operations. However, such generators are subjected to stress through the mechanical regulation of valve openings to compensate for FR [3]. For instance, in Korea, FR is performed using a governor-free control method with turbine governors responding within 10 s by providing power for 30 s, and the AGC responds within 30 s by providing power for 30 min. However, power plants using these approaches operate below their rated capacity to provide FR services until they are needed, making them inefficient [11,12]. Studies have been conducted concerning improving the frequency response characteristics in power systems, but little has been done to cater to the power response and SoC management of the ESS. In [13], ESS de-loading or curtailment of generation units, load-demand side management, and utilization of kinetic energy reserves have been highlighted as methods in which virtual inertia implementation can provide frequency regulation in relation to the inertial response. Reference [14] proposed a droop control strategy as a frequency regulation method for a microgrid using an ESS to regulate the ESS output power. Furthermore, ESS participation in primary frequency regulation uses both virtual droop control and virtual inertial control, which could increase the frequency nadir and effectively reduce the rate of change of frequency (RoCoF) [15]. Reference [9] proposed a control strategy in which virtual inertial and virtual negative inertia control methods were implemented to prevent frequency deterioration and accelerate frequency recovery, respectively, for ESS participation in primary frequency control (PFC). Another previous study [16] proposed droop control for battery energy storage system (BESS) participation in system-grid FR by adjusting the BESS output according to the fixed sagging coefficient. In a different study [17], a BESS based on virtual droop control was implemented to provide grid-frequency stability. However, these control strategies do not consider the limitations of the system characteristic dynamics and frequency variation requirements. A previous study demonstrated the advantage of using an ESS to replace the governor in a synchronous generator from the perspective of the SoC management scheme and FR performance. However, this study did not compare the proposed method with a conventional FR-ESS system [18].

In this study, to ensure the effect of contingency events on frequency regulation while also considering the importance of ESS-SoC management, an adaptive droop control strategy of the ESS instead of the governor is proposed. Considering the inertia value of the participating generators as the control quantity, an adaptive droop controller for the FR-ESS was designed using an algorithm to allocate the ESS output. Adjustment of the droop constant of the ESS improves the output power injected based on the frequency deviation rate for the ESS such that it may participate in primary frequency regulation. Therefore, the ESS reserves can be optimally utilized to improve the RoCoF and increase the frequency nadir, thereby improving the frequency stability of the power network. In addition, the mutual influence between the conventional generator controller and FR-ESS controller was investigated to evaluate the grid frequency response dynamics. To investigate the technical impact of this system on the overall system network, the complete test system was comprehensively modeled using electromagnetic transient analysis software (PSCAD/EMTDC). PSCAD/EMTDC is a widely used power system transient analysis tool that has intuitive simulation and modeling tools that are greatly enhanced by its state-of-the-art graphical user interface [19]. The effectiveness of the proposed control strategy was verified by simulation under the condition of a generator and load tripping disturbance.

2. Power System Architecture

2.1. Test System Model

The power system network shown in Figure 1 is considered to validate the effectiveness of the proposed method in accessing the frequency regulation response, generator output power, power grid output, and output active power of the ESS, and SoC limitations. The electrical architecture of the tested power system consists of two generators: Gen 1 (trip generator) and Gen 2 (with and without governor operation) with rated capacities of 100 and 612 MVA, respectively, and a utility grid connected to the main transformer (610.4 MVA) via a 345 kV AC bus. The ESS was installed on the 6.9 kV AC bus via an interconnection transformer (63 MVA) at the point of common coupling (PCC) with a rated capacity of 25 MW/6.25 MWh. At the PCC, the ESS terminal voltages synchronized with the system voltage can be measured appropriately. Generally, an ESS comprises a power conversion system (PCS) for DC-to-AC output conversion and a storage medium. The total load capacity of 1650 MW/315 MVAR, which consists of load A (1500 MW/300 MVAR) and load B (150 MW/15 MVAR), was connected to the 22 kV bus. Load B is the trip load. In this study, the generator was modeled as a synchronous machine in PSCAD, where the governor, exciter (ST4B), power system stabilizer (PSS2B), and turbine models were included for a more realistic simulation. Figure 1 shows the three-bus network, single-line model used for the simulation of the test results. The system parameters used for the design and simulation process are listed in Table 1.

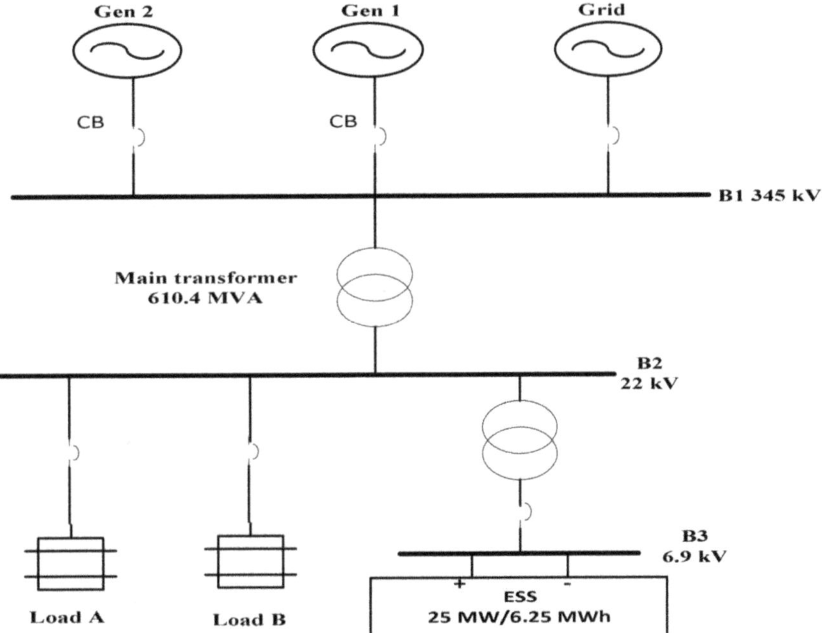

Figure 1. Single-line diagram of the power system.

Table 1. System Parameters.

System Parameters		
Parameters	**Value**	**Unit**
Generator rated power ($P_{Gen,rat}$)	612	MVA
Power conversion system (PCS) rating	25	MW
Inverter DC power rating	25	kW
Total system load	1650/315	MW/MVAR
System frequency	60	Hz
Offset frequency	59.8–60.2	Hz
ESS size/capacity $\left(P_{ESS,cap} \right)$	6.25	MWh
ESS rating ($P_{ESS,rat}$)	25	MW
State of Charge initial (SoC_o)	50	%
System sampling time	50	microsec
Droop rate (R_{droop})	4.62	%
Inertia constant (H)	3	s

2.2. ESS Modeling and Control Scheme

The ESS modeled in this study, which is an inverter-based reserve (IBR) system as described in Section 2, was designed such that it can inject and/or supply or absorb a certain amount of energy over a given period. The ESS structure shown in Figure 2 consists of a DC source composed of battery banks, DC link capacitors, a three-phase pulse-width modulation (PWM) inverter, inductors, and capacitor (LC) filters. The three-phase inverter is controlled using an active power/reactive power (P/Q) controller.

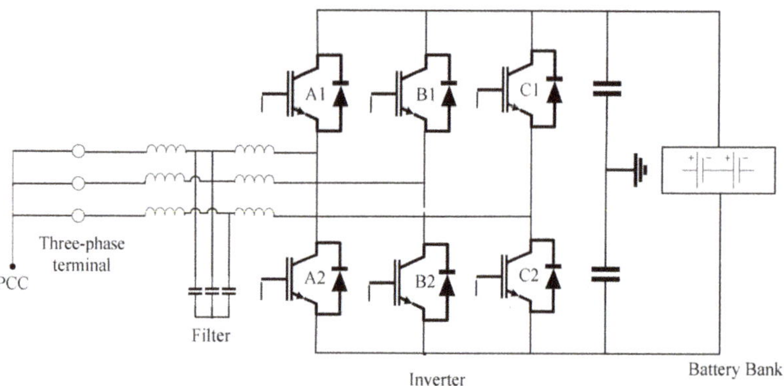

Figure 2. Energy storage system structure [15,20].

2.2.1. Voltage Source Inverter Model

The most commonly used inverter type is the VSI, where the AC power provided on the output side functions as a voltage source. The input DC source voltage is usually an independent source, such as a battery, which is referred to as a DC-link inverter. This structure is the most widely used because it naturally behaves as a voltage source and is employed in many industrial applications. Compared with single-phase VSIs used in low-range power applications, three-phase VSIs are implemented in medium- to high-power applications.

VSIs are required in island or autonomous operation to keep the voltage stable. In microgrid applications, VSIs have been found to be interesting because they do not require any external reference to remain synchronized. The model is convenient as it provides performances such as ride-through capability and power quality enhancement to distributed power generation systems. VSIs can change behavior from voltage to current sources when they operate in grid-connected mode. This source inverter is often connected to energy storage devices to regulate both frequency and voltage in low-inertia grids. Therefore, in this study, the use of VSI drives is more efficient than current source inverters (CSI); VSI drives are distinctive for their use of insulated gate bipolar transistors (IGBTs) with fast switching times that create a PWM voltage output with regulated frequency and voltage. By contrast, CSIs use gate turn-off thyristors (GTOs) or symmetrical gate commuted thyristors (SGCTs) that generate PWM output with regulated frequency output current with high harmonics, which necessitates filters on both input and output sides. VSIs are implemented in this study because the active and reactive power can be controlled independently, thereby reducing the need for reactive-power compensation. They contribute to the stabilization of the AC network at PCC. Hence, they have the capability of better sustaining the PCC voltage.

The voltage source inverter (VSI) in this study uses the classic active power (P) and reactive power (Q) control method, also referred to as P/Q control, which was developed based on IGBT semiconductor switches, as shown in Figure 2. Figure 2 shows a schematic of a three-phase VSI interacting with energy storage and an AC system. It is connected to the AC system through line filters composed of parallel inductors and capacitors. In this inverter, three-phase reference voltages are generated using the sinusoidal pulse-width modulation (PWM) technique, as depicted in Figure 3.

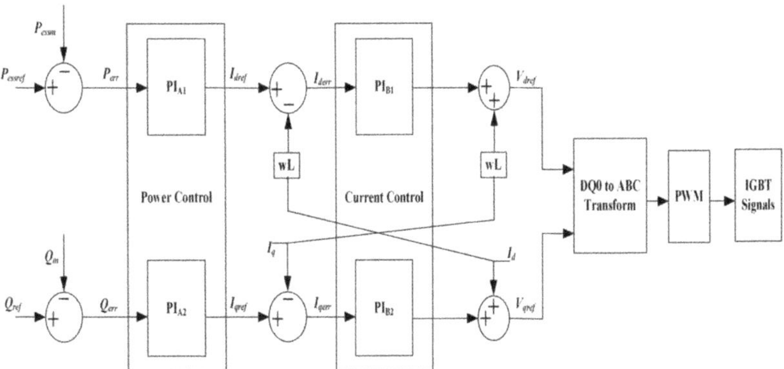

Figure 3. Active and reactive power (PQ) control scheme.

2.2.2. P/Q Control Scheme

The ESS control strategy proposed in this study is the P/Q control scheme designed in the PSCAD/EMTDC simulation program. This system, which is depicted in Figure 3, consisted of two cascaded control loops (outer slow power control and inner fast current control). The energy management system of the ESS provided both the ESS active power reference (P_{essref}) and reactive power (Q_{ref}), which were dependent on the ESS state and load balance in the outer control loop. By contrast, the inner control loop independently controlled the direct-axis (I_{dref}) and quadrature-axis (I_{qref}) current references. To implement these control loops, proportional and integral (PI) controllers were used. The P controller regulated ESS output power in accordance with the power reference generated from frequency controller (ΔP_{essref}). The ESS AC side output power (P_{essm}) measured was fed into the P controller to calculate the error (P_{err}), which the PI controller used in generating the reference for d-axis current $\left(I_{dref}\right)$ regulation. This controller output was regulated within the minimum (I_{dmin}) and maximum (I_{dmax}) d-axis current value through tuning of the PI gain parameters; this minimized frequency drop or rise and settling time. This PI controller gain parameters (K_p and K_i) values presented in the paper were obtained via the tuning rule of Ziegler and Nichols based on a measured step response to compensate single input, single output (SISO) process with time delay that satisfied both robustness and performance requirements by eliminating steady state error and reducing the overshoot with oscillations to obtain an improved transient response. The measured voltages (V_{dref}) and (V_{qref}) at each terminal of the control scheme were transformed from the DQ0 rotating frame to ABC using the Park transformation. The three-phase voltage reference signal of the PWM was determined. The reactive power reference (Q_{ref}) was set by considering the droop rate. The outputs P and Q of the inverter were adjusted using the droop coefficients [21]. The proportional and integral controller gains are presented in Table 2.

Table 2. PI Controllers' Gain.

PI Gain Values		
PI Controller	K_p	K_i
PI$_{A1}$	7.93651	0.0008521
PI$_{A2}$	8.52314	0.0006523
PI$_{B1}$	4.1746	0.000254
PI$_{B2}$	1.8574	0.000234

3. Proposed Control Strategy Implementation

In this section, the proposed ESS control strategy and its structure are discussed. The control implementation is based on the regulation of the active power of the ESS and generator without the governor based on the frequency deviation rate. The power-frequency control of the ESS in relation to the system inertia for improving the stability of the power system is discussed.

3.1. System Frequency Dynamics

The nominal frequency of the Republic of Korea is 60 Hz, with a deadband of 0.2 Hz. Therefore, a frequency variation within the 59.8 to 60.2 Hz range shows that the system is in steady state. Frequencies outside of this range are regarded as abnormal owing to a contingency event [22]. In this study, the deadband of the system frequency was set to 60 ± 0.2 Hz. The change in the power system frequency can be defined by the swing equation as follows [3]:

$$\frac{\Delta P_d}{S_{syst}} = \frac{2H_{syst}}{f_0} \times \frac{df}{dt} \tag{1}$$

where $(\Delta P_d = P_{gen} - P_{ld})$ represents the power deficit, which is the difference between the generation unit active power (P_{gen}) and load demand power (P_{ld}), S_{syst} is the rated capacity of the system, H_{syst} is the inertia constant of the system, f_0 is the nominal frequency, and $\frac{df}{dt}$ is the RoCoF.

The frequency nadir and RoCoF are emphasized in [15] as important elements to consider for system stability. Therefore, in this study, two essential elements related to the frequency response were examined: the RoCoF and inertia constant (H). The inertia constant of the synchronous generator (SG) is expressed in Equation (2) [3]:

$$H_{syst} = \frac{\sum_{i=1}^{n} H_i \times S_i}{S_{syst}} \tag{2}$$

where the inertia constant of the individual generator and nominal rating of the generator are H_i and S_i, respectively. A block diagram of the FR-ESS control strategy used in this study is shown in Figure 4. The three-phase ESS's current $(IL.abc)$ and voltage $(Vc.abc)$ at the PCC are measured by the measurement block. The d-axis and q-axis of the currents $(IL.dq)$ and voltages $(Vc.dq)$ are provided by the calculation block using the phase angle (θ) derived from the phase-locked loop (PLL). The $IL.dq$ and $Vc.dq$ are the inputs to the current and power controllers while the output of these controllers are $Vinv.dq^*$ and $ILdq^*$. The inverter voltages $(Vinv.abc)$ generated are sent to the PWM block as shown in Figure 4.

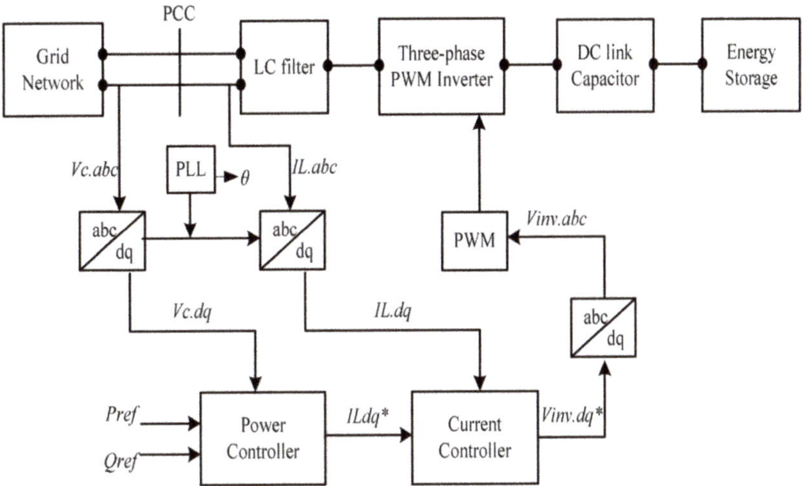

Figure 4. Block diagram of the ESS control in the PQ mode.

3.2. Generator Modeling Dynamics

In conventional power system networks, the role of the SG is to convert mechanical energy into electrical energy, which is coupled to the prime movers (steam turbines, as used in this study) that drive the rotor. The generator's rotor with a rotating mass contributes not only to the generator's active power output, but also adds an inertia property to the grid frequency via the supply of its stored kinetic energy when a contingency occurs. The dynamics of the SG rotor can be expressed as in Equations (3) and (4), where Equation (4) is similar to the swing equation described in Equation (1) [23].

$$\tau_m - \tau_e = J\alpha \tag{3}$$

$$\tau_m - \tau_e = J\frac{d\omega_m}{dt} \tag{4}$$

Here, τ_m is the mechanical torque exerted by the steam turbine; τ_e is the electrical torque exerted by the system load; J is the moment of inertia; α is the angular acceleration or retardation; and ω_m is the synchronous angular velocity.

To represent Equation (4) in real power terms rather than in torques using the relationship $P = \tau\omega$, it can be expressed as:

$$P_m - P_e = J\omega_m\frac{d\omega_m}{dt} \tag{5}$$

where P_m and P_e are the mechanical power input and electrical power output, respectively.

The inertia present in the SG also contributes to the effect of the RoCoF. The higher the inertia, the slower the RoCoF, and vice versa. Therefore, the mismatch between the active power and RoCoF of the grid network can be expressed as:

$$P_m - P_e = K_d\frac{df}{dt} \tag{6}$$

where K_d is the inertia coefficient.

An imbalance in the power network causes the rotor to speed up or down to offset the power mismatch, which is a characteristic of all SGs. This response, termed the inertial response (IR) if adopted only for FR, will support the system for a few seconds; thus, the stored kinetic energy will be consumed, resulting in system collapse. Therefore, a droop or PFC scheme mainly provided by SG governors was employed to adjust the generator output power in response to the grid frequency variation. In this study, the damping term was considered while modeling the SG, which was provided by the quadrature axis damper windings set at a value of two to resolve the spikes in voltage as a result of sudden disturbances. Therefore, with the incorporation of the damping term, the swing equation of the SG is:

$$\tau_m - \tau_e = J\frac{d\omega_m}{dt} + D\omega \tag{7}$$

where D is the damping coefficient of the generator.

3.3. Proposed Adaptive Control Method

Figure 5b,c depict the simplified and detailed proposed method of the FR-ESS with a generator, respectively. In the configuration shown in Figure 5b, the generator operated without a governor and the ESS was activated, whereas in Figure 5a, the governor regulated the angular speed ($\Delta\omega$) of the SG. The governor adjusted the mechanical power (P_m) with respect to the angular velocity variation of the generator rotor. As shown in the detailed representation in Figure 5c, the ESS was interfaced with the proposed adaptive control scheme, which functioned as PFC. During a contingency event, the generator output power changed and the ESS proposed in this study compensated for a power deficit by providing active power through the PCC to contribute to the FR. The main goal of the proposed adaptive control scheme is to enhance frequency regulation by reducing the RoCoF and frequency deviation.

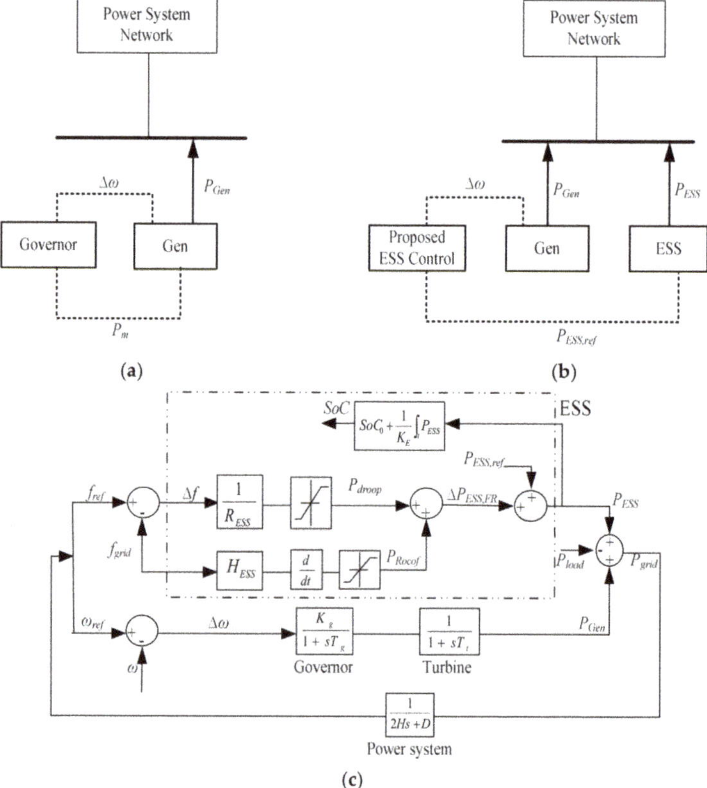

Figure 5. FR with/without the ESS and governor: (**a**) conventional and (**b**) proposed system. (**c**) Proposed adaptive control of ESS operation with a synchronous generator for FR and SoC calculation.

In the proposed configuration method, droop control of the FR-ESS, characterized as a power converter, was used by the generator without the governor owing to its fast action. The system frequency deviation Δf served as the input in the ESS control block, in which the ESS executed the droop control by adaptively adjusting the electrical power (P_e). The droop control block ($\frac{1}{R_{ESS}}$) generated an output signal P_{droop}. In other words, the adaptive droop control of the ESS was the product of the inverse droop rate and frequency deviation, which determined the amount of supporting power from the ESS, which was defined as in Equations (8) and (9) for both PFC and IR regulation, respectively, following the grid-frequency variation in Equation (10). In the conventional control scheme, the ESS droop rate was set to 5% (i.e., the control coefficient was 20), whereas that of the proposed control was 1% and the corresponding inertia constant value was 4 s.

Using this method, both the IR and PFC were provided to the power system through inertia constant adjustment and power-frequency droop rate adjustment, respectively.

$$P_{droop} = \frac{1}{R_{ESS}} \cdot \Delta f \tag{8}$$

$$P_{Rocof} = \frac{d}{dt} f_{grid} \cdot H_{ESS} \tag{9}$$

$$\Delta f = f_{grid} - f_{ref} \tag{10}$$

Here, P_{droop} is the change in the power output based on the droop characteristics, R_{ESS} is the ESS droop rate value, Δf expressed in (pu) is the frequency deviation from the contingency event, f_{grid} is the grid frequency, f_{ref} is the reference frequency, P_{Rocof} is the power required to regulate the RoCoF for the inertial response, and H_{ESS} is the inertia constant value of the ESS for the RoCoF.

Therefore, the power of the ESS based on the droop rate ($\Delta P_{ESS,FR}$) for frequency regulation is the summation of the power regulation of RoCoF (P_{Rocof}) and the power output based on the droop characteristics (P_{droop}), which can be defined as in Equation (11). The total output of the ESS (P_{ESS}) can then be derived as in Equation (12) for FR. This differs from the conventional droop method and enables the ESS to manage its energy optimally.

$$P_{droop} + P_{Rocof} = \Delta P_{ESS,FR} \tag{11}$$

$$\Delta P_{ESS,FR} + P_{ESS,ref} = P_{ESS} \tag{12}$$

Here, $\Delta P_{ESS,FR}$ is the ESS power based on the droop rate for FR, $P_{ESS,ref}$ is the ESS reference power, and P_{ESS} is the total ESS output power.

For the power system network to be stable after a disturbance occurs, the total power generation should match the load demand. Therefore, the amount of power flow to the grid P_{grid} can be calculated from Equation (13), which should match the load demand (P_{load}).

$$P_{ESS} + P_{Gen} - P_{load} = P_{grid} \tag{13}$$

Here, P_{Gen} is the SG active power output, P_{ESS} is the ESS active power output, P_{load} is the total load demand, and P_{grid} is the grid power.

Traditionally, the SoC is calculated by integrating the current (unit of current) [24,25]; however, it does not define the relationship between the battery power and SoC. In this study, we implemented the energy concept identified in [26] by integrating the power to calculate the SoC of the ESS in Equations (14) and (15) assuming that the battery's internal voltage was kept constant (i.e., power was proportional to current). As illustrated, the ESS power and SoC dynamics were employed to establish a relationship between the change in grid frequency and the SoC limit.

$$SoC(t) = SoC_o + \frac{1}{K_E} \int_{t_0}^{t} P_{ESS}(t) dt \tag{14}$$

$$K_E = E*h \tag{15}$$

Here, SoC is the ESS-SoC [%], SoC_o is the initial SoC value [%], K_E is the ESS energy [MWs], P_{ESS} is the ESS power [MW], E is the ESS size [MWh], and h is the constant used to convert hours to seconds.

The SoC limitation scheme has been developed for FR-ESS applications, which involve a reference value that the ESS is continually attempting to adapt to with a set nominal frequency (60 Hz). This reference value, termed the initial SoC (SoC_o), was 50% in this study, indicating the highest potential energy of charging and discharging [18]. The absorption or supply of ESS power (P_{ESS}) results in a change in the SoC. Similarly, when the ESS stops operating (i.e., it does not inject or absorb power), the SoC must remain constant.

3.4. ESS Control Algorithm

Figure 6 shows the control scheme for the FR-ESS based on the proposed methodology described in Section 3. The FR-ESS control algorithm in this study performed charging and discharging operations to reduce frequency variations by providing and/or absorbing power, which depended on the droop control signal it received. This was designed such that the proposed adaptive droop control scheme required a limit on the ESS capacity to prevent the excessive charging and overcharging of power. Therefore, the upper limit $P_{ESS,\ max}$ and lower limit $P_{ESS,\ min}$ were set appropriately. The control operation was based on the frequency deviation of the power system, which was initiated to implement droop control by adjusting the droop rate while ensuring that the ESS capacity limit was monitored when a contingency operation occurred. The ESS was activated depending on the grid-frequency deviation outside of the deadband in two modes of operation (charging and discharging). Therefore, the ESS operated in discharge mode when its active power output was greater than zero, providing power to the system, and vice versa. In idle mode, the active power was kept constant at zero. To avoid oscillation problems that may occur between the operation modes, a small deadband of 20% (the actual size of deadband can be adjusted based on the system requirements) was introduced for ESS control switching following Equations (16) and (17) for the charging and discharging modes, respectively. This deadband represented the offset frequencies f_{low} and f_{high}, which were 59.8 and 60.2 Hz, respectively. Therefore, a frequency variation within this range was regarded as a steady-state system. When the frequency was within the deadband, the ESS operated without charging or discharging to maintain the frequency. If the system frequency dropped below f_{low}, the ESS provided power to $P_{ESS,\ max}$. Otherwise, the ESS absorbed the power to charge up to $P_{ESS,\ min}$. The charging and discharging amounts of the power output can be calculated using the droop rate, as shown in Equation (18). The frequency deviation Δf can have both positive and negative values that define the ESS power injection and absorption, respectively, as expressed in Equation (10).

$$P_{ESS,\ min} \leq P_{ESS} < 0 \ ; \ and \ f_{grid} > f_{high} \tag{16}$$

$$0 < P_{ESS} \leq P_{ESS,\ max}; \ and \ f_{grid} < f_{low} \tag{17}$$

$$\Delta P_{ESS} = \frac{-\Delta f \ * \ P_{ESS,rat}}{R_{ESS} \ * \ f_o} \tag{18}$$

Here, $P_{ESS,\ min}$ (-25 MW) and $P_{ESS,\ max}$ ($+25$ MW) are the ESS power output charging and discharging limits, respectively; Δf is the frequency deviation [Hz]; ΔP_{ESS} is the ESS output power variation [MW]; $P_{ESS,rat}$ is the ESS rating [MW]; f_o is the nominal frequency [Hz]; R_{ESS} is the droop rate of the ESS.

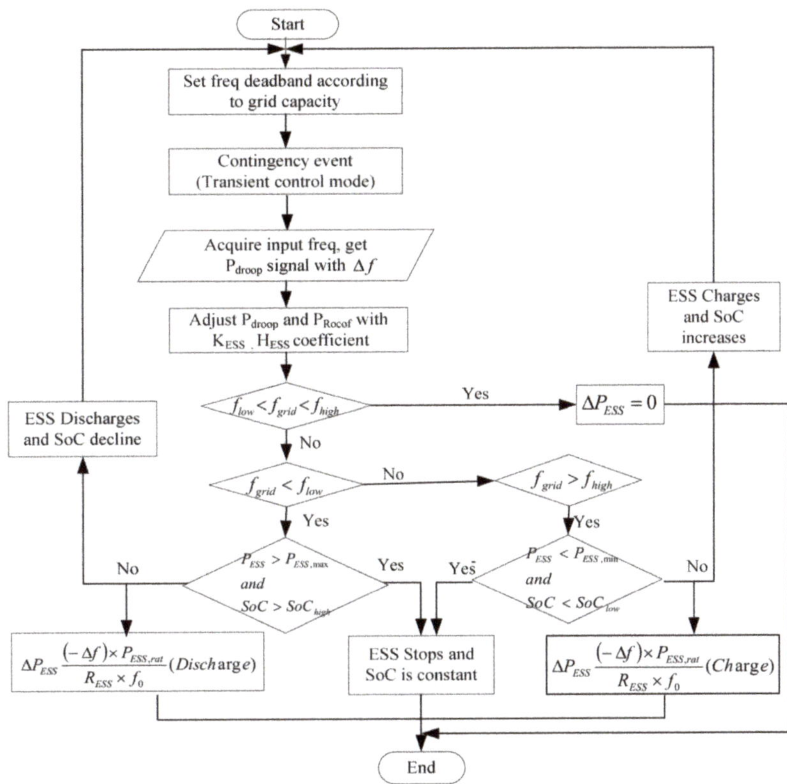

Figure 6. Flow chart of the control algorithm of the proposed method for FR.

In this study, the SoC limitation of the ESS was designed to operate based on the frequency deviation variation rate as well as the power output of the ESS. In most studies, an SoC reference value required to regulate the SoC of the ESS is set as a constant. This fixed value must be adjusted to avoid frequent charging and discharging of the ESS because it is not the same as the target value [27]. Therefore, in this study, the operation of the ESS was subject to the SoC range constraint to limit its output power according to the SoC and prevent the aforementioned issue when it participates in FR. The normal SoC operation range during the emergency case was (50, 80) for SoC_{low} and SoC_{high}, respectively. When the SoC exceeded the upper limit, the control scheme activated the discharging of the ESS, thereby reducing the SoC to prevent overcharging and vice versa when the SoC exceeded the lower limit. The operational constraints are defined in Equations (19) and (20) for normal operation and during an emergency, respectively. For the case considered in this study, a low SoC limit was investigated to compare the proposed method of ESS in PFC with the conventional method considering the rate of frequency deviation. The control mechanism involved using the droop control of the ESS to regulate the frequency variation. The control strategy can be summarized in Equation (21), which defines the charging, discharging, and idling powers (wherein both the charging and discharging powers are zero) of the ESS.

$$SoC = SoC_o \tag{19}$$

$$SoC_{low} \leq SoC \leq SoC_{high} \tag{20}$$

$$\Delta P_{ESS} = \begin{cases} P_{disch} \ of \ ESS; \ SoC < SoC_o \\ P_0 \ of \ ESS; \ SoC = Constant \\ P_{ch} \ of \ ESS; \ SoC_{limit} < SoC < SoC_o \end{cases} \tag{21}$$

4. Simulation Analysis and Results

To verify the effectiveness of the proposed control scheme using the test system shown in Figure 1, a case study was conducted during a contingency event by tripping both generator G1 and load B to compare three different control techniques via simulation. The case study conducted using PSCAD/EMTDC is described as follows.

Case 1: Generator G2 (with governor) while the ESS is deactivated.

Case 2: Generator G2 (without a governor) with the ESS and conventional droop control.

Case 3: Generator G2 (without a governor) with the ESS and adaptive droop control.

The simulations were carried out for a duration of 50 s and generator G1 was tripped at 5 s to cause a mismatch of the power imbalance, thereby causing the system frequency to deviate. Similarly, load B was tripped at 25 s, causing the system frequency to rise above the nominal value. This frequency deviation also has a significant effect on the grid active power, participating generators, and ESS in frequency regulation. Therefore, we investigated the effect of only the governor without an ESS (Case 1), the power compensation of the ESS using the conventional droop method (Case 2), and the proposed method (Case 3) to verify its advantages.

4.1. Frequency Variation

In the Korean power system, the minimum frequency deviation is 59.70 Hz after a disturbance occurs. Figure 7 shows a comparison of the grid frequency between the governor and generator, ESS compensation with a conventional (fixed) droop, and the proposed (adaptive droop) control scheme. As depicted in Figure 7, it can be deduced that the grid frequency is at a normal value (60 Hz) before the contingency events occur. When the tripping of the generator occurs at 5 s, the grid frequency curves of the three FR control techniques decreased rapidly. It can be seen that the minimum frequency deviations (nadir) for Cases 1 and 2 are the same (59.64 Hz); however, the frequency nadir for Case 3 is 59.71 Hz. Similarly, when the load was tripped at 25 s, the frequency response of the system increased. However, the maximum frequency increase was lower with the proposed method (60.15 Hz) than in Cases 1 and 2 (60.21 Hz). This shows that the proposed method can decrease the RoCoF with an improved frequency nadir compared with both Cases 1 and 2. Therefore, FR using the proposed method is improved compared with the other control methods. A rapid change in the rotating speed of the synchronous generator, which can result from a loss in a large generating unit, can lead to an unacceptable frequency decrease or sudden disconnect in load, which may in turn result in grid frequency instability. The existing or conventional generators supply or absorb their stored kinetic energy to adapt to frequency deviations in the inertia response (IR) stage [28]. However, the responses of the conventional methods were slow compared to that of the proposed method.

4.2. Active Power Output

Figures 8 and 9 show the results of the grid and generator active power responses compared to the three control methods (generator with a governor and ESS with the conventional proposed methods), respectively. It can be observed that the grid active power in Figure 8 exhibits slightly better oscillation damping with the proposed control compared to the conventional method, which exhibits higher oscillations in the transient stage (5 to 10 s). In the steady-state region (0 to 5 s and 35 to 50 s), the active power supplied to the grid is the same for all control strategies.

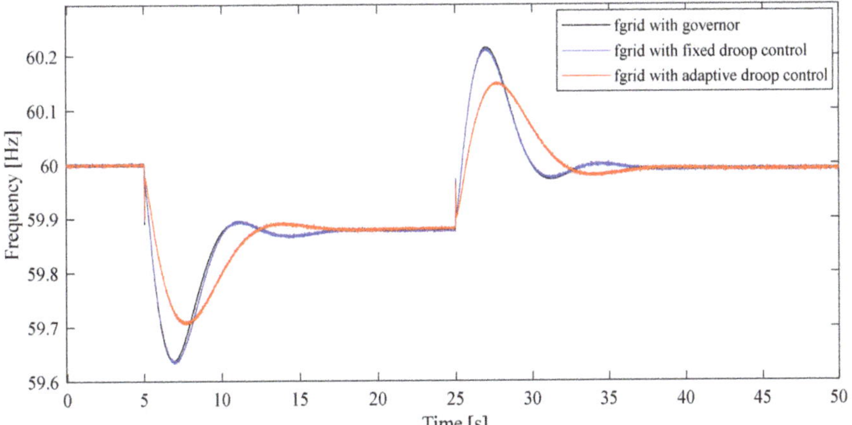

Figure 7. Comparison of the grid frequencies using only a governor, ESS with conventional control, and ESS with the proposed method.

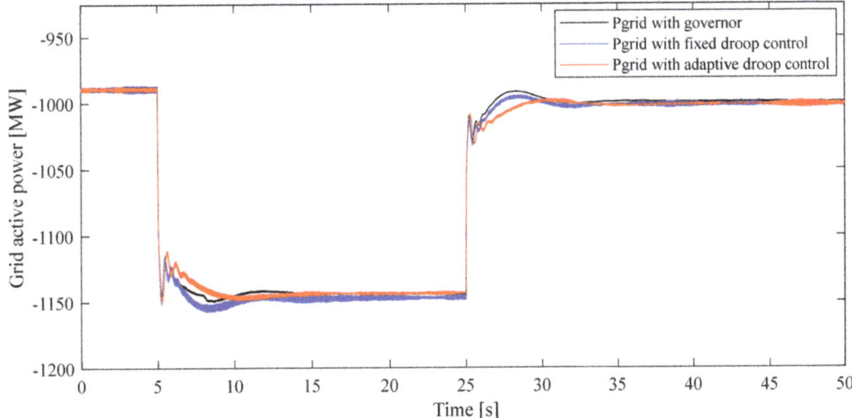

Figure 8. Comparison of the system's active power for the different control methods.

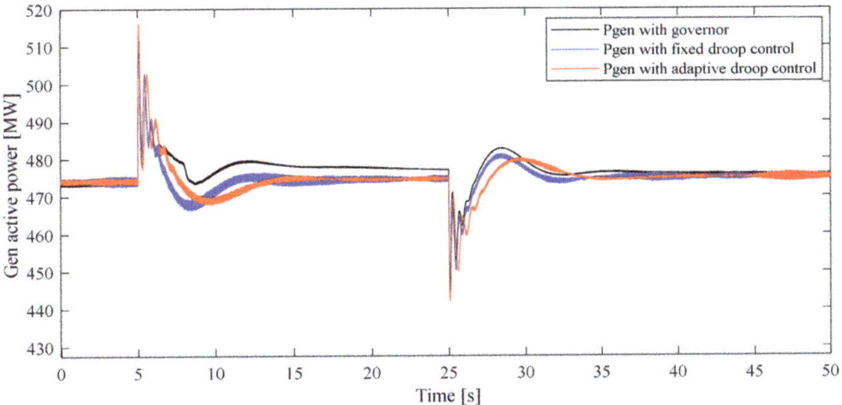

Figure 9. Comparison of the generator's active power for the different control methods.

The generator's active power in Case 1 was increased during the generator tripping event (5 to 25 s) from the steady state value of 474 MW at 0 s to 477.7 MW at 20 s to compensate for the frequency drop, as shown in Figure 9. However, for Cases 2 and 3, the active power output (P_{Gen}) was the same as the steady-state value (474 MW at 20 s). This shows the effect of the governor in Case 1 for FR compared to FR-ESS in Cases 2 and 3. Although the power fluctuated during the tripping events of both the generator and load, a new steady state was reached when the frequency stabilized. The variation in the output power of the generators in Case 1 occurred because the power compensation was limited to only the kinetic energy released by the SG through the governor with respect to the droop setting. Similarly, the amount of ESS power injected into the system with fixed droop control (Case 2) and adaptive droop control (Case 3) was limited by the droop rate constant.

4.3. ESS Active Power

According to the ESS power output shown in Figure 10, the operation of the ESS in terms of the injection and absorption of power to and from the grid can be observed when the grid frequency deviates. During the generator tripping event, the frequency drops, causing the ESS participating in FR to inject power into the system. However, the amount of power is dependent on the ESS droop-rate constant. Using the proposed method, more power can be injected into the system to minimize the frequency nadir. Likewise, when the frequency increases owing to load tripping, the ESS with the proposed method has the capability to absorb more power than the conventional method. Therefore, the ESS discharges its power by injecting a proportion defined by the droop rate value in Equation (8) such that a reasonable amount of active power provides compensation for stabilizing the system. In addition, this system charges in a similar manner. When there is no change in the grid frequency (no disturbance), the ESS remains in idle mode with a power equal to zero, as depicted in Figure 10. Using the proposed method, where the ESS uses a control scheme with a lower droop-rate value compared with the conventional control method, it can provide more power to the system during the transient period while ensuring that the maximum rated power is not exceeded. Therefore, the ESS can change its active power with respect to droop control to assist with FR.

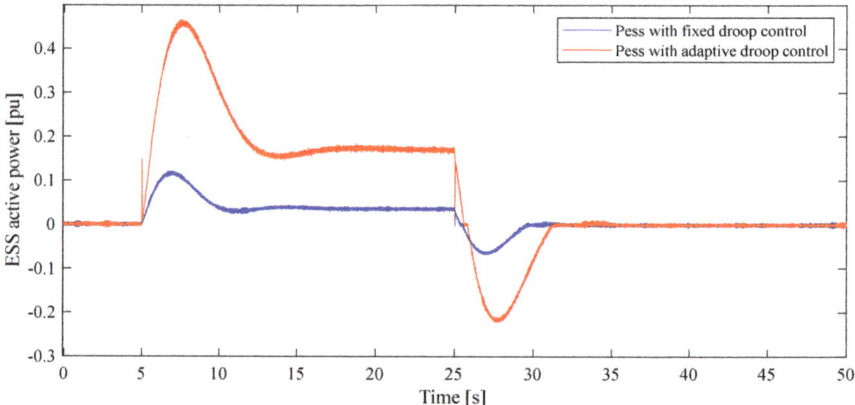

Figure 10. Comparison of both methods for ESS active power.

4.4. SoC Response

The control algorithm flow chart in Figure 6 shows the ESS active power limit and the SoC limit relation. When the power injected by ESS reached maximum, and the SoC was close to high limit, the ESS stopped to inject power and the SoC remained constant. Similarly, when the power absorbed by the ESS reached minimum, and the SoC was close to its lowest limit, the ESS stopped to absorb power and the SoC remains constant. When the ESS active power was within limit, the SoC of the ESS could increase and decrease rapidly with the proposed control to support FR. The SoC response of the ESS active power during charging, discharging, and idling is presented in Figure 11. It can be seen from this figure that at the steady state that occurred before the contingency event (0 to 5 s), the ESS-SoC was at the initial SoC (50%). However, when the generator was tripped, causing the frequency to drop, the ESS discharged owing to the power injection into the grid to provide FR (5 to 25 s). It could be observed that, at the duration when frequency drops rapidly below nominal value (5 to 10 s), the SoC curves in Case 3 was steeper than in Case 2. This effect in SoC difference corresponds to the amount of power injected by the ESS to improve the frequency nadir of the power system. At 10 s < t < 25 s, when the grid frequency was recovering to a new steady state, the SoC response in Case 2 declined not so rapidly as compared to that of Case 3. This signifies that a small amount of ESS active power is still injected into the grid as shown in Figure 10. At t = 25 s, when load trip occurred, the SoC curves for both the conventional and proposed method reached their low limit values of 49.84% and 49.22%, respectively. Similarly, as the frequency increased above the nominal value owing to load tripping, the ESS absorbed power, thereby charging (25 to 32 s). At this point, the SoC could be seen to have increased more rapidly in Case 3 than in Case 2. At the point when the frequency became stable and neared the nominal value (38 to 50 s), the ESS-SoC new limit was reached as a result of the ESS neither injecting nor absorbing power. From Figure 11, it can be deduced that the rate of SoC decrease was higher with the proposed method as compared to the conventional method, and the opposite is observed considering the increasing SoC rate. This shows that the proposed method can inject and absorb power to and from the grid rapidly to support FR through its fast charging and discharging rate without exceeding the limit.

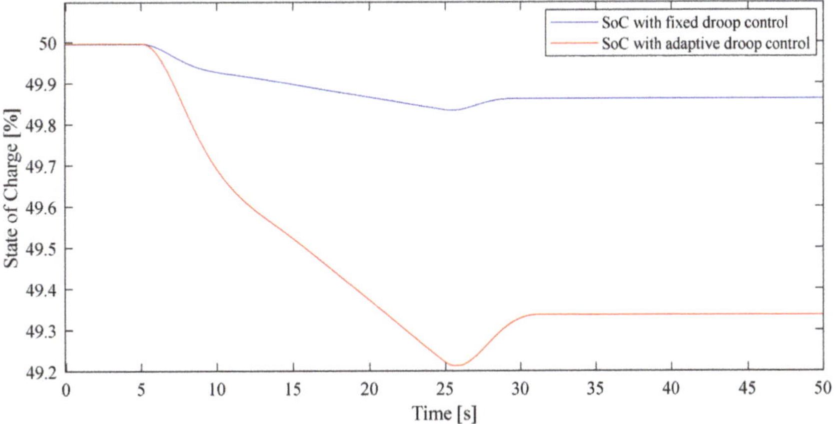

Figure 11. Comparison of SoC operations for the conventional and proposed methods.

5. Conclusions

In this study, we proposed an adaptive control strategy to coordinate a generator with an ESS in FR. The proposed method operated using an ESS to provide FR by employing both droop control and inertia constant adjustment to improve the frequency deviation, RoCoF, and frequency nadir of the system. An ESS control algorithm was developed to assist in managing the dynamic nature of the power system for an efficient power supply to the grid to protect the system from severe frequency deviation, thereby increasing system stability. We compared case studies involving only governor control, an ESS with the conventional fixed droop method, and an ESS with the proposed adaptive droop control scheme when the power system network was subjected to contingency events. To verify the performance of the proposed method, a dynamic simulation was performed to investigate the mutual influence between the conventional generator controller and the

FR-ESS controller with respect to grid frequency regulation and SoC limitation. It was observed that FR using an ESS can be more effective in contingency events (such as the non-functionality of transmission lines, generator trips, load trips, and three-phase faults) compared to the conventional generator. From the simulation results, we observed that, compared to the ESS with a fixed droop control scheme, the proposed FR-ESS controller provided the best results in regulating the frequency response by lowering the RoCoF and improving the minimum frequency deviation (higher frequency nadir) during the generator tripping event. Similarly, the frequency response amplitude was improved during the load tripping event using the proposed method. The SoC limitation was also investigated to compare the conventional and proposed methods. The result from the SoC of ESS shows that the proposed method has a faster action as compared to the conventional method in terms of charge and discharge rate without violating the SoC operating limits. The results indicated that inverter-based reserves such as ESSs respond faster to frequency deviations compared to the governor of a generating unit operating only with a conventional droop control scheme.

Considering the results presented in this study, system operators may use the proposed method to observe the effects of RoCoF based on the ESS power by changing the parameter settings according to the power plant conditions.

From the perspective of a transmission network, ESS has great potential to provide various grid stability supports. This study only focused on active power (P) and frequency control through the proposed generator with FR-ESS. In this kind of control system, the generator's reactive power and voltage are not considered. However, for future studies where the reactive power (Q) and voltage regulation is to be explored for performance evaluation, a harmonized optimal control method in terms of Q and voltage (V) will be required. Therefore, future studies of this kind of control for both P and Q for frequency and voltage regulation, respectively, is anticipated.

Author Contributions: Conceptualization, L.O.I., I.-Y.C., J.Y., J.W.S. and M.Y.; Methodology, L.O.I., J.W.S., M.Y. and J.S.; Investigation, M.Y. and J.S.; Supervision, Y.-M.S. and M.Y.; project administration, I.-Y.C. and J.Y.; Writing—original draft preparation, L.O.I.; Review and manuscript approval, L.O.I., Y.-M.S., J.W.S., M.Y. and J.S. All authors have read and agreed to the published version of the manuscript.

Funding: This work was supported by the Korea Electric Power Corporation grant (R21XO01-1) and National Research Foundation Grant (2020R1F1A1076638) funded by the Korean government.

Institutional Review Board Statement: Not applicable.

Informed Consent Statement: Not applicable.

Conflicts of Interest: The authors declare no conflict of interest.

References

1. Barelli, L.; Bidini, G.; Bonucci, F.; Castellini, L.; Castellini, S.; Ottaviano, A.; Pelosi, D.; Zuccari, A. Dynamic Analysis of a Hybrid Energy Storage System (H-ESS) Coupled to a Photovoltaic (PV) Plant. *Energies* **2018**, *11*, 396. [CrossRef]
2. IEC T120. Available online: http://www.iec.ch/dyn/www/f?p=103:7:11494349307735::::FSP_ORG_ID,FSP_LANG_ID:9463,25 (accessed on 4 April 2022).
3. Kundur, P.; Balu, N.J.; Lauby, M.G. *Power System Stability and Control*, 2nd ed.; The EPRI Power System Engineering Series; McGraw-Hill: New York, NY, USA, 1994; ISBN 0-07-035958-X.
4. Timur, Y.; Maximilian, J.Z.; William, H. Control of Energy Storage. *Energies* **2017**, *10*, 1010.
5. Aho, J.; Buckspan, A.; Laks, J.; Fleming, P.; Jeong, Y.; Dunne, F.; Churchfield, M.; Pao, L.; Johnson, K. A tutorial of wind turbine control for supporting grid frequency through active power control. In Proceedings of the 2012 American Control Conference (ACC), Montreal, QC, Canada, 27–29 June 2012; IEEE: Piscataway, NJ, USA, 2012; pp. 3120–3131.
6. Pan, X.; Xu, H.; Lu, C.; Song, J. Energy storage system control strategy in frequency regulation. In Proceedings of the 2016 IEEE International Conference on Automation Science and Engineering (CASE), Fort Worth, TX, USA, 21–26 August 2016; IEEE: Piscataway, NJ, USA, 2016; pp. 664–669.
7. Castillo, M.D.; Lim, G.P.; Yoon, Y.; Chang, B. Application of Frequency Regulation Control on the 4 MW / 8 MWh Battery Energy Storage System (BESS) in Jeju Island, Republic of Korea. *J. Energy* **2015**, *1*, 287–295.

8.	Kirby, B. *Frequency Regulation Basics and Trends, ORNL/TM-2004/291*; Technical Report; Oak Ridge National Laboratory: Oak Ridge, TN, USA, 2005.
9.	Fang, C.; Tang, Y.; Ye, R.; Lin, Z.; Zhu, Z.; Wen, B.; Ye, C. Adaptive Control Strategy of Energy Storage System Participating in Primary Frequency Regulation. *Processes* **2020**, *8*, 687. [CrossRef]
10.	Leitermann, O. Energy Storage for Frequency Regulation on the Electric Grid. Ph.D. Thesis, Massachusetts Institute of Technology, Cambridge, MA, USA, 2012.
11.	Federal Energy Regulatory Commission. Frequency Regulation Compensation in the Organized Wholesale Power Markets. 2011. Available online: https//www.jonesday.com/en/insights/2011/10 (accessed on 6 June 2022).
12.	Lim, G.P.; Park, C.W.; Labios, R.; Yoon, Y.B. Development of the control system for fast-responding frequency regulation in power systems using large-scale energy storage systems. *KEPCO J. Electr. Power Energy* **2015**, *1*, 9–13. [CrossRef]
13.	Tielens, P.; Van Hertem, D. Grid inertia and frequency control in power systems with high penetration of renewables. In Proceedings of the Young Researchers Symposium in Electrical Power Engineering, Delft, The Netherlands, 16–17 April 2012.
14.	Dreidy, M.; Mokhlis, H.; Mekhilef, S. Inertia response and frequency control techniques for renewable energy sources: A review. *Renew. Sustain. Energy Rev.* **2017**, *69*, 144–155. [CrossRef]
15.	Yoon, M.; Lee, J.; Song, S.; Yoo, Y.; Jang, G.; Jung, S.; Hwang, S. Utilization of Energy Storage System for Frequency Regulation in Large-Scale Transmission System. *Energies* **2019**, *12*, 3898. [CrossRef]
16.	Zhang, Y.J.; Zhao, C.; Tang, W.; Low, S.H. Profit-maximizing planning and control of battery energy storage systems for primary frequency control. *IEEE Trans. Smart Grid* **2016**, *9*, 712–723. [CrossRef]
17.	Li, Y.; He, L.; Liu, F.; Li, C.; Cao, Y.; Shahidehpour, M. Flexible voltage control strategy considering distributed energy storages for DC distribution network. *IEEE Trans. Smart Grid* **2017**, *10*, 163–172. [CrossRef]
18.	Nguyen-Hoang, N.-D.; Shin, W.; Lee, C.; Chung, I.-Y.; Kim, D.; Hwang, Y.-H.; Youn, J.; Maeng, J.; Yoon, M.; Hur, K.; et al. Operation Method of Energy Storage System Replacing Governor for Frequency Regulation of Synchronous Generator without Reserve. *Energies* **2022**, *15*, 798. [CrossRef]
19.	Manitoba HVDC Research Centre. *PSCAD/EMTDC User's Manual Guide v.4.5*; Manitoba HVDC Center: Winnipeg, MB, Canada, 2013.
20.	Ibrahim, L.O.; Sung, Y.M.; Hyun, D.; Yoon, M. A Feasibility Study of Frequency Regulation Energy Storage System Installation in a Power Plant. *Energies* **2020**, *13*, 5365. [CrossRef]
21.	Gkavanoudis, S.I.; Oureilidis, K.O.; Kryonidis, G.C.; Demoulias, C.S. A Control Method for Balancing the SoC of Distributed Batteries in Islanded Converter-Interfaced Microgrids. *Adv. Power Electron.* **2016**, *2016*, 8518769. [CrossRef]
22.	Oudalov, A.; Chartouni, D.; Ohler, C. Optimizing a battery energy storage system for primary frequency control. *IEEE Trans. Power Syst.* **2007**, *22*, 1259–1266. [CrossRef]
23.	Machowski, J.; Lubosny, Z.; Bialek, J.W.; Bumby, J.R. *Power System Dynamics: Stability and Control*; John Wiley & Sons: Hoboken, NJ, USA, 2020.
24.	Tremblay, O.; Dessaint, L.A. Experimental Validation of a Battery Dynamic Model for EV Applications. *World Electr. Veh. J.* **2009**, *3*, 289–298. [CrossRef]
25.	Shepard, C.M. Design of Primary and Secondary Cells. *J. Electrochem. Soc.* **1965**, *112*, 657. [CrossRef]
26.	Shim, J.W.; Verbič, G.; Kim, H.; Hur, K. On droop control of energy-constrained battery energy storage systems for grid frequency regulation. *IEEE Access* **2019**, *7*, 166353–166364. [CrossRef]
27.	Zhu, Z.; Ye, C.; Wu, S. Comprehensive control method of energy storage system to participate in primary frequency regulation with adaptive state of charge recovery. *Int. Trans. Electr. Energy Syst.* **2021**, *31*, e13220. [CrossRef]
28.	El-Bidairi, K.S.; Nguyen, H.D.; Mahmoud, T.S.; Jayasinghe SD, G.; Guerrero, J.M. Optimal sizing of Battery Energy Storage Systems for dynamic frequency control in an islanded microgrid: A case study of Flinders Island, Australia. *Energy* **2020**, *195*, 117059. [CrossRef]

 *sustainability*

Article

Maximum Power Point Tracking of a Grid Connected PV Based Fuel Cell System Using Optimal Control Technique

Muhammad Majid Gulzar

Control & Instrumentation Engineering Department & Center for Renewable Energy and Power Systems, King Fahd University of Petroleum & Minerals, Dhahran 31261, Saudi Arabia; muhammad.gulzar@kfupm.edu.sa

Abstract: The efficiency of renewable energy sources like PV and fuel cells is improving with advancements in technology. However, maximum power point (MPP) tracking remains the most important factor for a PV-based fuel cell power system to perform at its best. The MPP of a PV system mainly depends on irradiance and temperature, while the MPP of a fuel cell depends upon factors such as the temperature of a cell, membrane water content, and oxygen and hydrogen partial pressure. With a change in any of these factors, the output is changed, which is highly undesirable in real-life applications. Thus, an efficient tracking method is required to achieve MPP. In this research, an optimal salp swarm algorithm tuned fractional order PID technique is proposed, which tracks the MPP in both steady and dynamic environments. To put that technique to the test, a system was designed comprised of a grid-connected proton exchange membrane fuel cell together with PV system and a DC-DC boost converter along with the resistive load. The output from the controller was further tuned and PWM was generated which was fed to the switch of the converter. MATLAB/SIMULINK was used to simulate this model to study the results. The response of the system under different steady and dynamic conditions was compared with those of the conventionally used techniques to validate the competency of the proposed approach in terms of fast response with minimum oscillation.

Keywords: renewable energy sources; fuel cell; photovoltaic; maximum power point; fractional order PID

Citation: Gulzar, M.M. Maximum Power Point Tracking of a Grid Connected PV Based Fuel Cell System Using Optimal Control Technique. *Sustainability* 2023, *15*, 3980. https://doi.org/10.3390/su15053980

Academic Editor: Elena Cristina Rada

Received: 16 December 2022
Revised: 17 January 2023
Accepted: 15 February 2023
Published: 22 February 2023

1. Introduction

Conventional energy sources are being depleted at an alarming speed and becoming scarce; thus, the usage of unconventional energy sources is growing. Coal, petroleum, natural gas, and nuclear power are all major conventional sources. Because of their continued use, these resources have been exhausted to a great extent. Additionally, the usage of these sources contributes significantly to pollution, which contributes to global warming. Owing to these issues, scientists are forced to employ renewable energy sources (RES). Non-conventional/RES are sources of energy that are reproduced by natural processes regularly and do not deplete [1]. These sources do not damage our environment, are mostly cost-effective, and often do not require a huge investment, hence widely being accepted as more reliable. Moreover, these sources are called renewable because they are renewed or reproduced at an equal or greater rate with respect to the rate of their use.

Among all RES, solar PV, which utilizes the photovoltaic effect to produce electricity, is being widely used worldwide. Sunlight is absorbed using semiconductor materials—mostly silicon—and converted into electrical energy. The foremost drawback of solar energy is that a large area is required to install solar PV systems [2,3]. The fuel cell is, in fact, a device (electro-chemical) that uses a chemical reaction to produce electrical energy [4]. A fuel cell utilizes the energy of hydrogen (chemical energy) or other fuels to generate electricity in a clean and effective manner. When hydrogen is utilized as a fuel, the only things created are heat, electricity, and water. The prominent and unusual aspect of fuel cells is that they can be used for an extensive range of applications [5,6].

Fuel cells and batteries differ significantly from each other, as fuel cells do not require recharging since they do not run out of fuel. The fuel cell will keep producing heat and power until the fuel supply stops. A fuel cell is constructed using two electrodes sandwiched around an electrolyte. In a fuel cell, the anode is the negative $(-)$ electrode while the cathode is the positive $(+)$ electrode. Hydrogen is widely used as fuel in fuel cells and is provided to the anode [7], while the oxygen (from the air) is supplied to the cathode. A catalyst is needed to initiate the redox reaction. Platinum is used as a catalyst and, in some cases, enzymes are used. The catalyst converts hydrogen into electrons and protons (hydrogen ions). Here, electrons take an exterior circuit path and can be utilized to power a load, while protons cross through the electrolyte to combine with oxygen to form water [8].

Several classifications of fuel cells are currently under research, each using different fuels with different electrochemical reactions and construction. Each has a different catalyst requiring different operating conditions like temperature and has its own applications and drawbacks. Polymer Electrolyte Membrane Fuel Cells (PEMFC), also identified as proton exchange membrane fuel cells, provide a higher power density than other traditional fuel cells while being lighter in weight and smaller in size [9,10].

One vital aspect of the PV based fuel cell is tracking its maximum power point (MPP). For reliable and efficient use, it is important to use a PV based fuel cell at MPP, which depends on several factors, including irradiance, temperature, water content in membrane, and hydrogen and oxygen partial pressures. The Perturb and Observe (P&O) algorithm is the most popular to track MPP, owing to its simplicity. However, it may cause fluctuations across MPP due to excessive switching. To reduce these fluctuations, the step size can be reduced; but this will cause the tracking time to increase. The incremental conductance method is proposed in [11,12], which gives better results than the P&O, but still causes an overshoot. Sliding mode controller (SMC) is studied in [13], using a fuel cell stack with a boost converter, and performance is compared with incremental conductance and P&O. However, while it yields a significantly lower overshoot, the calculations are extensive and the design of the filter is difficult.

The Water cycle algorithm (WCA) is an effective algorithm to track MPP and it is inspired by the naturally occurring water cycle. The drawback of WCA is that it can trap in local optima [14]. The incremental conductance method has been implemented for MPP tracking, but its implementation is complex as it requires multiple sensors [15]. The author in [16] suggested a smart drive algorithm using a boost converter to track the MPP of the fuel cell, but the efficiency of the method turned out to be less than other metaheuristic techniques. Particle swarm optimization (PSO) is another technique used for MPP tracking, which is based on the natural process. Ref [17] discussed the PSO technique using a fuel cell stack with a boost converter. PSO is a metaheuristic approach, but the drawback is that it can also be stuck in local optima. Extremum seeking control is an efficient method but converges slowly [18].

A fuel cell model with a cuke converter is discussed in [19] and the firefly algorithm (FFA) is proposed to reach MPP. FFA is a metaheuristic technique, but it has a drawback, namely, in that it may be stuck in local optima. Backstepping techniques proposed in [20], show good efficiency, but the implementation requires great effort as it is very complex. Fuzzy logic control (FLC) is another important technique used to track MPP and is being used widely. FLC has been implemented using both Boost and Buck converters. The accuracy of FLC is low and one cannot be sure that the MPP calculated by the controller is accurate [21,22]. Convergence time for FLC, if used independently, can be very large. Another technique proposed in [23] is an artificial neural network (ANN); however, this technique requires an excessive amount of data.

A Jaya controller with cuke converter is implemented in [24] to improve MPP. The presented technique is metaheuristic but requires excessive computational time. The grey wolf optimization (GWO) method is also introduced in [25] to track MPP. GWO technique is motivated by the leadership hunting and hierarchy methodology of grey wolf packs. GWO is a metaheuristic technique, but its convergence rate is slow and it can be stuck in local optima. Another nature-inspired optimization technique is the salp swarm optimization (SSO) algorithm which is discussed in [26]. Despite being metaheuristic, it requires excessive computational time for processing. Anti-windup PID controllers are being used commonly in industries, due to their simplicity and fewer computational requirements [27]. However, PID is sensitive to excessive variations and can lead a system to instability. Moreover, it cannot be used for non-linear systems [28].

Higher-order sliding mode controllers tuned with a twisting algorithm (HOSM-TA) are implemented in [29]. They show high robustness against disturbances and uncertainties. The drawback of this technique is that it is very complex and there is no guarantee that the solution is accurate. Furthermore, it cannot be used for 1st order systems. Chattering is a phenomenon that decreases the efficiency of SMC and also causes heat loss in the system. To overcome this, a Quasi-continuous (QC) algorithm is proposed in [30]. This proposed technique shows considerable improvement against chattering and is also robust. Nonetheless, one of the major drawbacks is that it is complex in design with no guarantee of accuracy, and cannot be used for 1st order systems. Higher-order prescribed convergence law technique (PCL) is used to track MPP using a DC-DC boost converter, which is a robust technique and has a finite convergence time, but it is also complex with low accuracy [31]. Another MPP tracking technique is model predictive control (MPC), which offers multiple variable control and predicts upcoming disturbances and upcoming control actions. It is better than many other techniques in terms of energy savings and has enhanced transient response, but it requires specific background knowledge of the method to be implemented [32,33]. Tuning of PID with SSO technique shows good results with reasonable execution time and good accelerated convergence, and requires few parameters to be tuned [34]. However, it can suffer from premature convergence.

The integral fast terminal sliding mode control (IFTSMC) technique has advantages, e.g., robustness against uncertainties and disturbances, ability to reduce chattering, and high speed of convergence [35]. The golden section search technique is another technique for MPP. Although this technique is faster than many heuristic methods, the implementation of the same can be costly; furthermore, it requires knowledge of fuel cell plant specifications [36]. The forensic-based investigation algorithm (FBI) has been used for proportional integral derivative, which requires multiple sensors and, hence, can be costly [37]. The equilibrium optimizer algorithm is adopted to optimize FLC for MPPT. The algorithm itself is complex and also FLC lacks in accuracy [38,39].

Table 1 lists the key characteristics and provides a comparison of the various approaches previously employed.

Table 1. Summary of Fuel Cell-based MPPT Techniques.

Sr. #	Reference #	Algorithm/ Approach	Converter Type	Nature/Remarks/Notes
1	[5]	IC	Boost	Multi-sensors are required
2	[6]	PSO	Boost	Easily trapped in local optimum
3	[7,8]	ANFIS	Boost	ANN requires excessive data
4	[9]	P&O	High step ratio	Oscillations/fluctuations near MPP with large tracking time
5	[10]	MPC	Boost	Requires plant model and specific knowledge
6	[11,12]	P&O/InC	Buck	Oscillations/fluctuations near MPP multi-sensors needed

Table 1. *Cont.*

Sr. #	Reference #	Algorithm/Approach	Converter Type	Nature/Remarks/Notes
7	[13]	SMC	Boost	The design of the filter circuit is cumbersome
8	[14]	WCA	Boost	It may become stuck in local optima
9	[15]	INR	Boost	Multiple sensors are needed
10	[16]	Smart drive algorithm	Boost	Low accuracy
11	[17]	PSO	Boost	Easily stuck in local optima
12	[18]	Extremum seeking control	-	Slow convergence rate
13	[19]	Firefly algorithm	Cuke	Easily trapped in local optimum
14	[20]	Backstepping	Boost	Complex/Excessive effort in implementation
15	[21]	Fuzzy logic	Boost	Lacks precision
16	[22,23]	ANN	Boost	ANN requires excessive data
17	[24]	Jaya	Cuke	Requires Excessive computational time
18	[25]	GWO	Boost	Sluggish convergence and stuck in local optimum
19	[26]	SSA	Boost	Excessive computational time required
20	[27]	AW-PID	Buck-Boost	Inefficient and sensitive toward large load changings and not suitable for Nonlinear systems
21	[28]	FPID	Four switch Buck-Boost	Complex to implement
22	[29]	TA	Boost	Complex and accurate results not guaranteed
23	[30]	PCL	Boost	High complexity and low accuracy
24	[31]	QC	Boost	Cannot be used for 1st order systems, complex and less accurate
25	[32]	MPC	Two-level inverter	Requires plant model and specific knowledge
26	[33]	MPC	Boost	Requires plant model and specific knowledge
27	[34]	SSA-PID	Boost	Can become stuck in the local maximum

Table 1. *Cont.*

Sr. #	Reference #	Algorithm/Approach	Converter Type	Nature/Remarks/Notes
28	[35]	IFTSMC	Boost	Knowledge of system boundary uncertainty is required, also convergence issues when states are not near equilibrium.
29	[36]	GSS	Boost	Implementation cost high and knowledge of plant specification of the fuel cell required
30	[37]	FBI-PID	Boost	Multiple sensors required, hence costly
31	[38]	EO-FLC	Boost	Fuzzy logic may lack in accuracy

Novelty and Contribution

The literature review reveals specific areas that need further improvement; thus, the proposed technique has focused its utility on these areas. This main contribution of the proposed work is as follows:

1. This work presented an optimum salp swarm algorithm tuned fractional order PID controller for MPPT to modify input and output during transient operating conditions to attain an ideal duty ratio.
2. Compared to other traditional MPPT algorithms utilized in the literature, it offers high-power tracking capability, quick convergence speed, fewer controlling parameters, and ease of implementation.
3. The PV-based fuel cell grid connected technology offers a guarantee for steady and practical operation under varying load situations.

The paper is organized as follows. Section 2 provides the design and modeling of the PV and Fuel Cell. The proposed control strategy is described in Section 3. Section 4 is designated for the attained results and discussions. Section 5 is dedicated to the conclusion.

2. System Modeling

The system under study is designed to comprise a grid-connected proton exchange membrane fuel cell (PEM) together with PV system, a DC-DC boost converter along with resistive load, and a robust power point tracking controller. The output from the controller is further tuned and PWM is generated, which is fed to the switch of the converter. Figure 1 presents an overview of the model under study.

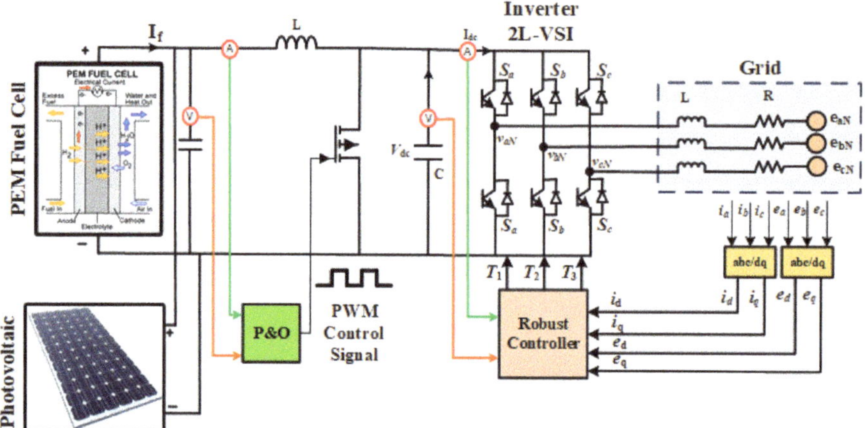

Figure 1. Fuel cell and PV to Grid stages.

2.1. Fuel Cell System Modelling

Fuel cells, first conceived by Sir William Grove in 1839, are now a viable source of energy. Fuel cells may be thought of as generators in their most basic form. Unlike traditional generators, which utilize internal combustion engines to turn an alternator, fuel cells create electricity by directly creating electrons with no moving components. As a result, they are quite effective and dependable. They are also almost silent, producing only water vapor in addition to energy and heat. As a result, they are suitable for indoor usage.

The voltage-current characteristics of fuel cells are intricate and nonlinear. A polarization curve illustrates the non-linear connection between a fuel cell's current density and voltage. The fuel cell output voltage is controlled by current density, which is affected by operational parameters. A PEMFC, a boost DC/DC converter, as well as a resistive load make up the system. The controlling variable within that system is the duty cycle of the boost converter; which is the driving variable for achieving MPPT, where C and L stand for the capacitance and inductance of the boost converter, respectively.

The fuel cell output is given by Equation (1)

$$V_{cell} = E_{Nernst} - V_{act} - V_{ohm} - V_{conc} \tag{1}$$

where E_{Nernst} is reversible thermos-dynamic potential which is defined by Nernst Equation (2)

$$E_{Nernst} = 1.229 - 8.5 \times 10^{-4}(T - 298.15) + 4.308 \times 10^{-5}T\left(\ln\left(P_{H_2} + 0.5\ln(P_{O_2})\right)\right) \tag{2}$$

where T indicates the absolute temperature in kelvins, P_{H_2} is hydrogen partial pressure (atmospheric) and P_{O_2} is the oxygen partial pressure. Activation voltage drop is given by Tafel Equation (3)

$$V_{act} = \zeta_1 + \zeta_2 T + \zeta_3 T \ln(C_{O_2}) + \zeta_4 T \ln(I_{FC}) \tag{3}$$

Here $i = 1, \ldots, 4$ are parametric coefficients for every cell model, and C_{O_2} denotes the dissolved-oxygen concentration in the interface of the cathode catalyst, as mentioned in Equation (4)

$$C_{O_2} = \frac{P_{O_2}}{(5.08 \times 10^6) \times \exp\left(-\frac{498}{T}\right)} \tag{4}$$

The overall ohmic voltage drop is calculated as Equation (5)

$$V_{ohm} = I_{FC}R_M \tag{5}$$

where R_M is the resistance (ohmic) and is made up of the electrode resistances as well as the resistances of the polymer membrane and electrodes. Here, R_M is provided by Equation (6)

$$R_M = \frac{r_m\, t_m}{A} \tag{6}$$

where, t_m is the membrane thickness, which is in centimeters, A is the activation area in micro centimeters, and r_m is the membrane resistivity Ωcm to proton conductivity. Membrane humidity and temperature have a significant impact on membrane resistivity, which can be computed as Equation (7)

$$r_m = \frac{181.6\left[1 + 0.03\left(\frac{I_{FC}}{A}\right) + 0.0062\left(\frac{T}{303}\right)^2\left(\frac{I_{FC}}{A}\right)^{2.5}\right]}{\left[\lambda_m - 0.634 - 3\left(\frac{I_{FC}}{A}\right)\right] \times \exp\left(4.18\left(T - \frac{303}{T}\right)\right)} \tag{7}$$

where, water content is represented by λ_m of the membrane and is an input of the PEMFC model. In addition, it is a function of the average water activity a_m as represented in Equation (8)

$$\lambda_m = \left\{ \begin{array}{l} 0.043 + 17.81 a_m - 39.85 a_m^2 + 36 a_m^3 \; 0 < a_m < 1 \\ 14 + 1.4(a_m - 1) \; 1 < a_m \leq 3 \end{array} \right\} \tag{8}$$

The relationship between the average water activity and the anode and cathode water vapor partial pressures, ($P_{v,an}$, $P_{v,ca}$ respectively) is given by Equation (9)

$$a_m = \frac{1}{2}(a_{an} + a_{ca}) = \frac{1}{2}\left[\frac{P_{v,an} + P_{v,ca}}{P_{sat}}\right] \tag{9}$$

The saturation pressure of water P_{sat} can be calculated with the subsequent empirical expression as mentioned in Equation (10)

$$\mathrm{lpg}_{10}P_{sat} = -2.1794 + 0.02953T - 9.1813 \times 10^{-5}T^2 + 1.4454 \times 10^{-7}T^3 \tag{10}$$

The values (real-time) of λ_m can vary from 0 to 14. The concentration voltage drop is expressed as Equation (11)

$$V_{conc} = -\frac{RT}{nF}\ln\left(1 - \frac{i_{FC}}{i_L A}\right) \tag{11}$$

where, i_L is the limiting current and it is the maximum rate at which the reactant may be given to an electrode.

Fuel cells are linked together in a series to produce the desired voltage. Thus, the N_{FC} series cells per string have nonlinear $V - I$ characteristics, as mentioned in Equation (12)

$$V_{FC} = N_{FC}V_{cell} \tag{12}$$

2.2. PV System Modelling

The PV system is one of the most extensively used RES. The current source is parallel to the diode and precisely converts solar energy into electrical energy by accelerating the flow of holes and electrons inside the photovoltaic cell. It is required to create the PV source to unavoidably operate on its MPPT in order to get maximum power, since it is a non-linear current source [40].

A PV array needs to go through a number of processes to connect with a thermal power supply. As demonstrated in Figure 2, the design of a PV system includes a number of components, including a converter, an inverter, modeling, and a computation of the average power that is actually sent to the grid.

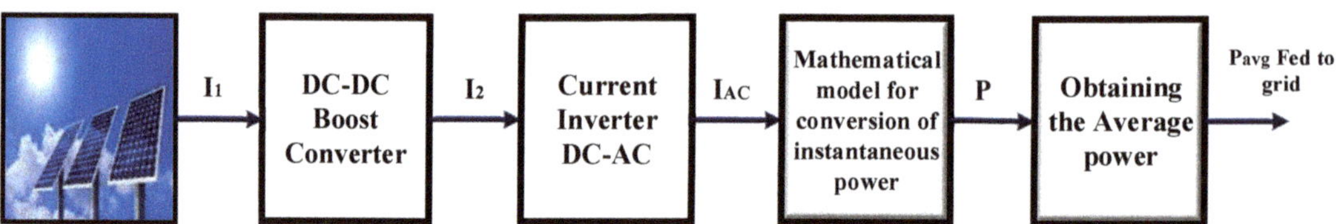

Figure 2. Grid Connected PV Farm Prototype.

AC voltage of PV can be calculated, by using Equation (13).

$$q = \frac{v_{dc}}{v_{ac}} \tag{13}$$

Here q is the gain between AC-DC voltage. The boost converter transfer function (TF) can be projected using Equations (14) and (15)

$$m_1 = \frac{v_2}{v_1} = \frac{i_1}{i_2} \tag{14}$$

$$g_1(s) = \frac{1}{m_1} \tag{15}$$

where $\frac{1}{m_1}$ is the boost converter gain. The inverter TF is mentioned in Equation (16)

$$g_2(s) = \frac{I_{ac}(s)}{I_2(s)} = \frac{s^2}{s^2 + \omega^2} \tag{16}$$

Here, $\omega = 2\pi f = 2\pi(50) = 314.12\,\text{rad/s}$. For instantaneous power, the TF is mentioned in Equation (17), where $\frac{v_m}{i_m}$ is the impedance.

$$P(s) = \frac{v_m i_m}{2s} + \frac{v_m i_m}{2} \frac{s}{s^2 + (2\omega)^2} \tag{17}$$

The instantaneous power gain is given in Equation (18)

$$g_3(s) = \frac{p(s)}{I_{ac}(s)} = v_m \left(\frac{(s^2 + \omega^2)(s^2 + (2\omega)^2)}{s^2(s^2 + (4\omega)^2)} \right) \tag{18}$$

The average power is mentioned in Equation (19)

$$p_{avg}(s) = \frac{v_m i_m}{2s} \tag{19}$$

The average power gain is shown in Equation (20)

$$g_4(s) = \frac{p_{avg}(s)}{p(s)} \tag{20}$$

3. Proposed Robust Controller

The proposed robust controller is a combination of the Salp Swarm Algorithm tuned Fractional order PID controller to achieve the MPPT of the hybrid PV based fuel cell system, as shown in Figure 3.

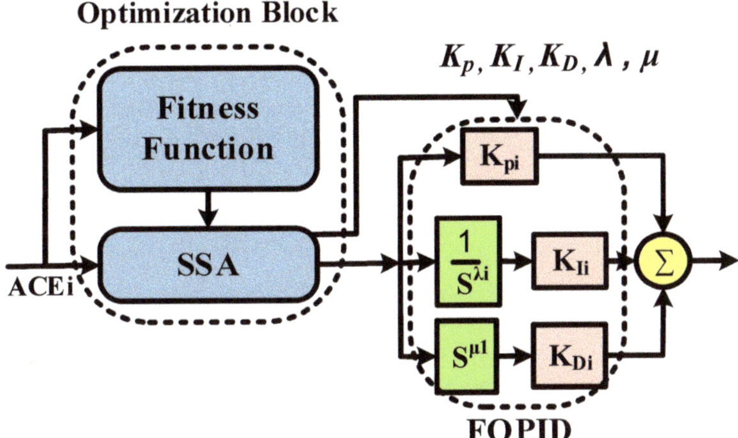

Figure 3. Proposed Robust Controller.

3.1. Salp Swarm Algorithm

The method is inspired by transparent body salp vertebrates, which are famous for generating spiral chains when they travel to find food [41]. The leader and followers are the two basic divisions of the salp swarm. The leader's role is to direct the group as they look for food, and they update their position using Equation (21)

$$K_j^i = \begin{cases} M_i + C_1\left(\left(ub_j - lb_j\right)C_2 + lb_j\right) & C_3 \geq 0 \\ M_i - C_1\left(\left(ub_j - lb_j\right)C_2 + lb_j\right) & C_3 < 0 \end{cases} \tag{21}$$

Here, lb_j and ub_j stands for the lower and upper limits of the jth dimension, while M stands for the target food and K is the 2D salp position. The variables C_2 and C_3 are uniform coefficients. The C_1, which is shown in Equation (22), is utilized to balance the exploitation of food in search space.

$$C_1 = 2e^{-\left(\frac{4t}{t_{max}}\right)^2} \tag{22}$$

where t and t_{max} denotes the current and maximum iterations, respectively. Equation (23) is used to update the position of the follower salp.

$$K_j^i = \frac{1}{2}at^2 + v_0 t \quad i \geq 2 \tag{23}$$

The flow chart of the complete Salp Swarm Algorithm is depicted in Figure 4.

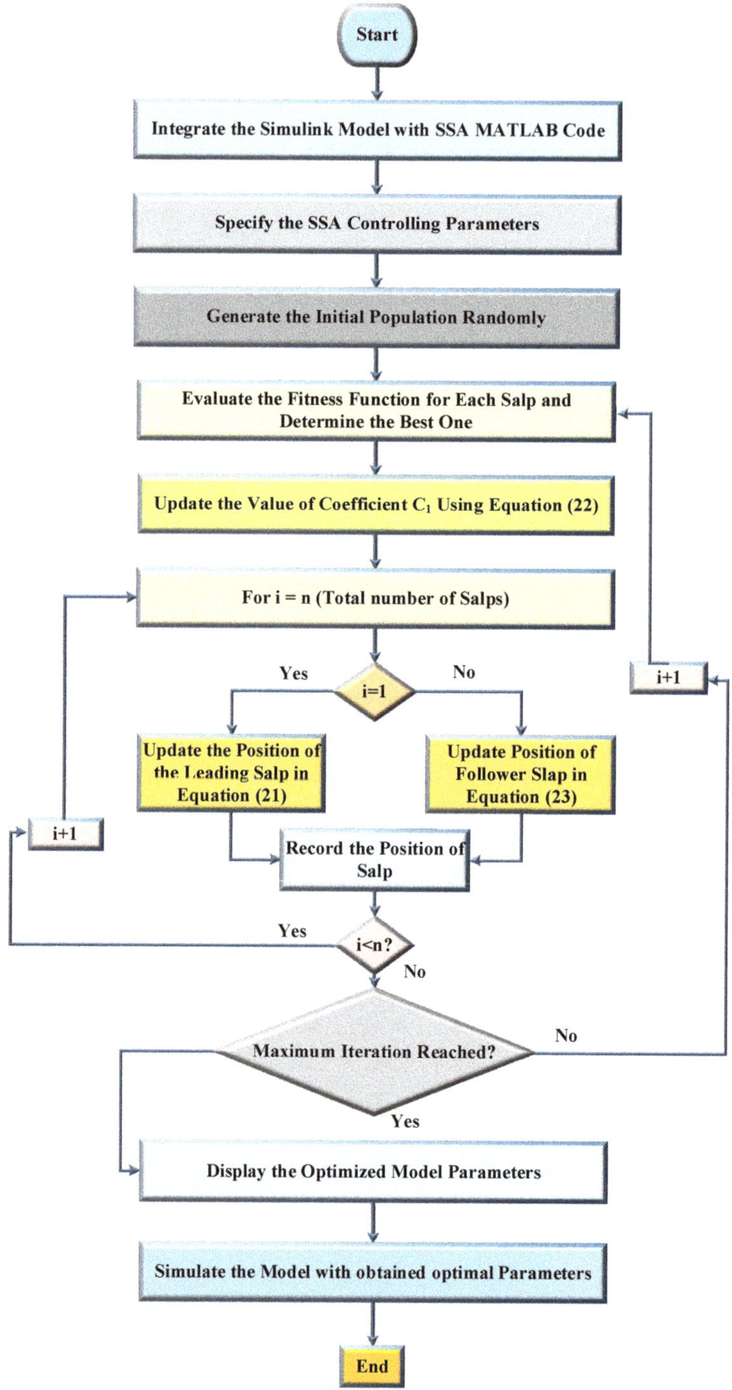

Figure 4. Flowchart of Salp Swarm Algorithm.

3.2. Fractional Order PID Controller

The use of fractional calculus was developed/increased when Podlubny proposed the $PI^{\lambda}D^{\mu}$ controller in 1999. This generality of the typical PID controller included a fractional integration of order λ and a fractional derivation of order μ, and it has since led numerous scholars to a new area of study called the modification of the fractional-order controller $PI^{\lambda}D^{\mu}$. Fractional order controllers are described using fractional calculus, where the calculus of proportional α—derivative is well-defined by the basic operator $_{\alpha}D_t^{\alpha}$, as mentioned in Equation (24).

$$_{\alpha}D_t^{\alpha} = \begin{cases} \frac{d^{\alpha}}{dt^{\alpha}} & \alpha > 0 \\ 1 & \alpha = 0 \\ \int\limits_{\alpha}^{t} (d\tau)^{-\alpha} & \alpha < 0 \end{cases} \tag{24}$$

Upper and lower bounds are determined by α and t, while $\alpha \in \mathbb{R}$ and α operator can be substituted in the frequency domain as $F(s) = \frac{1}{s^{\alpha}}$. The output equation of the fractional-order controller in the time domain is given by Equation (25)

$$u = k_p e(t) + k_i D_t^{-\lambda} e(t) + k_d D_t^{\mu} e(t) \tag{25}$$

where, k_d is the differentiating constant, whereas k_i is the integration constant, and k_p is the proportional constant, μ is the fractional order of the differentiating action and λ is the fractional order of the integrating action.

In contrast to standard PID controllers, fractional-order controllers include two extra parameters that represent the order of integrating and derivative values, respectively. Based on the modification of these two factors, one can discover a wide range of fractional order controller choices.

As can be seen in Figure 5, the fractional order PI D controller expands the traditional PID controller from a point to a plane. The design of PID control may benefit greatly from this expansion's increased flexibility. Clearly, by selecting $[\lambda, \mu] = [1,\ 1]$, a traditional PID corrector can be regained; and by selecting $[\lambda, \mu] = [1,\ 0]$ and $[\lambda, \mu] = [0,\ 1]$, one can get traditional PI and PD controllers, respectively.

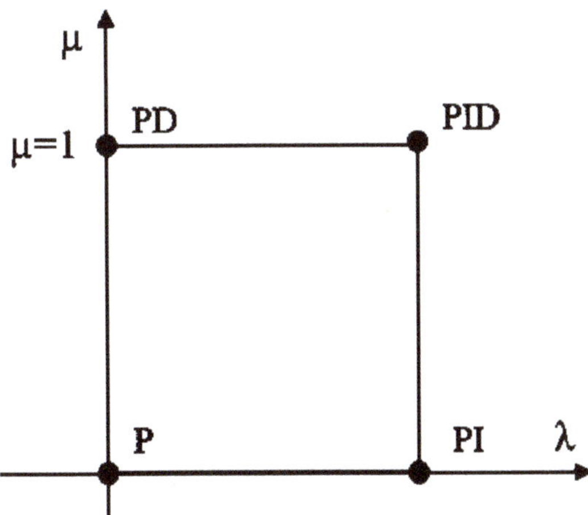

Figure 5. Types of Controllers According to Coefficients.

The FOPID executes much better than the traditional PID, since it uses discretized values. Moreover, as the stability region of the FOPID is wider than that of the PID controller, it is evident from Figure 6 that it enables more flexibility to the controller.

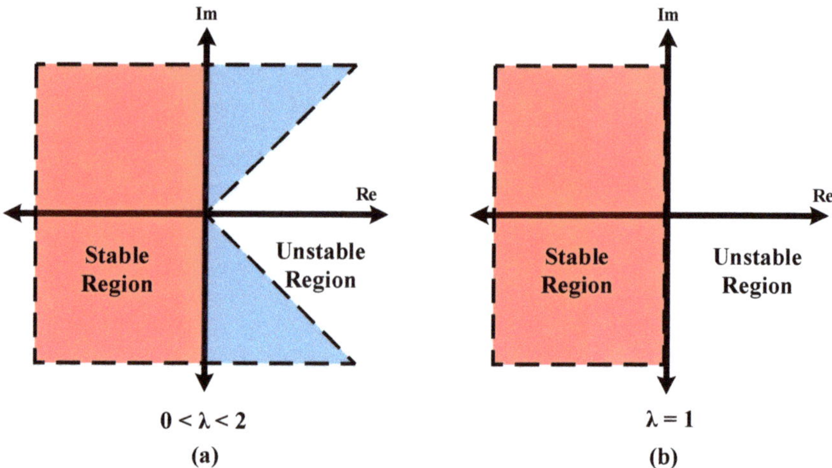

Figure 6. Stability Region of Controllers (**a**) FOPID (**b**) PID.

4. Results and Discussion

The hybrid system constituted of grid tied PEMFC and PV is simulated on MAT-LAB/Simulink for dynamic operation. The proposed FOPID controller is tested under different load conditions. Results are then compared with the conventional PI controller which substantiates the efficiency of the proposed (FOPID) controller.

Figure 7 depicts the change in irradiance, applied as input to PV to test the response of the controller, while Figure 8 shows the consumption of oxygen and hydrogen in the fuel cell. After the initial disturbance, the fuel consumption attains a constant value.

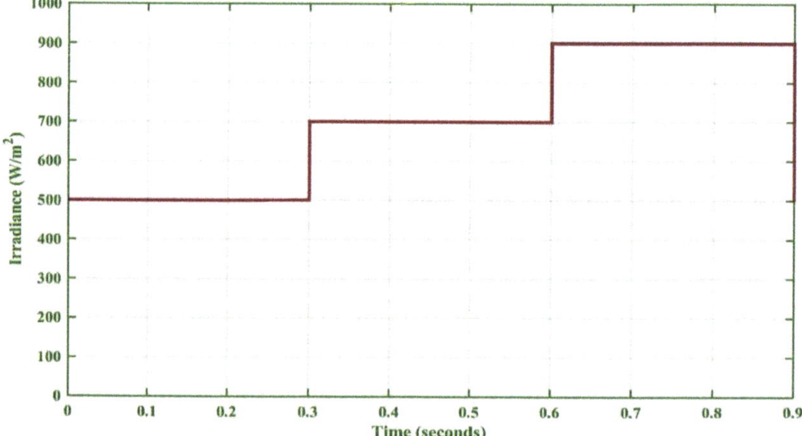

Figure 7. Abrupt Irradiance Change.

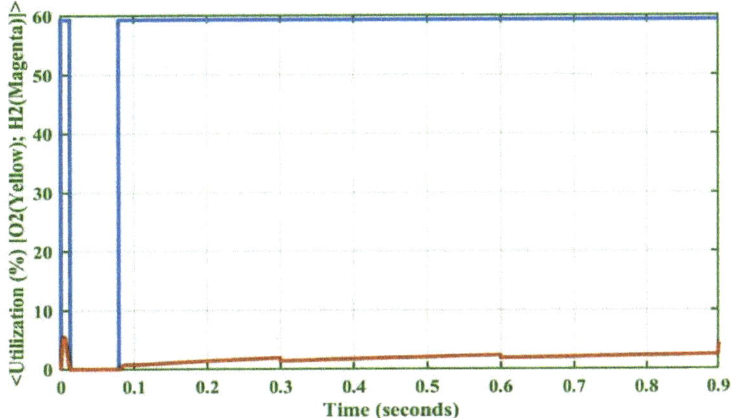

Figure 8. Oxygen and Hydrogen Consumption.

Figures 9 and 10 indicate the fuel cell output voltage and output current, respectively.

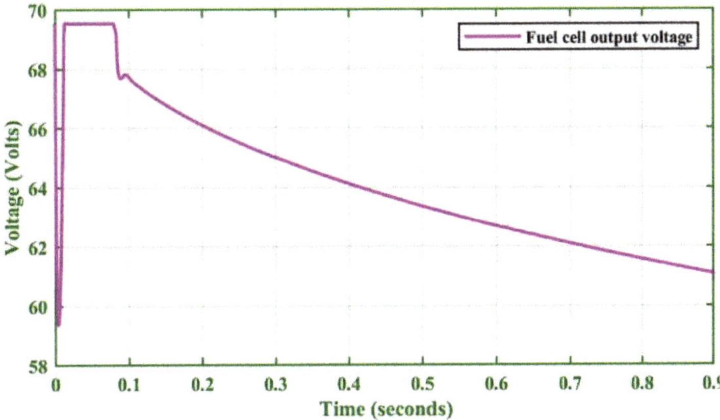

Figure 9. Fuel Cell Output Voltage.

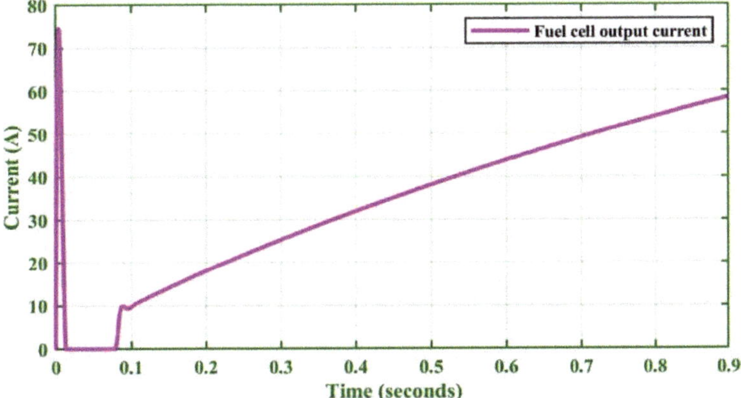

Figure 10. Fuel Cell Output Current.

The purpose of this research was to propose and test a robust control system that is effective under varying circumstances. The same was put to the test. Figure 11 shows the output current of the hybrid system using the conventional PI controller. Nevertheless, Figure 12 depicts the output current of the hybrid system using the proposed FOPID controller and a comparison is displayed with the conventional PI controller in Figure 13.

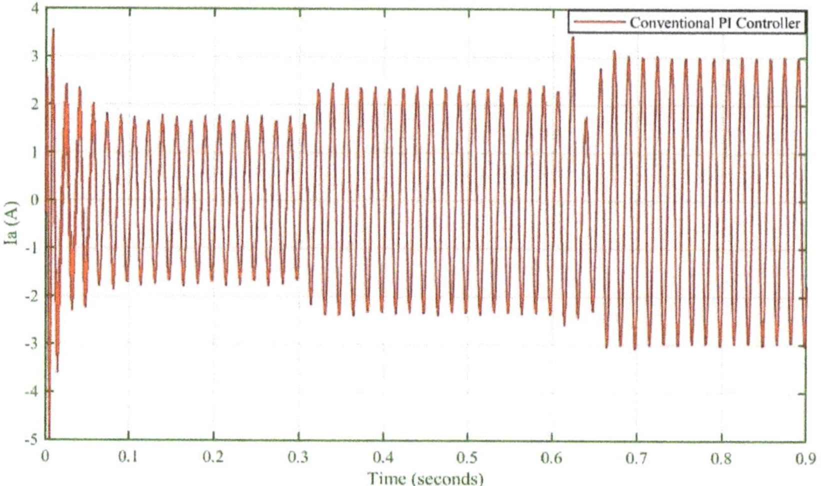

Figure 11. Output Current−Using Conventional PI.

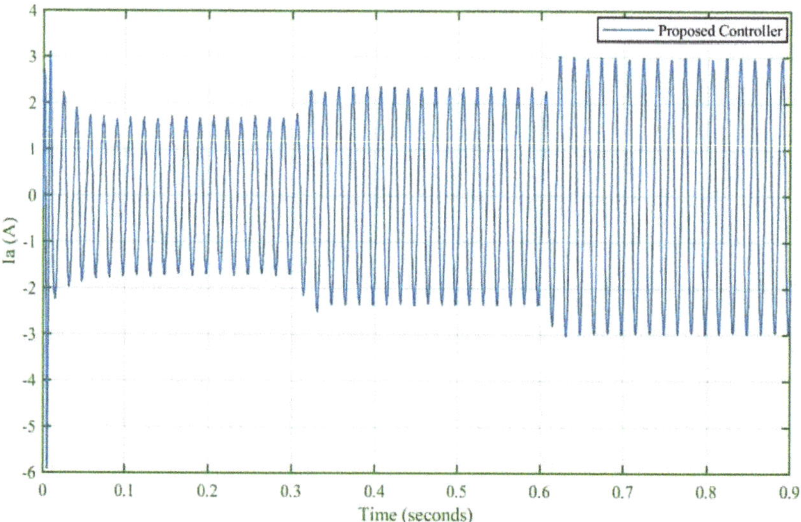

Figure 12. Output Current−Using Proposed Controller.

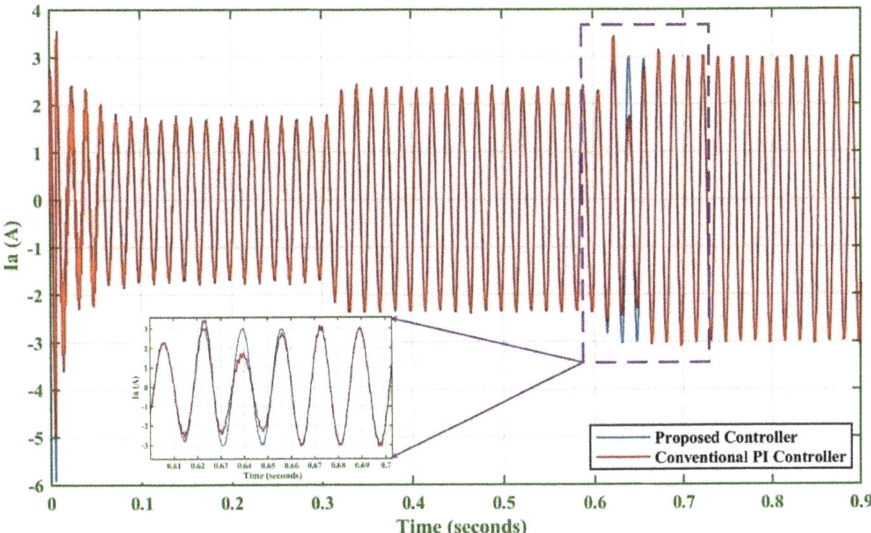

Figure 13. Output Current−Controller Comparison.

Figure 14 reveals the output power, for both controllers, and it is noticeable that the proposed FOPID controller shows significantly fewer oscillations and less settling time as compared to the conventional PI controller. DC output voltage (V_{dc}) is shown in Figure 15, which clearly indicates the superior performance of the projected controller and is more optimum against the uncertainties in the system.

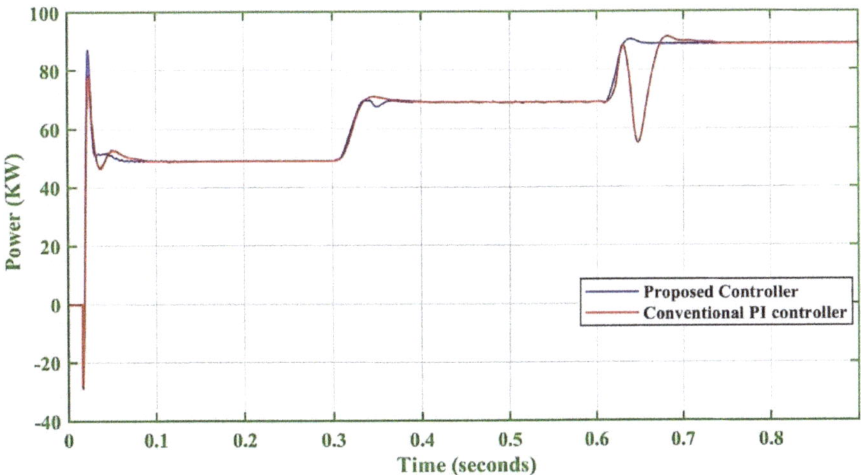

Figure 14. Output Power−Controller Comparison.

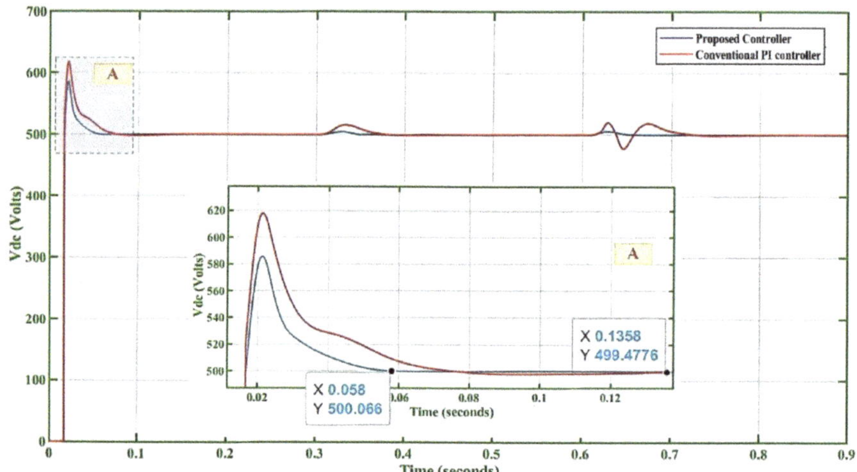

Figure 15. DC Output Voltage—Controller Comparaison.

5. Conclusions

This research work has presented a robust control strategy using optimum salp swarm algorithm tuned fractional order PID controller for the tracking of MPP of grid tied PEMFC along with PV. The proposed controller tracks the MPP whenever uncertainty of fluctuation occurs. Conventional P&O is used to control the duty cycle of the DC-DC converter, while FOPID controls the output of the DC-AC inverter. The overall capability of the suggested controller is significantly improved over the typical/conventional PI controller; and it offers high-power tracking capability, quick convergence speed, fewer controlling parameters, and ease of implementation. In the given test bench for abrupt irradiance change, the settling time is observed just 0.058 s with minimum overshoot, as compared to the conventional PI controller. Moreover, the overall suggested regulating technique adapts to the unforeseen power system scenario fairly successfully, with minimal oscillation.

Funding: This research was funded by the King Fahd University of Petroleum & Minerals, Dhahran 31261, Saudi Arabia.

Institutional Review Board Statement: Not applicable.

Informed Consent Statement: Not applicable.

Data Availability Statement: Not applicable.

Acknowledgments: This publication is based upon work supported by King Fahd University of Petroleum & Minerals. Author at KFUPM acknowledge the Interdisciplinary Research Center for Renewable Energy and Power Systems for the support received under Grant no. EC221008.

Conflicts of Interest: The author declares no conflict of interest.

References

1. Zheng, J.; Du, J.; Wang, B.; Klemeš, J.J.; Liao, Q.; Liang, Y. A hybrid framework for forecasting power generation of multiple renewable energy sources. *Renew. Sustain. Energy Rev.* **2023**, *172*, 113046. [CrossRef]
2. Harrabi, N.; Souissi, M.; Aitouche, A.; Chaabane, M. Modeling and control of photovoltaic and fuel cell based alternative power systems. *Int. J. Hydrogen Energy* **2018**, *43*, 11442–11451. [CrossRef]
3. Gulzar, M.M.; Iqbal, M.; Shahzad, S.; Muqeet, H.A.; Shahzad, M.; Hussain, M.M. Load Frequency Control (LFC) Strategies in Renewable Energy-Based Hybrid Power Systems: A Review. *Energies* **2022**, *15*, 3488. [CrossRef]
4. Mohamed, N.; Aymen, F.; Altamimi, A.; Khan, F.A.; Lassaad, S. Power Management and Control of a Hybrid Electric Vehicle Based on Photovoltaic, Fuel Cells, and Battery Energy Sources. *Sustainability* **2022**, *14*, 2551. [CrossRef]

5.   Javed, M.Y.; Gulzar, M.M.; Rizvi, S.T.H.; Arif, A. A hybrid technique to harvest maximum power from PV systems under partial shading conditions. In Proceedings of the 2016 International Conference on Emerging Technologies (ICET), Islamabad, Pakistan, 18–19 October 2016; IEEE: Manhattan, NY, USA, 2016; pp. 1–5.

6.   Mohamed, A.; Peer, K.R.M.; Chandrakala, V.; Saravanan, S. Comparative study of maximum power point tracking techniques for fuel cell powered electric vehicle. In Proceedings of the IOP Conference Series: Materials Science and Engineering, Wuhan, China, 10–12 October 2019; IOP Publishing: Bristol, UK, 2019; Volume 577, No. 1. p. 012031.

7.   Luta, D.N.; Raji, A.K. Fuzzy rule-based and particle swarm optimisation MPPT techniques for a fuel cell stack. *Energies* **2019**, *12*, 936. [CrossRef]

8.   Reddy, K.J.; Sudhakar, N. ANFIS-MPPT control algorithm for a PEMFC system used in electric vehicle applications. *Int. J. Hydrogen Energy* **2019**, *44*, 15355–15369. [CrossRef]

9.   Rezk, H.; Fathy, A. Performance improvement of PEM fuel cell using variable step-size incremental resistance MPPT technique. *Sustainability* **2020**, *12*, 5601. [CrossRef]

10.   Derbeli, M.; Charaabi, A.; Barambones, O.; Napole, C. High-performance tracking for proton exchange membrane fuel cell system pemfc using model predictive control. *Mathematics* **2021**, *9*, 1158. [CrossRef]

11.   Saleh, B.; Yousef, A.M.; Abo-Elyousr, F.K.; Mohamed, M.; Abdelwahab, S.A.M.; Elnozahy, A. Performance Analysis of Maximum Power Point Tracking for Two Techniques with Direct Control of Photovoltaic Grid-Connected Systems. *Energy Sources Part A Recovery Util. Environ. Eff.* **2022**, *44*, 413–434. [CrossRef]

12.   Hanan, M.; Ai, X.; Javed, M.Y.; Gulzar, M.M.; Ahmad, S. A two-stage algorithm to harvest maximum power from photovoltaic system. In Proceedings of the 2018 2nd IEEE Conference on Energy Internet and Energy System Integration (EI2), Beijing, China, 20–22 October 2018; IEEE: Manhattan, NY, USA, 2018; pp. 1–6.

13.   Wang, M.H.; Huang, M.; Jiang, W.; Liou, K.-J. Maximum power point tracking control method for proton exchange membrane fuel cell. *IET Renew. Power Gener.* **2016**, *10*, 908–915. [CrossRef]

14.   Nasiri, A.I.; Sarvi, M. A new maximum power point tracking method for PEM fuel cells based on water cycle algorithm. *J. Renew. Energy Environ.* **2016**, *3*, 35–42.

15.   Rezk, H. Performance of incremental resistance MPPT based proton exchange membrane fuel cell power system. In Proceedings of the 2016 Eighteenth International Middle East Power Systems Conference (MEPCON), Cairo, Egypt, 27–29 December 2016; IEEE: Manhattan, NY, USA, 2016; pp. 199–205.

16.   Derbeli, M.; Sbita, L.; Farhat, M.; Barambones, O. Proton exchange membrane fuel cell—A smart drive algorithm. In Proceedings of the 2017 International Conference on Green Energy Conversion Systems (GECS), Hammamet, Tunisia, 23–25 March 2017; IEEE: Manhattan, NY, USA, 2017; pp. 1–5.

17.   Ahmadi, S.; Abdi, S.; Kakavand, M. Maximum power point tracking of a proton exchange membrane fuel cell system using PSO-PID controller. *Int. J. Hydrogen Energy* **2017**, *42*, 20430–20443. [CrossRef]

18.   Liu, J.; Zhao, T.; Chen, Y. Maximum power point tracking with fractional order high pass filter for proton exchange membrane fuel cell. *IEEE/CAA J. Autom. Sin.* **2017**, *4*, 70–79. [CrossRef]

19.   Priyadarshi, N.; Sharma, A.K.; Azam, F. A hybrid firefly-asymmetrical fuzzy logic controller based MPPT for PV-wind-fuel grid integration. *Int. J. Renew. Energy Res.* **2017**, *7*, 1546–1560.

20.   Derbeli, M.; Barambones, O.; Sbita, L. A robust maximum power point tracking control method for a PEM fuel cell power system. *Appl. Sci.* **2018**, *8*, 2449. [CrossRef]

21.   Fan, L.; Ma, X. Maximum power point tracking of PEMFC based on hybrid artificial bee colony algorithm with fuzzy control. *Sci. Rep.* **2022**, *12*, 4316. [CrossRef]

22.   Sibtain, D.; Gulzar, M.M.; Shahid, K.; Javed, I.; Murawwat, S.; Hussain, M.M. Stability Analysis and Design of Variable Step-Size P&O Algorithm Based on Fuzzy Robust Tracking of MPPT for Standalone/Grid Connected Power System. *Sustainability* **2022**, *14*, 8986.

23.   Attia, H.; Hossin, K. Efficient maximum power point tracker based on neural network and sliding-mode control for buck converters. *Clean Energy* **2022**, *6*, 716–725. [CrossRef]

24.   Priyadarshi, N.; Padmanaban, S.; Bhaskar, M.S.; Blaabjerg, F.; Holm-Nielsen, J.B.; Azam, F.; Sharma, A.K. A hybrid photovoltaic-fuel cell-based single-stage grid integration with Lyapunov control scheme. *IEEE Syst. J.* **2019**, *14*, 3334–3342. [CrossRef]

25.   Rana, K.P.S.; Kumar, V.; Sehgal, N.; George, S. A novel dPdI feedback based control scheme using GWO tuned PID controller for efficient MPPT of PEM fuel cell. *ISA Trans.* **2019**, *93*, 312–324. [CrossRef]

26.   Fathy, A.; Abdelkareem, M.A.; Olabi, A.; Rezk, H. A novel strategy based on salp swarm algorithm for extracting the maximum power of proton exchange membrane fuel cell. *Int. J. Hydrogen Energy* **2021**, *46*, 6087–6099. [CrossRef]

27.   Belhaj, F.Z.; El Fadil, H.; El Idrissi, Z.; Koundi, M.; Gaouzi, K. Modeling, Analysis and Experimental Validation of the Fuel Cell Association with DC-DC Power Converters with Robust and Anti-Windup PID Controller Design. *Electronics* **2020**, *9*, 1889. [CrossRef]

28.   Qi, Z.; Tang, J.; Pei, J.; Shan, L. Fractional controller design of a DC-DC converter for PEMFC. *IEEE Access* **2020**, *8*, 120134–120144. [CrossRef]

29.   Derbeli, M.; Barambones, O.; Farhat, M.; Ramos-Hernanz, J.A.; Sbita, L. Robust high order sliding mode control for performance improvement of PEM fuel cell power systems. *Int. J. Hydrogen Energy* **2020**, *45*, 29222–29234. [CrossRef]

30.   Silaa, M.Y.; Derbeli, M.; Barambones, O.; Cheknane, A. Design and Implementation of High Order Sliding Mode Control for PEMFC Power System. *Energies* **2020**, *13*, 4317. [CrossRef]

31.   Derbeli, M.; Barambones, O.; Silaa, M.Y.; Napole, C. Real-Time Implementation of a New MPPT Control Method for a DC-DC Boost Converter Used in a PEM Fuel Cell Power System. *Actuators* **2020**, *9*, 105. [CrossRef]

32. Gulzar, M.M.; Sibtain, D.; Ahmad, A.; Javed, I.; Murawwat, S.; Rasool, I.; Hayat, A. An Efficient Design of Adaptive Model Predictive Controller for Load Frequency Control in Hybrid Power System. *Int. Trans. Electr. Energy Syst.* **2022**, *2022*, 7894264. [CrossRef]
33. Gulzar, M.M.; Rizvi, S.T.H.; Javed, M.Y.; Sibtain, D.; Din, R.S.U. Mitigating the load frequency fluctuations of interconnected power systems using model predictive controller. *Electronics* **2019**, *8*, 156. [CrossRef]
34. Fathy, A.; Alharbi, A.G.; Alshammari, S.; Hasanien, H.M. Archimedes optimization algorithm based maximum power point tracker for wind energy generation system. *Ain Shams Eng. J.* **2022**, *13*, 101548. [CrossRef]
35. Silaa, M.Y.; Derbeli, M.; Barambones, O.; Napole, C.; Cheknane, A.; De Durana, J.M.G. An efficient and robust current control for polymer electrolyte membrane fuel cell power system. *Sustainability* **2021**, *13*, 2360. [CrossRef]
36. Bahri, H.; Harrag, A. Ingenious golden section search MPPT algorithm for PEM fuel cell power system. *Neural Comput. Appl.* **2021**, *33*, 8275–8298. [CrossRef]
37. Fathy, A.; Rezk, H.; Alanazi, T.M. Recent approach of forensic-based investigation algorithm for optimizing fractional order PID-based MPPT with proton exchange membrane fuel cell. *IEEE Access* **2021**, *9*, 18974–18992. [CrossRef]
38. Kraiem, H.; Flah, A.; Mohamed, N.; Messaoud, M.H.B.; Al-Ammar, E.A.; Althobaiti, A.; Alotaibi, A.A.; Jasiński, M.; Suresh, V.; Leonowicz, Z.; et al. Decreasing the Battery Recharge Time if Using a Fuzzy Based Power Management Loop for an Isolated Micro-Grid Farm. *Sustainability* **2022**, *14*, 2870. [CrossRef]
39. Rezk, H.; Aly, M.; Fathy, A. A novel strategy based on recent equilibrium optimizer to enhance the performance of PEM fuel cell system through optimized fuzzy logic MPPT. *Energy* **2021**, *234*, 121267. [CrossRef]
40. Xie, M.; Gulzar, M.M.; Tehreem, H.; Xie, M.; Javed, M.Y.; Rizvi, S.T.H. Automatic Voltage Regulation of Grid Connected Photovoltaic System Using Lyapunov Based Sliding Mode Controller: A Finite-Time Approach. *Int. J. Control Autom. Syst.* **2020**, *18*, 1550–1560. [CrossRef]
41. Iqbal, M.; Gulzar, M.M. Master-slave design for frequency regulation in hybrid power system under complex environment. *IET Renew. Power Gener.* **2022**, *16*, 3041–3057. [CrossRef]

sustainability  

Review

A Review on Biohydrogen Sources, Production Routes, and Its Application as a Fuel Cell

Antony V. Samrot [1,*], Deenadhayalan Rajalakshmi [2], Mahendran Sathiyasree [2], Subramanian Saigeetha [3], Kasirajan Kasipandian [4], Nachiyar Valli [2], Nellore Jayshree [2], Pandurangan Prakash [2] and Nagarajan Shobana [2]

[1] School of Bioscience, Faculty of Medicine, Bioscience and Nursing, MAHSA University, Jalan SP2, Bandar Saujana Putra, Jenjarom 42610, Malaysia

[2] Department of Biotechnology, School of Bio and Chemical Engineering Sathyabama Institute of Science and Technology, Chennai 600119, Tamil Nadu, India

[3] Department of Biotechnology, School of Biosciences and Technology, Vellore Institute of Technology, Vellore 632014, Tamil Nadu, India

[4] Faculty of Engineering, Built Environment and IT, MAHSA University, Jalan SP2, Bandar Saujana Putra, Jenjarom 42610, Malaysia

* Correspondence: antonysamrot@gmail.com

Abstract: More than 80% of the energy from fossil fuels is utilized in homes and industries. Increased use of fossil fuels not only depletes them but also contributes to global warming. By 2050, the usage of fossil fuels will be approximately lower than 80% than it is today. There is no yearly variation in the amount of CO_2 in the atmosphere due to soil and land plants. Therefore, an alternative source of energy is required to overcome these problems. Biohydrogen is considered to be a renewable source of energy, which is useful for electricity generation rather than relying on harmful fossil fuels. Hydrogen can be produced from a variety of sources and technologies and has numerous applications including electricity generation, being a clean energy carrier, and as an alternative fuel. In this review, a detailed elaboration about different kinds of sources involved in biohydrogen production, various biohydrogen production routes, and their applications in electricity generation is provided.

Keywords: biohydrogen; gasification; feedstocks; biohydrogen production; dark fermentation

Citation: Samrot, A.V.; Rajalakshmi, D.; Sathiyasree, M.; Saigeetha, S.; Kasipandian, K.; Valli, N.; Jayshree, N.; Prakash, P.; Shobana, N. A Review on Biohydrogen Sources, Production Routes, and Its Application as a Fuel Cell. *Sustainability* 2023, 15, 12641. https://doi.org/10.3390/su151612641

Academic Editor: Muhammad Khalid

Received: 24 December 2022
Revised: 27 February 2023
Accepted: 7 March 2023
Published: 21 August 2023

1. Introduction

Over the past decade, the growth of industries has been increasing enormously, which has resulted in the requirement for alternate energy sources. At the beginning of human history, wood biomass was used for heating, cooking, and shelter, which made it an ideal energy source for man. However, fossil fuels were exploited to meet the energy demands due to the growth of the human population [1]. Depletion and the inability to replenish the energy sources due to increasing industries resulted in the usage of fossil fuels. Increased usage of fossil fuels not only depletes them but also causes significant global warming by emitting harmful greenhouse gases [2]. In recent years, emissions of carbon dioxide and other harmful gases by human activities have been rising more recently than in previous years. Pollution due to fossil fuels can be controlled by the transition from fossil fuels to alternative renewable resources [3]. Sustainable development requires energy as a main component, which must be available constantly at an affordable range for a long period. The conversion of wastes into useful forms is the best way for sustainable development, for example, biohydrogen, biogas, and biofuel, which release less greenhouse gas than fossil fuels [4]. Electricity plays a major role in everyday life, of which 32.9% is produced from fossil fuels supplying approximately 213 Terawatt per hour (TWh) worldwide [5]. In India, the most contributing source of fossil fuel is coal, which contributes approximately 69.5% to power generation [6]. The balance between the preservation of the environment and economic growth is considered "sustainability" [7].

The demand for hydrogen is increasing rapidly nowadays as hydrogen is considered a clean source of energy and a valuable gas. It is used as a feedstock in many industries [8,9]. The ionic form of hydrogen is present abundantly in the universe. It is odorless, colorless, tasteless, and non-toxic [10]. Hydrogen draws prominent attention as a future fuel because of its versatility and efficiency. It can be used as the best and most efficient fuel for transportation as the combustion of this fuel produces only water vapor and eliminates the release of hydrocarbons, carbon monoxide, carbon dioxide, and other micro particles that cause environmental pollution [11]. Hydrogen is used in the hydrogenation of coal, oil, petroleum, and shale oil and is also used in the production of ammonia. Hydrogen can be produced by oil and natural gas using the steam reforming process and other methods such as coal gasification and water electrolysis. However, these processes are considered non-renewable and do not draw much attention. Therefore, the eco-friendly production of bio hydrogen gas using renewable sources such as agricultural waste, inorganic waste, and microorganisms is highly encouraged [12]. The production of hydrogen becomes more interesting when produced from renewable sources because it can be operated at ambient pressure and temperature with a lower amount of energy consumption. Energy production from hydrogen is 122 kJ/g, which is 2.75 times greater than hydrocarbon fuels so it acts as a potential energy carrier [10]. Obtaining hydrogen from biomass is rather challenging as the amount of hydrogen present in biomass is nearly 6% versus 25% for methane, and the lower energy content is due to the 40% of oxygen present in biomass [13].

2. Definition

The term biohydrogen in Greek refers to Bio- or life, hydro- or water, and gen- or genes, which indicates non-degradable organic fuel obtained from biological sources such as plants, microorganisms, animals, etc. [13]. Hydrogen produced biologically is termed "Biohydrogen". It draws much attention because it is a clean, non-degradable, non-condensable fuel with higher efficiency, high energy density, and a lack of pollution [14]. Biohydrogen is a natural or transient byproduct of several microbial-mediated biochemical reactions. It can be produced either by a biological process or the thermochemical treatment of biomass [2]. Biohydrogen has the ability to be converted into usable power at a higher efficiency. However, the lower yields, storage, and rate of production remain barriers to biohydrogen production [15].

3. Feedstocks of Biohydrogen Production

The sources selected for the production of hydrogen gas should be low cost and biodegradable and must have a high level of carbohydrate content with the presence of simple sugars such as glucose, lactose, and sucrose, which can be used as reliable biodegradable substrates for biohydrogen production [8]. The production of biohydrogen via bio photolysis of water using cyanobacteria, microalgae, and photosynthetic anoxygenic bacteria is most suitable as it utilizes major natural resources such as sunlight, water, etc. [16]. These microorganisms either supply electrons as an alternate source for the sake of survival in minimal optimum conditions or the need to prevent the reduction of the electron transport chain and act as a security valve. In addition to these biochemical reactions, hydrogen gas can also be produced during nitrogen fixation by the nitrogenases enzyme, which is a major mechanism in the heterocyst forming blue-green algae [17].

3.1. Agricultural Waste

Over the last decade, many research works have been carried out focusing on the findings of alternate sources of green, clean, and renewable energy. However, the production of biofuels from food sources such as corn and sugar has served as an alternate source but has indirectly increased food prices, which has resulted in a global food crisis. Hence, nowadays, the production of biofuels from agricultural wastes has gained much attention from researchers [18]. The production of hydrogen gas from agricultural waste, which consists of lignocellulose material, contributes to the global energy conversion process.

Agricultural waste is rich in hemicellulose and cellulose after conversion into mono or disaccharides and can be used in dark fermentation, photo fermentation, and bio photolysis (direct and indirect) [19].

3.1.1. Lignocellulose Waste

Lignocellulose waste is considered a macromolecule consisting of lignin, cellulose, and hemicellulose. Lignin is a highly insoluble, irregular polymer that bonds with hemicellulose with a covalent bond, and in the cell wall, cellulose is enwrapped in a complex containing lignin and cellulose. This complex nature causes the barriers to transform into lignocellulosic waste. Waste such as residues of plants, agricultural waste, and the logging of wood is considered to be lignocellulosic waste and they are degraded slowly as they are difficult to degrade [20]. Around 180 million tons per year of lignocellulose materials are produced as byproducts or in the form of agricultural residues, which can be used as an inexpensive source for the production of biofuels [21,22]. These materials, due to their low fiber porosity, heterogeneity, and crystallinity are not readily fermentable, and pre-treatment is required for the process of forming fermentable sugars [23]. Nowadays, researchers are focused on the next-generation organic matter, which includes lignocellulose, rather than using first-generation products, as lignocellulose is a rich source of fermentable sugars and can be used for the production of biohydrogen. Some of the steps to be followed when lignocellulose materials are used for biohydrogen production are as follows:

- Lignocellulose materials consist of a hetero polymeric substance, and in order to break the complex, the raw materials must be pre-treated.
- A large number of monomeric sugars was obtained by hydrolysis of cellulose and hemicellulose.
- The obtained monomers liberated from the fractions were converted into the respective biofuel by the utilization of a microorganism using techniques involved in the bioprocess [24].

High yields of biohydrogen are obtained by following the aforementioned steps [25]. The production of biohydrogen from lignocellulose waste has attracted the attention of many researchers due to its efficiency. Several researchers proved the efficiency and positive response of biohydrogen production by utilizing various lignocellulosic substrates and also identified the sources responsible for the inhibition [26]. The production of biohydrogen from lignocellulosic biomass after the pre-treatment, hydrolysis, and utilization of different microbial cultures via the process of dark fermentation has improved the yield and rates of biohydrogen production [27]. The production of biohydrogen from various substrates of lignocellulose via the process of dark fermentation is considered to be effective. The next most effective process used for the production of biohydrogen after dark fermentation is photo fermentation [28]. Taguchi et al. [29] isolated *Clostridium* sp. strain no. 2 from termites and produced biohydrogen with 18.6 mmol/g of the substrate using xylan from oat spelts. Taguchi et al. [30] used the same Clostridium sp. for the hydrolysis of cellulose and observed that the bacterium consumed 0.92 mmol of glucose per h and produced 4.1 mmol of hydrogen per h. The increase in the concentration of cellulose (12.5 g/L to 50 g/L) decreased the yield (2.18 mmol/g of cellulose to 0.42 mmol/g of cellulose). At high temperatures, high conversion of cellulose into hydrogen took place (43 mL of hydrogen/g of cellulose at 37 °C to 69 mL of hydrogen/g of cellulose at 55 °C; 567 mL of hydrogen was produced from 1 g of cellulose) [31]. The production of biohydrogen from lignocellulosic biomass is described in Figure 1. Some of the lignocellulose biomass used and its composition, types of monomers present, and the amount of hydrogen produced is tabulated below (Table 1).

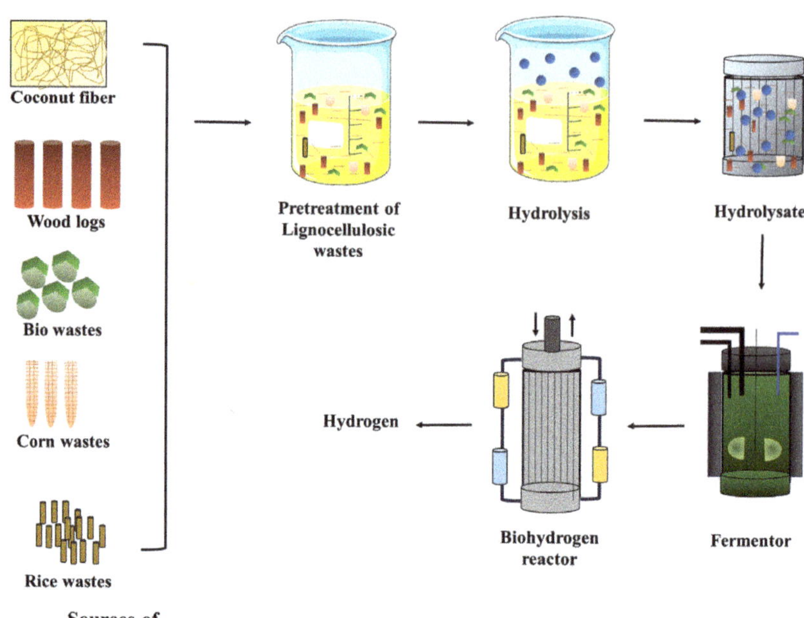

Figure 1. Biohydrogen production from lignocellulosic wastes.

Table 1. Lignocellulose biomass, its composition, and hydrogen production.

LCB	Monomer Composition	Composition of LCB	Hydrogen Production	Reference
Beer less	Not mentioned	Not mentioned	68.6 mL of biohydrogen per gram of total volatile solids	[32]
Corn stover	1.5 g/L-Xylose 10 g/L-glucose 0.2 g/L-Arabinose	37.6%-cellulose 21.5%-hemicellulose 19.1%-Lignin	12.9 mmol/L in an hour	[33]
Grass	Not mentioned	Not mentioned	4.9 mol hydrogen gas per gram of total solid	[34]
Soy bean straw	3.6% of TRS	39.6%-cellulose 14.6%-hemicellulose 23.4%-lignin	60.2 mL of hydrogen per gram of dry straw	[35]
Wheat bran	Not mentioned	8.27%-cellulose 33.7%-hemicellulose	128.2 mL of hydrogen per gram of total volatile solid	[36]

LCB—Lignocellulose Biomass.

3.1.2. Livestock Waste

After the depletion of fossil fuels, solid wastes have become a promising factor for the production of renewable sources [37]. The terrestrial surface is occupied by livestock, and it plays a major role as a significant global asset. Livestock is considered to be an important provider of nutrition for growing crops in a small area. In recent days, in developing countries, livestock is considered to be the fastest-growing agricultural subsector [38]. Livestock serves as an important factor in increasing food security and contributing to rural and agricultural development [39]. Nowadays, the waste generated by livestock from cattle, swine buildings, and poultry is a major source of contamination of underground water systems due to its odor, gases, and dust. Due to the contamination caused by these wastes, many researchers proposed the idea of generating useful products from these wastes. These livestock wastes include fodder, manure, and slaughterhouse and poultry farm wastes. The improper maintenance of these wastes is harmful to both human health and the environment. From these polluting substances, a renewable non-polluting energy source is produced, named biohydrogen [37,40]. However, the production of biohydrogen is inhibited due to the presence of ammoniacal nitrogen (NH_3-N) in chicken manure and the presence of high sulphate content in swine manure. In order to produce biohydrogen, the high sulphate content can be treated with a rich carbohydrate source such as lignocellulose materials, which provide the perfect C/N ratio and enhance the buffering capacity and provide nutritional manure [37]. Livestock waste can be used as a substrate along with the carbohydrate source for the efficient production of biohydrogen [41].

Lateef et al. [42] produced biohydrogen with cow manure as a source along with waste milk as a co-substrate. After adding the organic load, which is obtained from the co-digestion of cow manure, the production of biohydrogen increased. Tenca et al. [43] produced biohydrogen with a yield of 126 ± 22 mL H_2/g $_{VS-added}$ when swine manure was used, along with fruit and vegetable waste. Marone et al. [44] produced biohydrogen with a maximum yield of 117 mL H_2/g $_{VS-added}$ through the co-fermentation of buffalo slurry with cheese whey and crude glycerol using a mixed microbial culture. Bari et al. [45] produced biohydrogen from organic waste by fermentation process and had various industrial applications like steel making, ammonia production, Glass making etc. Fan et al. [32] produced biohydrogen with a yield of 68.6 mL H_2/g TVS when beer-less wastes were converted into biohydrogen via cow dung compost. The hydrogen yield and hydrogen production rate were higher (30.00 mL/g $_{VS-added}$ and 1.00 L/L/d, respectively) when the biohydrogen was produced via the co-digestion of cattle manure and food wastes with an optimal mixing ratio of 47 to 51%, a hydraulic retention time of 2 days, and a substrate concentration of 76 to 86 g/L [46]. The production of biohydrogen using the co-digestion of cattle manure with specified risk materials has been reported by Gilroyed et al. [47]. The maximum hydrogen production rate and hydrogen yield was 109.55 mL H_2/L per day and 0.84 mol H_2/mol of total sugar consumed, respectively, when elephant dung was used as the inoculum for sugarcane bagasse hydrolysate [48]. The maximum hydrogen production rate and hydrogen yield were 215.4 ($\pm$62.1) mL H_2/L/d 152.2 and ($\pm$43.9) mL H_2/g $_{VS-added}$, respectively, achieved at an organic loading rate of 2.1 g VS/L/d of cheese whey via the dark fermentation method using buffalo manure as a buffering agent [49]. The co-digestion of cassava wastewater along with buffalo dung for biohydrogen production gave a maximum hydrogen production rate and hydrogen yield of 839 mL H_2/L/d and 16.90 mL H_2/g $_{COD-added}$, respectively [50]. Perera et al. [51] produced a maximum hydrogen yield of 2.9–5.3 M hydrogen/M sucrose when sucrose along with dairy cattle manure was used for production. Biohydrogen was produced when the liquid swine manure was co-fermented with molassesm of which the hydrogen production rate and hydrogen yield of 31.9 L/d and 1.52 L/g sugar, respectively, was obtained [52]. Zhu et al. [53] produced biohydrogen with swine manure co-fermented with glucose as a substrate. Biohydrogen production from livestock waste is illustrated and tabulated in Figure 2 and Table 2, respectively.

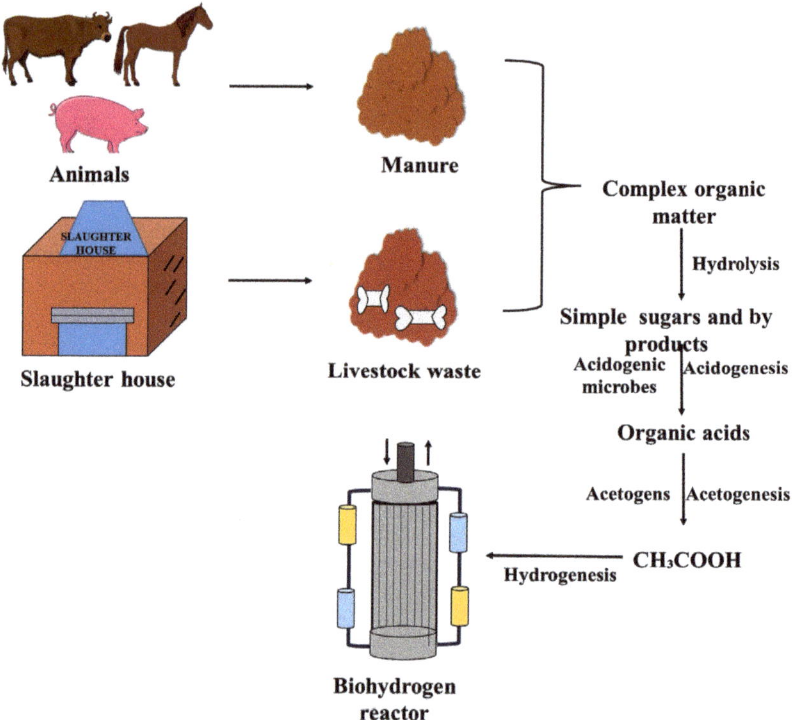

Figure 2. Biohydrogen production from livestock wastes.

Table 2. Livestock waste as a source for hydrogen production.

Livestock Wastes	Hydrogen Yield	Reference
Swine manure with glucose	5 L H_2/L d	[54]
Dairy manure with mixed cultures	31.5 mL/g-TVS	[40]
Pig slurry with inoculum (Mesophilic methanogenic sludge)	3.65 (mL H_2/g $_{VS\text{-added}}$)	[55]
Bovine manure	58.48 mL of H_2/g of manure;	[53]
Cow dung compost	68.6 Ml of H_2/g TVS	[45]
Swine manure	126 ± 22 mL of H_2/g $_{VS\text{-added}}$	[43]

3.2. Industrial Waste

The growth of the world relies mostly on industrialization. Pollution is caused by these industries by utilizing more water and the excessive production of effluents [56]. Industrial wastes are substances that cause severe environmental pollution as they are non-biodegradable. The application of these industrial wastes in road construction has attracted many researchers in recent days [57]. The waste materials generated from these industries are renewable biomass and can be used for the production of biohydrogen. Industrial wastewater and biodiesel industry wastes are some examples of industrial wastes that can be used for biohydrogen production. Many reports on biohydrogen production from electrolysis and other chemical processes have been reported, but the biological conversion of wastes into hydrogen can be the best alternative method and also the most cost-efficient method [58]. Many starch- and cellulose-based materials are present in the waste products from the food and agricultural industries. These waste products are rich in carbohydrate content. It is easier to process the starch waste content by hydrolyzing it into maltose or glucose via enzymatic or acid hydrolysis followed by conversion into carbohydrates and then into hydrogen gas, but cellulose-containing wastes are difficult to process as they require the pre-treatment of wastes, and hydrolyses, followed by conversion into carbohydrates and then into biohydrogen production [8,23,59].

Biohydrogen was produced using the waste from food industries by Alexandropoulou et al. [60] using the continuous-type reactor under different pH and hydraulic retention times. The obtained hydrogen yield was 96.27 ± 3.36 and 101.75 ± 213.7 L H_2/kg FIW for 12 and 6 h, respectively. Moreno-Andrade et al. [61] produced hydrogen using different industrial wastes as feedstocks. The feedstocks used were tequila vinasses, sugar vinasses, wastewater from the plastic industry, aircraft wastewater, and physio-chemically treated wastewater from the plastic industry. The tequila vinasses produced the maximum amount of hydrogen followed by wastewater from the plastic industry, aircraft wastewater, physio-chemically treated wastewater from the plastic industry, and sugar vinasses, and it was observed that the hydrogen production in aircraft wastewater increased when an anaerobic sequencing batch reactor was used. Moreno-Dávila et al. [62] produced hydrogen with 60.75 mmol/h$*$g volatile solids when pre-treated wastes of paper industries were used as the source. The process followed for biohydrogen production was simultaneous saccharification and fermentation. Oceguera-Contreras et al. [63] produced biohydrogen with a yield of 1246.36, 1571.81, and 232.72 mL H_2/L from the bagasse, molasses, and vinasses agro-industry wastes when vermihumus-associated microorganisms as inoculum were used as a source and found that these microbes not only produce biohydrogen but also help in the degradation of lignocellulosic waste material.

Lopez-Hidalgo et al. [64] produced hydrogen from agro-industrial wastes such as cheese whey and wheat straw hydrolysate. The authors reported that both the wheat straw hydrolysate and cheese whey produced hydrogen efficiently as both an individual substrate and even when mixed together. Lucas et al. [65] produced biohydrogen using cassava wastewater, dairy wastewater, and citrus processing wastewater as sources and the production of hydrogen was found to be 31.41, 28.95, and 37.25 mL/g. Gomez-Romero et al. [66] utilized fruit and vegetable wastes and crude cheese whey for the production of biohydrogen. The yield of produced hydrogen was 813.3 mL H_2 g COD^{-1} and was determined at 17.5 h (Hydraulic Retention Time) with an organic loading rate of 80.02 g COD L^{-1} d^{-1}. The usage of agro-industrial wastes such as starch wastes produces biohydrogen efficiently and is a cheaper process. A variety of raw materials from agro-industries can be used for the production of biohydrogen [67]. Biohydrogen production from industrial wastes is illustrated and tabulated in Figure 3 and Table 3, respectively.

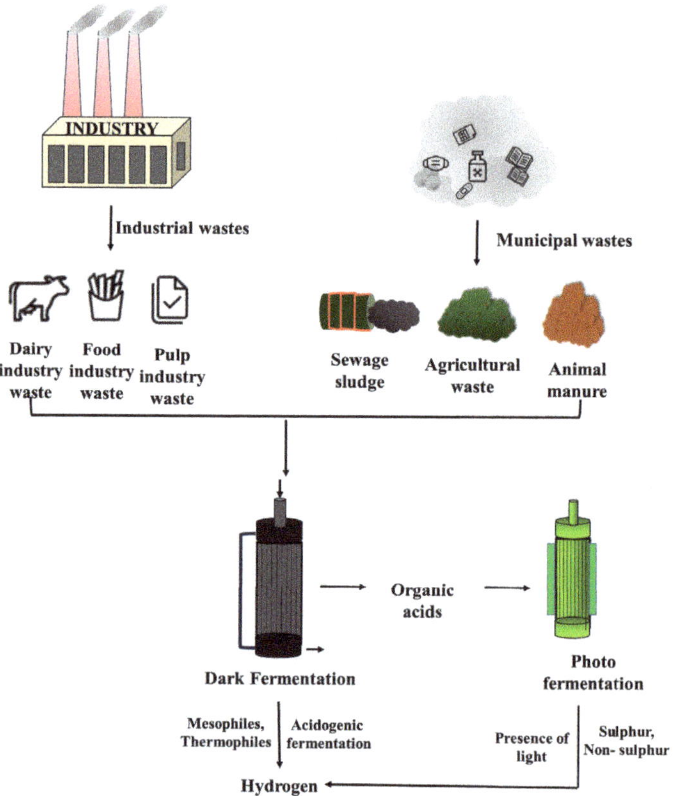

Figure 3. Biohydrogen production from industrial wastes.

Table 3. Production of biohydrogen from industrial wastes.

Industrial Waste	Hydrogen Production Rate	Reference
Molasses	700 mL H_2/L/D	[68]
Paper and pulp industry	2.03 mol H_2/mol sugar	[69]
Palm oil mill effluent	0.41 mmol H_2/g COD	[70]
Textile designing wastewater	1.52 mol/mol hexose	[71]
Palm oil mill effluent	0.66 mol H_2/mol total monomeric sugar	[72]
Textile wastewater	1.37 mol H_2/mol hexose	[73]
Rice mill wastewater	1.97 mol H_2/mol of sugar	[74]

3.3. Municipal Wastes—Waste Sludge

The management and generation of waste products are becoming a global challenge and causing environmental problems. The management and recycling of waste are the best way to avoid pollution. Waste sludge causes much environmental pollution and also affects human health in many ways. Waste sludges are used for the generation of many renewable resources in order to maintain the quality of the environment, reduce many risk factors, produce sustainable energy, and serve as a reliable source of energy production [75,76]. Biohydrogen was produced via the co-digestion of food waste and sewage sludge, and the maximum hydrogen production rate was observed to be 111:2 mL H_2/g VSS/h [77]. Cai et al. [78] produced biohydrogen from sewage sludge and reported that the hydrogen yield of alkali pre-treated sludge was higher than dry sludge. The yield increased from 9.1 mL of H_2/g of dry solids (DS) to 16.6 mL of H_2/g of DS when alkali-pre-treated sludge was used. Yin and Wang [79] produced hydrogen using waste sludge and reported that the irradiation and gamma irradiation combined with the alkali pretreatment was able to produce biohydrogen by dissolving the waste-activated sludge. The co-fermentation of sewage sludge and fallen leaves produced biohydrogen. The mixing ratio of 20:80 of fallen leaves and sewage sludge produced biohydrogen with a yield of 37.8 mL/g $_{VS\text{-added}}$ [80]. Natural sludge was used as an inoculum to produce biohydrogen using corn stalks via anaerobic fermentation, and the maximum hydrogen yield was observed to be 126.22 mL g^{-1}-CS [81]. A Continuous Mixed Immobilized Sludge Reactor (CMISR) using activated carbon as a support carrier was used for hydrogen production via dark fermentation from enzymatically hydrolyzed food waste. The maximum hydrogen production rate of 353.9 mL/h/L was obtained under the conditions of a packing ratio of 15% and an organic loading rate of 40 kg/m^3/d [82]. Yang and Wang [83] reported that the combined sodium citrate and ultrasonic pretreatment disrupted the sludge floc structure and promoted biohydrogen fermentation performance. Yang and Wang [84] produced biohydrogen from waste-activated sludge in which the sludge consisted of a complex structure due to the presence of an extracellular polymeric substance, which had to be pre-treated. The maximum hydrogen yield of 38.8 mL/g $_{VS\text{-added}}$ was obtained after the combined pre-treatment of sodium citrate pre-treatment and ultrasonic pre-treatment.

Yang and Wang [84] produced biohydrogen from waste-activated sludge with the addition of a cationic binding agent (sodium citrate) to disintegrate the extracellular polymeric substance present in the sludge. The addition of the binding agent improved biohydrogen production from 3.7 to 18.8 mL/g $_{VS\text{-added}}$ when 0.3 g of sodium citrate/of SS was added. Biohydrogen was produced via the dark fermentation method by using waste-activated sludge from fructose processing manufacturing and the maximum hydrogen yield obtained was 7.8 mmol [85]. Biohydrogen production from municipal wastes is illustrated and tabulated in Figure 3 and Table 4, respectively.

Table 4. Production of biohydrogen from municipal wastes.

Municipal Wastes	Substrate	Hydrogen Production Rate (L/L/d)	Hydrogen Yield (mol H$_2$/mol Glucose, Hexose Equivalent)	Reference
Anaerobic digested sludge	Sucrose	Nd	3.06	[86]
Suspended & granular anaerobic sludge	Ground wheat	Nd	25.7	[87]
Anaerobic digested sludge	Glucose	120.4 mL H$_2$/h	1.9	[88]
Cassava stillage	Nd	Nd	53.8 (mL H$_2$/g VS)	[89]
Cattle wastewater	Nd	Nd	319 mL H$_2$/g COD consumed.	[90]
Wastewater sludge	Nd	Nd	2.1 mmol-H$_2$/g-COD	[91]
Distillery wastewater	Nd	Nd	3.35 (mol/mol glucose)	[92]

3.4. Microbial Routes

The production of biohydrogen on a large scale came into thought after the rapid depletion of fossil fuels. It has been known for more than 70 years that algae can make bio-hydrogen under illumination. The evolution of hydrogen was induced in the cells when pre-incubation in the dark was performed on the cells. Hydrogen production is due to the hydrogenase enzyme expressed during the period of incubation [93]. The fermentative hydrogen production depends on the type of inoculum used, the reactor type, and its temperature settings. Many types of inoculums are used for hydrogen production and must be pure cultures of hydrogen-producing bacteria, mixed cultures of anaerobic bacteria obtained from compost piles, and anaerobic sludges [94–96]. The metabolic shifts in pure cultures are easily visible, and the utilization of pure cultures enables us to understand the conditions that promote a high hydrogen production rate and yield [97]. Biohydrogen production from municipal wastes is tabulated in Table 5.

Table 5. Production of biohydrogen from microbial routes wastes.

Culture	Substrate Type	Hydrogen Production Rate (L/L/d)	Hydrogen Yield (mol H_2/mol Glucose, Hexose Equivalent)	Reference
3-Clostridium DMHC-10	Glucose	2.14	3.35	[92]
5-Clostridium beijerinckii L9	Glucose	1.9	2.81	[98]
Clostridium butyricum and *Enterobacter aerogenes* HO-39	Sweet potato starch residue	0.977	2.7	[99]
14-Escherichia coli S3	Glucose	0.33	1.45	[100]
Clostridium butyricum	Glucose	0.41	2.09	[100]
Escherichia coli	Glucose	0.33	1.45	[100]
Clostridium butyricum and *Escherichia coli*	Glucose	0.52	1.65	[99]
Clostridium tyrobutyricum FYa102	Glucose (0.36 g/L, 1.4 g/L peptone and ammonium chloride respectively, were added with the substrate)	1.6	1.47	[98]
Caldicellulosiruptor saccharolyticus	Carrot pulp hydrolysate	7	2.8	[59]
Clostridium thermocellum and *Clostridium thermosaccharolyticum*	Corn stalk waste	0.34	ND	[101]
Klebsiella pneumoniae DSM2026	Glycerol	12.2	0.53	[102]
Clostridium butyricum TISTR 1032	Sugarcane juice	3	1.33	[103]
Clostridium acetobutylicum X9	Microcrystalline cellulose	21.33	0.59	[104]
Clostridium acetobutylicum ATCC 824	Cassava wastewater	1.32	2.41	[105]
Enterobacter cloacae IIT-BT08	Glucose	Not mentioned	2.2	[106]
Caldicellulosiruptor saccharolyticus DSM 8903	Hydrolyzed potato steam peels	Not mentioned	3.4	[107]
Thermotoga neapolitana DSM 4349	Hydrolyzed potato steam peels	Not mentioned	3.3	[107]

4. Biohydrogen Production

4.1. Bio Photolysis

Light-dependent production of hydrogen from water is a biological process that converts sunlight into chemical energy [108]. The enzymes are responsible for catalyzing chemical reactions such as nitrogenase, Ni-Fe- hydrogenase, and Fe- hydrogenase. The bio photolysis process makes use of the Fe- hydrogenase enzyme [109].

$$2H_2O \xrightarrow{\text{Light energy}} 2H_2 + O_2 \tag{1}$$

The various routes of biohydrogen production are illustrated in Figure 4.

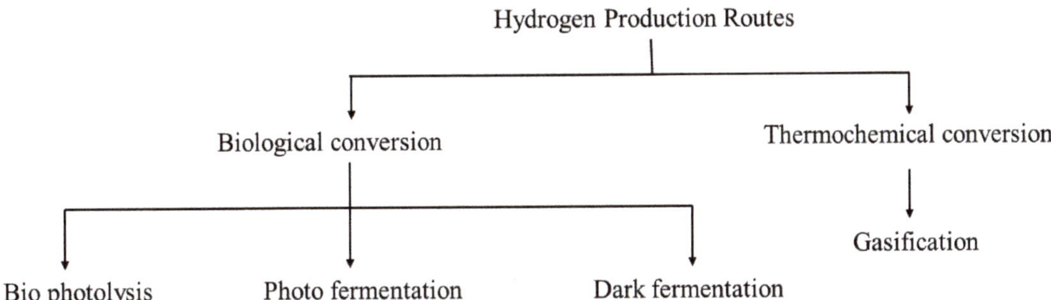

Figure 4. Routes of hydrogen production.

4.1.1. Direct Bio Photolysis

There are many advantages of direct bio photolysis of hydrogen production. This reaction can be observed in laboratory conditions and is self-limited by the oxygen that builds up in the cellular environment and takes place during the initial transition to conventional photosynthesis. Photosystem I and Photosystem II are involved in photosynthesis, where photosystem I reduces carbon dioxide and photosystem II splits H_2O and produces oxygen [110]. The photosynthetic apparatus absorbs sunlight directly and uses photoenergy for the splitting of water, and the resulting low-potential reductant reduces the hydrogenase enzyme system. Thus, photo energy could convert the readily available substrate and H_2O into O and H molecules [109].

Photoautotrophic organisms do produce hydrogen from water using the hydrogenase enzyme under anaerobic conditions in the presence of light energy. Cyanobacteria and green algae produce hydrogen via direct bio-photolysis through chlorophyll and other pigments that have the ability to absorb photons at a wavelength of less than 680 nm [111]. Green algae can produce hydrogen when exposed to light or uptake hydrogen via the CO_2-fixation process when exposed to darkness in anaerobic conditions. The unicellular green algae *Chlamydomonas reinhardtii* has gained a great deal of attention in recent decades for its direct bio-photolysis production of hydrogen molecules. Cyanobacteria are prokaryotes that can perform oxygenic photosynthetic reactions [112,113].

When plants are used as a source for biohydrogen production, only CO_2 reduction takes place as plants cannot undergo the process of producing hydrogen as it does not have the hydrogenase enzyme, but green macroalgae and cyanobacteria can produce hydrogen as they do have the hydrogenase enzyme [114]. *Synechocystis* sp. PCC 6803 was used for the production of hydrogen by direct bio photolysis, and 0.037 mmol H_2/mg Chl/h of hydrogen was produced in the dark within 120 h [115–117].

4.1.2. Indirect Bio Photolysis

Cyanobacteria and microalgae are employed to produce hydrogen from water, where photosynthesis occurs and solar energy is transformed into electrical energy [118]. In indirect bio photolysis, the hydrogen and oxygen evolution takes place at separate stages linked to carbon dioxide fixation, where CO_2 is used for the production of the cellular substance, and these are used for the production of hydrogen. Primarily cyanobacteria are used during indirect bio photolysis as it has the property of using carbon dioxide in the air as a carbon source and the energy source is provided by solar energy [108].

An alternate process for direct photolysis is indirect bio photolysis, where carbon dioxide acts as an electron carrier between photosynthesis and hydrogen production. The reason for the wide usage of nitrogen-fixing cyanobacteria in this process for hydrogen production is that it can produce hydrogen using the nitrogenase enzyme present in it, even in the absence of nitrogen, which is also possible under laboratory conditions [118]. The most commonly used cyanobacteria in indirect bio photolysis are *Oscillatoria* sp.,

Gloeocapsa sp., *Anabaena* sp., and *Calothrix* sp. [119]. Generally, four steps are involved in the production of biohydrogen via indirect bio photolysis [110]:

i.   Photosynthesis for the production of biomass.
ii.   The concentration of biomass.
iii.   Dark fermentation in aerobic conditions, which produces 4 mol hydrogen/mol glucose along with 2 mol of acetates.
iv.   The production of hydrogen.

Indirect bio photolysis is a two-step process that starts with photosynthesis and sugar reduction, followed by the induction of light. The maximum efficiency for the conversion of light is 16.3%. Better conversion of light takes place at the lowest illumination, and at the highest illumination, the efficiency is less [118]. To date, reports regarding indirect bio photolysis are fewer, and more studies must be conducted in order to obtain a better understanding of this process.

4.2. Dark Fermentation

The production of biohydrogen via dark fermentation involves the use of anaerobic or facultative anaerobic bacteria in anaerobic conditions. Even for the estimation of fermentative hydrogen production, various substances can be used such as carbohydrates, proteins, sugar molecules, and lipids. Glucose biotransformation toward acetate is widely preferred [1]. The bacteria are responsible for producing biohydrogen from organic waste during dark fermentation. The substrate primarily used is lignocellulose biomass, but other raw materials such as municipal waste and wastewater from industries are also able to be used as a substrate for the production of biohydrogen. Compared with photo fermentation, dark fermentation is considered to be the most promising method for biohydrogen production [120].

Anaerobic bacteria are responsible for using the organic substance as the source of electrons and the energy required for converting it into hydrogen. The reactions taking place during dark fermentation occur as a rapid process as there are no requirements for solar radiation. Large quantities of biomass are treated using a large fermenter [121]. Under anaerobic conditions, protons can act as electron acceptors to accept the electrons generated and bacteria reduce the protons in hydrogen by using hydrogenase, which maintains the electrical neutrality for the uninterrupted and continuous supply of ATP [122]. This hydrogenase enzyme can be divided into many types depending on the metal-binding capacity, and microbial hydrogen metabolism greatly depends on the hydrogenase enzyme [123]. Dark fermentation can take place using both mixed and pure cultures, but there is an advantage of using a pure culture over a mixed culture as the metabolic changes can be monitored easily [124].

$$C_6H_{12}O_6 + 2H_2O \rightarrow 2CH_3COOH + 2CO_2 + 4H_2 \tag{2}$$

$$C_6H_{12}O_6 + 2H_2O \rightarrow 2H_2 + CH_3CH_2CH_2COOH + 2CO_2 \tag{3}$$

From Equations (2) and (3), it is evident that 4 mol H_2/mol glucose can be produced if acetic acid is the volatile fatty acid (VFA) product. Moreover, 2 mol H_2/mol glucose can be produced if butyric acid is the Volatile Fatty Acid (VFA) product [125]. The advantages of this method include that it can produce hydrogen even for a day without light, various carbon sources can be used as the substrate, there is no oxygen limitation problem as it is an anaerobic reaction, and the byproducts produced during dark fermentation are valuable byproducts such as acetic acid, lactic acid, etc. [126]. Dark fermentation for the production of biohydrogen is illustrated in Figure 3.

4.3. Photofermentation

Photosynthetic and Non-Sulfur (PNS) bacteria have the ability to convert the volatile fatty acid into carbon dioxide and hydrogen under anoxygenic conditions [127]. PNS bacteria is a non-taxonomic group that is capable of growing as photoautotrophs, photoheterotrophs, and chemoheterotrophs, depending on the availability of carbon, oxygen, and light sources [128]. The optimum growth conditions for PNS bacteria are pH 7 and temperatures ranging between 30 and 35 °C [8]. This method is considered to be an effective process for producing hydrogen without the generation of oxygen. Organic components are decomposed under the presence of light by anaerobic or photosynthetic bacteria via the nitrogenase-catalyzed reaction [129]. *Rhodobacter capsulatus, Rhodobacter sphaeroides, Rhodopseudomonas palustris,* and *Rhodovulum sulfidophilum* are some of the PNS bacteria responsible for photo fermentation. Photo fermentation can be performed in both batch and continuous systems by supplying an artificial light source or illumination. Various physical parameters such as the temperature, pH, medium composition, and intensity of light affect the productivity of hydrogen by bacteria [130].

PNS bacteria have the ability to reduce H^+ ions to hydrogen in the gaseous phase by extracting power from the oxidation of certain compounds such as fatty acids of low molecular weight and light energy [131]. For the PNS organism to grow and produce hydrogen, photo heterotrophy is generally preferred. This photo fermentation is carried out via the catalytic action of two enzymes involving hydrogenase and nitrogenase via the Tricarboxylic Acid (TCA) cycle [112,132]. The production of hydrogen gas by PNS bacteria is possible as a result of one of the important enzymes: Nitrogenase. It is highly sensitive to oxygen as it is an iron sulfur molybdenum enzyme. The main source for photo fermentation is light, which is most required for developing a photobioreactor with a greater illumination facility for industrial purposes [133].

The production of hydrogen under dark fermentation is usually lower compared to photo fermentation, but a 14 h light and 10 h dark cycle can improve the rate of hydrogen production [8].

4.4. Gasification

After biological conversion, gasification became the most widely studied field. More studies on gasification have been performed by China and the United States of America, while the UK, Italy, Malaysia, Canada, and Japan have also contributed many findings in the field of producing hydrogen using gasification. At high temperatures and high pressures, organic feedstock undergoes partial oxidation, which is termed gasification. During this process, several byproducts can also be produced such as tar, biochar, light hydrocarbon, etc. [134]. Gasification is not a biological process but it is still used for the conversion of organic wastes into biohydrogen. The optimization of operating parameters helps in improving hydrogen production [135].

$$2C + O_2 \rightarrow 2CO_2 C + O_2 \rightarrow 2CO \tag{4}$$

$$C + O_2 \rightarrow CO_2 C + O_2 \rightarrow CO_2 \tag{5}$$

$$C + H_2O \rightarrow CO + H_2 C + H_2O \rightarrow CO + H_2 \tag{6}$$

$$C + CO_2 \rightarrow 2CO C + CO_2 \rightarrow 2CO \tag{7}$$

$$C + 2H_2 \rightarrow CH_4 C + 2H_2 \rightarrow CH_4 \tag{8}$$

$$CO + H_2O \rightarrow CO_2 + H_2 CO + H_2O \rightarrow CO_2 + H_2 \tag{9}$$

$$CH_4 + H_2O \rightarrow CO + 3H_2 \tag{10}$$

Biomass is considered to be a very good source for gasification because of its low sulfur content, and if the moisture content is less than 35% for any kind of biomass, then it can be converted into fuel gas [136]. Gasification is considered to be a biological process that converts biomass into carbon monoxide, carbon dioxide, hydrogen, and methane with controlled amounts of steam and oxygen used at high temperatures [137].

Biomass gasification usually takes place between 700 and 1200 °C using oxygen, air, and other gasifying agents. Steam employment during gasification enhances the production of hydrogen and produces high-heating-value gas with no N_2. Some of the major steps involved in steam gasification are pyrolysis, the homogenous reaction by volatiles produced during pyrolysis, and heterogenous char gasification [138]. For the production of hydrogen, a good-quality gas from the gasifier should consist of a low tar content and a high hydrogen content, but the gas quality can be affected by various parameters such as the pressure, temperature, equivalence ratio, gasifier design, and characteristics of biomass [139]. The advantages and disadvantages of the above-mentioned techniques are summarized in Table 6.

Table 6. Advantages and disadvantages of various biohydrogen production techniques.

S. No.	Techniques		Advantages	Disadvantages	References
1	Biological conversion	Bio photolysis	Ability to produce hydrogen from water in mild conditions, such as those involving moderate temperatures and pressures; anaerobic conditions can be maintained more easily	High energy costs; high oxygen sensitivity of hydrogenase; low light conversion efficiency; Need lighting; Need for ATP	[140–142]
		Photo fermentation	Utilizes light as a source of energy instead of sugar; PNS bacteria are capable of producing hydrogen in a variety of light energies	poor solar energy conversion efficiency; requires anaerobic photobioreactors with a lot of solar exposure	[142–144]
		Dark fermentation	Light independent method; bioremediation; No oxygen limitation	Thermodynamically unfavorable due to limited production of hydrogen; accumulation of oxygen causing inhabitation of biohydrogen	[144]
2	Thermochemical conversion	Gasification	Higher conversion can be achieved	Gas conditioning and tar removal are to be performed	[145]

4.5. Applications of Biohydrogen in Fuel Cells

Fuel cells convert chemical energy for the production of electricity. It is considered an electrochemical conversion device. Hydrogen and microbial fuel cells, when coupled together, produce electricity without the emission of water and other toxic elements as byproducts. H_2 is produced effectively in MFC and can also be used in the generation of electricity and aids in the purification of wastewater [146]. Hydrogen produced by the biohydrogen separation system is used as fuel in fuel cells. The Proton Exchange Membrane Fuel Cell (PEMFC) has received much attention due to its portable nature and ability to work in low-temperature conditions [147]. Wei et al. [148] produced biohydrogen through anaerobic fermentation by using the starch in wastewater as a source, and it was transferred immediately to PMEFC for the effective generation of electricity.

Biohydrogen was produced from dairy wastewater and was transferred to PMEFC for electricity generation. Contaminants such as carbon monoxide, carbon dioxide, ammonia, and hydrogen sulfide present/produced in the fuel cells affect the performance of the fuel cells as CO_2 poisons the surface of the catalyst by damaging the electrochemically active surface area and blocks the hydrogen from reaching the active platinum sites [149]. Biohydrogen is produced by a *C. sorokiniana* strain under sulphur-deprived conditions.

This produced hydrogen is transferred into PEMFC and converted into electricity, and when 27.09 mL of hydrogen was injected, 8.9 mA of current was generated [150]. Hydrogen was produced photobiologically by *Chlamydomonas reinhardtii* and was integrated with PEMFC for electricity generation, and 1.81 mA cm^{-2} of current density was produced for approximately 50 h and 0.23 mA cm^{-2} for approximately 80 h [151]. Hydrogen was produced using the marine microalgae *Tetraselmis subcordiformis* and was coupled with an alkaline fuel cell for the production of electricity [152].

The heat-treated microbial population (HTMP) and the Acid-Treated Microbial Population (ATMP) produced higher H$_2$ yields at 35 °C, where HTMP was *Clostridium* sp. and ATMP was a mixed microbial population. With a flux of 0.9 L/h hydrogen, a PEMFC was operated successfully [153]. Using a single-chambered MFC and pre-fermented wastewater, biohydrogen was produced and simultaneously biohydrogen production was linked to electricity generation. MFC was used to treat wastewater and for bioenergy production [154].

5. Discussion

From a green perspective, biohydrogen adheres to the green chemistry concept because the wastes produced by food, vegetables, and manure are not released into the environment but are instead treated and used to generate hydrogen gas. MFCs offer a cutting-edge, versatile alternative method of producing hydrogen. Over the past few decades, a number of technological developments have improved the yield of the product. However, this technology is still far from being able to serve as a profitable real-world application [155]. The idea of a hydrogen economy is gaining popularity, and technologists are working to obtain methods of producing H$_2$ with a zero-emission plan. More studies on the sustainability of this process must be carried out to understand its efficiency.

6. Conclusions

The use of biohydrogen is an alternative source of energy, as it is a renewable source of energy. It can be produced by various sources such as agricultural waste, industrial waste, municipal waste, and microbial routes. Dark fermentation is considered the most effective method for producing biohydrogen with a higher yield and purity, even though it is not currently feasible for large-scale implementation. It is a very reliable source of energy for electricity generation around the world, but it has its own limitations. The greatest challenge is to ensure that the process is sustainable, considering the low level of substrate conversion, production rate, and yield. Cost-wise and yield-wise, current biohydrogen technologies are not yet competitive with conventional H$_2$ production methods. It is essential to conduct extensive research in order to reduce costs and maximize H$_2$ yield with the current production technologies. As a result, future research should focus on increasing the sustainability and measuring the economic feasibility of biohydrogen production in order to enable its scalability.

Author Contributions: A.V.S.: Conceptualization, writing—original draft, editing, supervision; D.R.: Writing—original draft, editing; M.S.: Writing—original draft, editing; S.S.: Writing—original draft, editing; K.K.: Writing—original draft, editing; N.V.: Writing—original draft, editing; N.J.: Writing—original draft, editing; P.P.: Writing—original draft, editing; N.S.: Writing—original draft, editing. All authors have read and agreed to the published version of the manuscript.

Funding: This research received no external funding.

Institutional Review Board Statement: Not applicable.

Informed Consent Statement: Not applicable.

Data Availability Statement: The data used to support the findings of this study are included in the article. Should further data or information be required, these are available from the corresponding author upon request.

Acknowledgments: The authors thank the Sathyabama Institute of Science and Technology, India and MAHSA University, Malaysia for providing facilities and support to complete this review.

Conflicts of Interest: The authors declare no conflict of interest.

References

1. Ntaikou, I.; Antonopoulou, G.; Lyberatos, G. Biohydrogen production from biomass and wastes via dark fermentation: A review. *Waste Biomass Valorization* **2010**, *1*, 21–39. [CrossRef]
2. Mohan, S.V.; Pandey, A. Biohydrogen production: An introduction. In *Biohydrogen*; Elsevier: Amsterdam, The Netherlands, 2013; pp. 1–24.
3. Chandrasekhar, K.; Lee, Y.J.; Lee, D.W. Biohydrogen production: Strategies to improve process efficiency through microbial routes. *Int. J. Mol. Sci.* **2015**, *16*, 8266–8293. [CrossRef] [PubMed]
4. Kothari, R.; Tyagi, V.V.; Pathak, A. Waste-to-energy: A way from renewable energy sources to sustainable development. *Renew. Sustain. Energy Rev.* **2010**, *14*, 3164–3170. [CrossRef]
5. Rahman, M.M.; Tan, J.H.; Fadzlita, M.T.; Muzammil, A.W.K. A Review on the development of Gravitational Water Vortex Power Plant as alternative renewable energy resources. In *IOP Conference Series: Materials Science and Engineering*; IOP Publishing: Miri, Malaysia, 2017; Volume 217, p. 012007.
6. Jha, S.K.; Puppala, H. Prospects of renewable energy sources in India: Prioritization of alternative sources in terms of Energy Index. *Energy* **2017**, *127*, 116–127. [CrossRef]
7. Demirtas, O. Evaluating the best renewable energy technology for sustainable energy planning. *Int. J. Energy Econ. Policy* **2013**, *3*, 23.
8. Kapdan, I.K.; Kargi, F. Bio-hydrogen production from waste materials. *Enzym. Microb. Technol.* **2006**, *38*, 569–582. [CrossRef]
9. Yasin, N.H.M.; Mumtaz, T.; Hassan, M.A. Food waste and food processing waste for biohydrogen production: A review. *J. Environ. Manag.* **2013**, *130*, 375–385. [CrossRef]
10. Azwar, M.Y.; Hussain, M.A.; Abdul-Wahab, A.K. Development of biohydrogen production by photobiological, fermentation and electrochemical processes: A review. *Renew. Sustain. Energy Rev.* **2014**, *31*, 158–173. [CrossRef]
11. Elbeshbishy, E.; Dhar, B.R.; Nakhla, G.; Lee, H.S. A critical review on inhibition of dark biohydrogen fermentation. *Renew. Sustain. Energy Rev.* **2017**, *79*, 656–668. [CrossRef]
12. Ghimire, A.; Frunzo, L.; Pirozzi, F.; Trably, E.; Escudie, R.; Lens, P.N.; Esposito, G. A review on dark fermentative biohydrogen production from organic biomass: Process parameters and use of by-products. *Appl. Energy* **2015**, *144*, 73–95. [CrossRef]
13. Demirbas, A. *Biohydrogen*; Springer: London, UK, 2009; pp. 163–219.
14. Percival Zhang, Y.-H. Hydrogen Production from Carbohydrates: A Mini-Review. In *Sustainable Production of Fuels, Chemicals, and Fibers from Forest Biomass*; ACS Symposium Series; Oxford University Press: Oxford, UK, 2011; Volume 1067, pp. 203–216.
15. Hallenbeck, P.C.; Ghosh, D. Advances in fermentative biohydrogen production: The way forward? *Trends Biotechnol.* **2009**, *27*, 287–297. [CrossRef]
16. Oh, Y.K.; Raj, S.M.; Jung, G.Y.; Park, S. Current status of the metabolic engineering of microorganisms for biohydrogen production. *Bioresour. Technol.* **2011**, *102*, 8357–8367. [CrossRef]
17. Esper, B.; Badura, A.; Rögner, M. Photosynthesis as a power supply for (bio-)hydrogen production. *Trends Plant Sci.* **2006**, *11*, 543–549. [CrossRef]
18. Guo, X.M.; Trably, E.; Latrille, E.; Carrere, H.; Steyer, J.P. Hydrogen production from agricultural waste by dark fermentation: A review. *Int. J. Hydrogen Energy* **2010**, *35*, 10660–10673. [CrossRef]
19. Li, Y.C.; Liu, Y.F.; Chu, C.Y.; Chang, P.L.; Hsu, C.W.; Lin, P.J.; Wu, S.Y. Techno-economic evaluation of biohydrogen production from wastewater and agricultural waste. *Int. J. Hydrogen Energy* **2012**, *37*, 15704–15710. [CrossRef]
20. Cheng, C.L.; Lo, Y.C.; Lee, K.S.; Lee, D.J.; Lin, C.Y.; Chang, J.S. Biohydrogen production from lignocellulosic feedstock. *Bioresour. Technol.* **2011**, *102*, 8514–8523. [CrossRef]
21. Saratale, G.D.; Saratale, R.G.; Chang, J.S. *Biohydrogen from Renewable Resources*; Pandey, A., Chang, J.S., Hallenbeck, P., Larroche, C., Eds.; Elsevier: Amsterdam, The Netherlands, 2013; pp. 185–221.
22. Saratale, G.D.; Kshirsagar, S.D.; Sampange, V.T.; Saratale, R.G.; Oh, S.E.; Govindwar, S.P.; Oh, M.K. Cellulolytic enzymes production by utilizing agricultural wastes under solid state fermentation and its application for biohydrogen production. *Appl. Biochem. Biotechnol.* **2014**, *174*, 2801–2817. [CrossRef]
23. Kang, S.W.; Park, Y.S.; Lee, J.S.; Hong, S.I.; Kim, S.W. Production of cellulases and hemicellulases by *Aspergillus niger* KK2 from lignocellulosic biomass. *Bioresour. Technol.* **2004**, *91*, 153–156. [CrossRef]
24. Kumar, G.; Bakonyi, P.; Periyasamy, S.; Kim, S.H.; Nemestóthy, N.; Bélafi-Bakó, K. Lignocellulose biohydrogen: Practical challenges and recent progress. *Renew. Sustain. Energy Rev.* **2015**, *44*, 728–737. [CrossRef]
25. Monlau, F.; Barakat, A.; Trably, E.; Dumas, C.; Steyer, J.P.; Carrère, H. Lignocellulosic materials into biohydrogen and biomethane: Impact of structural features and pretreatment. *Crit. Rev. Environ. Sci. Technol.* **2013**, *43*, 260–322. [CrossRef]
26. Sivagurunathan, P.; Kumar, G.; Mudhoo, A.; Rene, E.R.; Saratale, G.D.; Kobayashi, T.; Xu, K.; Kim, S.-H.; Kim, D.-H. Fermentative hydrogen production using lignocellulose biomass: An overview of pre-treatment methods, inhibitor effects and detoxification experiences. *Renew. Sustain. Energy Rev.* **2017**, *77*, 28–42. [CrossRef]
27. Soares, J.F.; Confortin, T.C.; Todero, I.; Mayer, F.D.; Mazutti, M.A. Dark fermentative biohydrogen production from lignocellulosic biomass: Technological challenges and future prospects. *Renew. Sustain. Energy Rev.* **2020**, *117*, 109484. [CrossRef]

28. Singh, A.; Sevda, S.; Abu Reesh, I.M.; Vanbroekhoven, K.; Rathore, D.; Pant, D. Biohydrogen production from lignocellulosic biomass: Technology and sustainability. *Energies* **2015**, *8*, 13062–13080. [CrossRef]

29. Taguchi, F.; Mizukami, N.; Yamada, K.; Hasegawa, K.; Saito-Taki, T. Direct conversion of cellulosic materials to hydrogen by *Clostridium* sp. strain no. 2. *Enzym. Microb. Technol.* **1995**, *17*, 147–150. [CrossRef]

30. Taguchi, F.; Yamada, K.; Hasegawa, K.; Taki-Saito, T.; Hara, K. Continuous hydrogen production by *Clostridium* sp. strain no. 2 from cellulose hydrolysate in an aqueous two-phase system. *J. Ferment. Bioeng.* **1996**, *82*, 80–83. [CrossRef]

31. Liu, H.; Zhang, T.; Fang, H.H. Thermophilic H_2 production from a cellulose-containing wastewater. *Biotechnol. Lett.* **2003**, *25*, 365–369. [CrossRef]

32. Fan, Y.T.; Zhang, G.S.; Guo, X.Y.; Xing, Y.; Fan, M.H. Biohydrogen-production from beer lees biomass by cow dung compost. *Biomass Bioenergy.* **2006**, *30*, 493–496. [CrossRef]

33. Ren, N.Q.; Cao, G.L.; Guo, W.Q.; Wang, A.J.; Zhu, Y.H.; Liu, B.F.; Xu, J.F. Biological hydrogen production from corn stover by moderately thermophile *Thermoanaerobacterium thermosaccharolyticum* W16. *Int. J. Hydrogen Energy* **2010**, *35*, 2708–2712. [CrossRef]

34. Sigurbjornsdottir, M.A.; Orlygsson, J. Combined hydrogen and ethanol production from sugars and lignocellulosic biomass by *Thermoanaerobacterium* AK54, isolated from hot spring. *Appl. Energy.* **2012**, *97*, 785–791. [CrossRef]

35. Han, H.; Wei, L.; Liu, B.; Yang, H.; Shen, J. Optimization of biohydrogen production from soybean straw using anaerobic mixed bacteria. *Int. J. Hydrogen Energy.* **2012**, *37*, 13200–13208. [CrossRef]

36. Pan, C.; Fan, Y.; Hou, H. Fermentative production of hydrogen from wheat bran by mixed anaerobic cultures. *Ind. Eng. Chem. Res.* **2008**, *47*, 5812–5818. [CrossRef]

37. Keskin, T.; Abubackar, H.N.; Arslan, K.; Azbar, N. Biohydrogen production from solid wastes. In *Biohydrogen*; Elsevier: Amsterdam, The Netherlands, 2019; pp. 321–346.

38. Thornton, P.K. Livestock production: Recent trends, future prospects. *Philos. Trans. R. Soc. B Biol. Sci.* **2010**, *365*, 2853–2867. [CrossRef]

39. Solomon, A.; Authority, E.L.M. *Livestock Marketing in Ethiopia: A Review of Structure, Performance, and Development Initiatives*; ILRI (aka ILCA and ILRAD): Addis Ababa, Ethiopia, 2003; Volume 52.

40. Xing, Y.; Li, Z.; Fan, Y.; Hou, H. Biohydrogen production from dairy manures with acidification pretreatment by anaerobic fermentation. *Environ. Sci. Pollut. Res.* **2010**, *17*, 392–399. [CrossRef]

41. Saratale, G.D.; Saratale, R.G.; Banu, J.R.; Chang, J.S. Biohydrogen production from renewable biomass resources. In *Biohydrogen*; Elsevier: Amsterdam, The Netherlands, 2019; pp. 247–277.

42. Lateef, S.A.; Beneragama, N.; Yamashiro, T.; Iwasaki, M.; Ying, C.; Umetsu, K. Biohydrogen production from co-digestion of cow manure and waste milk under thermophilic temperature. *Bioresour. Technol.* **2012**, *110*, 251–257. [CrossRef]

43. Tenca, A.; Schievano, A.; Perazzolo, F.; Adani, F.; Oberti, R. Biohydrogen from thermophilic co-fermentation of swine manure with fruit and vegetable waste: Maximizing stable production without pH control. *Bioresour. Technol.* **2011**, *102*, 8582–8588. [CrossRef]

44. Marone, A.; Varrone, C.; Fiocchetti, F.; Giussani, B.; Izzo, G.; Mentuccia, L.; Rosa, S.; Signorini, A. Optimization of substrate composition for biohydrogen production from buffalo slurry co-fermented with cheese whey and crude glycerol, using microbial mixed culture. *Int. J. Hydrogen Energy* **2015**, *40*, 209–218. [CrossRef]

45. El Bari, H.; Lahboubi, N.; Habchi, S.; Rachidi, S.; Bayssi, O.; Nabil, N.; Mortezaei, Y.; Villa, R. Biohydrogen production from fermentation of Organic Waste, storage and applications. *Clean. Waste Syst.* **2022**, *3*, 100043. [CrossRef]

46. Liu, S.; Li, W.; Zheng, G.; Yang, H.; Li, L. Optimization of Cattle Manure and Food Waste Co-Digestion for Biohydrogen Production in a Mesophilic Semi-Continuous Process. *Energies* **2020**, *13*, 3848. [CrossRef]

47. Gilroyed, B.H.; Li, C.; Hao, X.; Chu, A.; McAllister, T.A. Biohydrogen production from specified risk materials co-digested with cattle manure. *Int. J. Hydrogen Energy* **2010**, *35*, 1099–1105. [CrossRef]

48. Fangkum, A.; Reungsang, A. Biohydrogen production from sugarcane bagasse hydrolysate by elephant dung: Effects of initial pH and substrate concentration. *Int. J. Hydrogen Energy* **2011**, *36*, 8687–8696. [CrossRef]

49. Ghimire, A.; Luongo, V.; Frunzo, L.; Pirozzi, F.; Lens, P.N.; Esposito, G. Continuous biohydrogen production by thermophilic dark fermentation of cheese whey: Use of buffalo manure as buffering agent. *Int. J. Hydrogen Energy* **2017**, *42*, 4861–4869. [CrossRef]

50. Wadjeam, P.; Reungsang, A.; Imai, T.; Plangklang, P. Co-digestion of cassava starch wastewater with buffalo dung for bio-hydrogen production. *Int. J. Hydrogen Energy* **2019**, *44*, 14694–14706. [CrossRef]

51. Perera, K.R.J.; Nirmalakhandan, N. Modeling fermentative hydrogen production from sucrose supplemented with dairy manure. *Int. J. Hydrogen Energy* **2011**, *36*, 2102–2110. [CrossRef]

52. Wu, X.; Lin, H.; Zhu, J. Optimization of continuous hydrogen production from co-fermenting molasses with liquid swine manure in an anaerobic sequencing batch reactor. *Bioresour. Technol.* **2013**, *136*, 351–359. [CrossRef] [PubMed]

53. Zhu, J.; Li, Y.; Wu, X.; Miller, C.; Chen, P.; Ruan, R. Swine manure fermentation for hydrogen production. *Bioresour. Technol.* **2009**, *100*, 5472–5477. [CrossRef]

54. Wu, X.; Zhu, J.; Dong, C.; Miller, C.; Li, Y.; Wang, L. Continuous biohydrogen production from liquid swine manure supplemented with glucose using an anaerobic sequencing batch reactor. *Int. J. Hydrogen Energy* **2009**, *34*, 6636–6645. [CrossRef]

55. Kotsopoulos, T.A.; Fotidis, I.A.; Tsolakis, N.; Martzopoulos, G.G. Biohydrogen production from pig slurry in a CSTR reactor system with mixed cultures under hyper-thermophilic temperature (70 C). *Biomass Bioenergy* **2009**, *33*, 1168–1174. [CrossRef]

56. Ahmad, T.; Aadil, R.M.; Ahmed, H.; Rahman, U.; Soares, B.C.V.; Souza, S.L.Q.; Pimentel, T.C.; Scudino, H.; Guimarães, J.T.; Esmerino, E.A.; et al. Treatment and utilization of dairy industrial waste: A review. *Trends Food Sci. Technol.* **2019**, *88*, 361–372. [CrossRef]

57. Sen, T.; Mishra, U. Usage of Industrial Waste Products in Village Road Construction. *Int. J. Environ. Sci. Dev.* **2010**, *1*, 122. [CrossRef]

58. Chong, M.L.; Sabaratnam, V.; Shirai, Y.; Hassan, M.A. Biohydrogen production from biomass and industrial wastes by dark fermentation. *Int. J. Hydrogen Energy* **2009**, *34*, 3277–3328. [CrossRef]

59. De Vrije, T.; De Haas, G.G.; Tan, G.B.; Keijsers, E.R.P.; Claassen, P.A.M. Pretreatment of Miscanthus for hydrogen production by Thermotoga elfii. *Int. J. Hydrogen Energy* **2002**, *27*, 1381–1390. [CrossRef]

60. Alexandropoulou, M.; Antonopoulou, G.; Trably, E.; Carrere, H.; Lyberatos, G. Continuous biohydrogen production from a food industry waste: Influence of operational parameters and microbial community analysis. *J. Clean. Prod.* **2018**, *174*, 1054–1063. [CrossRef]

61. Moreno-Andrade, I.; Moreno, G.; Kumar, G.; Buitrón, G. Biohydrogen production from industrial wastewaters. *Water Sci. Technol.* **2015**, *71*, 105–110. [CrossRef]

62. Moreno-Dávila, I.M.M.; Ríos-González, L.J.; Rodríguez-de la Garza, J.A.; Morales-Martínez, T.K.; Garza-García, Y. Biohydrogen production from paper industry wastes by SSF: A study of the influence of temperature/enzyme loading. *Int. J. Hydrogen Energy* **2019**, *44*, 12333–12338. [CrossRef]

63. Oceguera-Contreras, E.; Aguilar-Juárez, O.; Oseguera-Galindo, D.; Macías-Barragán, J.; Bolaños-Rosales, R.; Mena-Enríquez, M.; Arias-García, A.; Montoya-Buelna, M.; Graciano-Machuca, O.; De León-Rodríguez, A. Biohydrogen production by vermihumus-associated microorganisms using agro industrial wastes as substrate. *Int. J. Hydrogen Energy* **2019**, *44*, 9856–9865. [CrossRef]

64. Lopez-Hidalgo, A.M.; Alvarado-Cuevas, Z.D.; De Leon-Rodriguez, A. Biohydrogen production from mixtures of agro-industrial wastes: Chemometric analysis, optimization and scaling up. *Energy* **2018**, *159*, 32–41. [CrossRef]

65. Lucas, S.D.M.; Peixoto, G.; Mockaitis, G.; Zaiat, M.; Gomes, S.D. Energy recovery from agro-industrial wastewaters through biohydrogen production: Kinetic evaluation and technological feasibility. *Renew. Energy* **2015**, *75*, 496–504. [CrossRef]

66. Gomez-Romero, J.; Gonzalez-Garcia, R.A.; Chairez, I.; Torres, L.; García-Peña, E.I. Continuous two-staged co-digestion process for biohydrogen production from agro-industrial wastes. *Int. J. Energy Res.* **2016**, *40*, 257–272. [CrossRef]

67. Vendruscolo, F. Starch: A potential substrate for biohydrogen production. *Int. J. Energy Res.* **2015**, *39*, 293–302. [CrossRef]

68. Lu, L.; Ren, N.; Xing, D.; Logan, B.E. Hydrogen production with effluent from an ethanol–H$_2$-coproducing fermentation reactor using a single-chamber microbial electrolysis cell. *Biosens. Bioelectron.* **2009**, *24*, 3055–3060. [CrossRef]

69. Lakshmidevi, R.; Muthukumar, K. Enzymatic saccharification and fermentation of paper and pulp industry effluent for biohydrogen production. *Int. J. Hydrogen Energy* **2010**, *35*, 3389–3400. [CrossRef]

70. Mohammadi, P.; Ibrahim, S.; Annuar, M.S.M.; Law, S. Effects of different pretreatment methods on anaerobic mixed microflora for hydrogen production and COD reduction from palm oil mill effluent. *J. Clean. Prod.* **2011**, *19*, 1654–1658. [CrossRef]

71. Lin, C.Y.; Chiang, C.C.; Nguyen, M.L.T.; Lay, C.H. Enhancement of fermentative biohydrogen production from textile desizing wastewater via coagulation-pretreatment. *Int. J. Hydrogen Energy* **2017**, *42*, 12153–12158. [CrossRef]

72. Taifor, A.F.; Zakaria, M.R.; Yusoff, M.Z.M.; Toshinari, M.; Hassan, M.A.; Shirai, Y. Elucidating substrate utilization in biohydrogen production from palm oil mill effluent by Escherichia coli. *Int. J. Hydrogen Energy* **2017**, *42*, 5812–5819. [CrossRef]

73. Li, Y.C.; Chu, C.Y.; Wu, S.Y.; Tsai, C.Y.; Wang, C.C.; Hung, C.H.; Lin, C.Y. Feasible pretreatment of textile wastewater for dark fermentative hydrogen production. *Int. J. Hydrogen Energy* **2012**, *37*, 15511–15517. [CrossRef]

74. Ramprakash, B.; Muthukumar, K. Comparative study on the production of biohydrogen from rice mill wastewater. *Int. J. Hydrogen Energy* **2014**, *39*, 14613–14621. [CrossRef]

75. Tyagi, V.K.; Lo, S.L. Sludge: A waste or renewable source for energy and resources recovery? *Renew. Sustain. Energy Rev.* **2013**, *25*, 708–728. [CrossRef]

76. Johnson, O.A.; Napiah, M.; Kamaruddin, I. Potential uses of waste sludge in construction industry: A review. *Res. J. Appl. Sci. Eng. Technol.* **2014**, *8*, 565–570. [CrossRef]

77. Kim, S.H.; Han, S.K.; Shin, H.S. Feasibility of biohydrogen production by anaerobic co-digestion of food waste and sewage sludge. *Int. J. Hydrogen Energy* **2004**, *29*, 1607–1616. [CrossRef]

78. Cai, M.; Liu, J.; Wei, Y. Enhanced biohydrogen production from sewage sludge with alkaline pretreatment. *Environ. Sci. Technol.* **2004**, *38*, 3195–3202. [CrossRef]

79. Yin, Y.; Wang, J. Biohydrogen production using waste activated sludge disintegrated by gamma irradiation. *Appl. Energy* **2015**, *155*, 434–439. [CrossRef]

80. Yang, G.; Hu, Y.; Wang, J. Biohydrogen production from co-fermentation of fallen leaves and sewage sludge. *Bioresour. Technol.* **2019**, *285*, 121342. [CrossRef] [PubMed]

81. Wang, Y.; Wang, H.; Feng, X.; Wang, X.; Huang, J. Biohydrogen production from cornstalk wastes by anaerobic fermentation with activated sludge. *Int. J. Hydrogen Energy* **2010**, *35*, 3092–3099. [CrossRef]

82. Han, W.; Liu, D.N.; Shi, Y.W.; Tang, J.H.; Li, Y.F.; Ren, N.Q. Biohydrogen production from food waste hydrolysate using continuous mixed immobilized sludge reactors. *Bioresour. Technol.* **2015**, *180*, 54–58. [CrossRef] [PubMed]

83. Yang, G.; Wang, J. Biohydrogen production from waste activated sludge pretreated by combining sodium citrate with ultrasonic: Energy conversion and microbial community. *Energy Convers. Manag.* **2020**, *225*, 113436. [CrossRef]

84. Yang, G.; Wang, J. Enhancing biohydrogen production from waste activated sludge disintegrated by sodium citrate. *Fuel* **2019**, *258*, 116177. [CrossRef]

85. Lin, Y.H.; Zheng, H.X.; Juan, M.L. Biohydrogen production using waste activated sludge as a substrate from fructose-processing wastewater treatment. *Process Saf. Environ. Prot.* **2012**, *90*, 221–230. [CrossRef]

86. Zhu, H.; Béland, M. Evaluation of alternative methods of preparing hydrogen producing seeds from digested wastewater sludge. *Int. J. Hydrogen Energy* **2006**, *31*, 1980–1988. [CrossRef]

87. Argun, H.; Kargi, F. Effects of sludge pre-treatment method on bio-hydrogen production by dark fermentation of waste ground wheat. *Int. J. Hydrogen Energy* **2009**, *34*, 8543–8548. [CrossRef]

88. Wang, J.; Wan, W. Comparison of different pretreatment methods for enriching hydrogen-producing bacteria from digested sludge. *Int. J. Hydrogen Energy* **2008**, *33*, 2934–2941. [CrossRef]

89. Luo, G.; Xie, L.; Zou, Z.; Zhou, Q.; Wang, J.Y. Fermentative hydrogen production from cassava stillage by mixed anaerobic microflora: Effects of temperature and pH. *Appl. Energy* **2010**, *87*, 3710–3717. [CrossRef]

90. Tang, G.L.; Huang, J.; Sun, Z.J.; Tang, Q.Q.; Yan, C.H.; Liu, G.Q. Biohydrogen production from cattle wastewater by enriched anaerobic mixed consortia: Influence of fermentation temperature and pH. *J. Biosci. Bioeng.* **2008**, *106*, 80–87. [CrossRef]

91. Wang, C.C.; Chang, C.W.; Chu, C.P.; Lee, D.J.; Chang, B.V.; Liao, C.S. Producing hydrogen from wastewater sludge by *Clostridium bifermentans*. *J. Biotechnol.* **2003**, *102*, 83–92. [CrossRef]

92. Kamalaskar, L.B.; Dhakephalkar, P.K.; Meher, K.K.; Ranade, D.R. High biohydrogen yielding Clostridium sp. DMHC-10 isolated from sludge of distillery waste treatment plant. *Int. J. Hydrogen Energy* **2010**, *35*, 10639–10644. [CrossRef]

93. Shaishav, S.; Singh, R.N.; Satyendra, T. Biohydrogen from algae: Fuel of the future. *Int. Res. J. Environ. Sci.* **2013**, *2*, 44–47.

94. Hafez, H.; Baghchehsaraee, B.; Nakhla, G.; Karamanev, D.; Margaritis, A.; El Naggar, H. Comparative assessment of decoupling of biomass and hydraulic retention times in hydrogen production bioreactors. *Int. J. Hydrogen Energy* **2009**, *34*, 7603–7611. [CrossRef]

95. Hafez, H.; Nakhla, G.; El Naggar, H. An integrated system for hydrogen and methane production during landfill leachate treatment. *Int. J. Hydrogen Energy* **2010**, *35*, 5010–5014. [CrossRef]

96. Elsharnouby, O.; Hafez, H.; Nakhla, G.; El Naggar, M.H. A critical literature review on biohydrogen production by pure cultures. *Int. J. Hydrogen Energy* **2013**, *38*, 4945–4966. [CrossRef]

97. Hafez, H.; Nakhla, G.; El Naggar, H. Biological hydrogen production. In *Handbook of Hydrogen Energy*; Sherif, S.A., Ed.; CRC Press: Boca Raton, FL, USA; Taylor and Francis: Abingdon, UK, 2012.

98. Lin, P.; Whang, L.; Wu, Y.; Ren, W.; Hsiao, C.; Li, S.; Chang, J. Biological hydrogen production of the genus *Clostridium*: Metabolic study and mathematical model simulation. *Int. J. Hydrogen Energy* **2007**, *32*, 1728–1735. [CrossRef]

99. Yokoi, H.; Maki, R.; Hirose, J.; Hayashi, S. Microbial production of hydrogen from starch-manufacturing wastes. *Biomass Bioenerg* **2002**, *22*, 389–395. [CrossRef]

100. Seppälä, J.J.; Puhakka, J.A.; Yli-Harja, O.; Karp, M.T.; Santala, V. Fermentative hydrogen production by *Clostridium butyricum* and *Escherichia coli* in pure and cocultures. *Int. J. Hydrogen Energy* **2011**, *36*, 10701–10708. [CrossRef]

101. Li, Q.; Liu, C. Co-culture of *Clostridium thermocellum* and *Clostridium thermosaccharolyticum* for enhancing hydrogen production via thermophilic fermentation of cornstalk waste. *Int. J. Hydrogen Energy* **2012**, *37*, 10648–10654. [CrossRef]

102. Liu, F.; Fang, B. Optimization of bio-hydrogen production from biodiesel wastes by *Klebsiella pneumoniae*. *Biotechnol. J.* **2007**, *2*, 374–380. [CrossRef] [PubMed]

103. Plangklang, P.; Reungsang, A.; Pattra, S. Enhanced biohydrogen production from sugarcane juice by immobilized *Clostridium butyricum* on sugarcane bagasse. *Int. J. Hydrogen Energy* **2012**, *37*, 15525–15532. [CrossRef]

104. Wang, A.; Ren, N.; Shi, Y.; Lee, D. Bioaugmented hydrogen production from microcrystalline cellulose using cocultured *Clostridium acetobutylicum* and *Ethanoigenens harbinense*. *Int. J. Hydrogen Energy* **2008**, *33*, 912–917. [CrossRef]

105. Cappelletti, B.M.; Reginatto, V.; Amante, E.R.; Antonio, R.V. Fermentative production of hydrogen from cassava processing wastewater by *Clostridium acetobutylicum*. *Renew. Energy* **2011**, *36*, 3367–3372. [CrossRef]

106. Kumar, N.; Das, D. Enhancement of hydrogen production by *Enterobacter cloacae* IIT-BT 08. *Process Biochem.* **2000**, *35*, 589–593. [CrossRef]

107. Mars, A.E.; Veuskens, T.; Budde, M.A.; Van Doeveren, P.F.; Lips, S.J.; Bakker, R.R.; de Vrije, T.; Claassen, P.A. Biohydrogen production from untreated and hydrolyzed potato steam peels by the extreme thermophiles *Caldicellulosiruptor saccharolyticus* and *Thermotoga neapolitana*. *Int. J. Hydrogen Energy* **2010**, *35*, 7730–7737. [CrossRef]

108. Manish, S.; Banerjee, R. Comparison of biohydrogen production processes. *Int. J. Hydrogen Energy* **2008**, *33*, 279–286. [CrossRef]

109. Hallenbeck, P.C.; Benemann, J.R. Biological hydrogen production: Fundamentals and limiting processes. *Int. J. Hydrogen. Energy* **2002**, *27*, 1185–1193. [CrossRef]

110. Ni, M.; Leung, D.Y.; Leung, M.K.; Sumathy, K. An overview of hydrogen production from biomass. *Fuel Process. Technol.* **2006**, *87*, 461–472. [CrossRef]

111. Saifuddin, N.; Priatharsini, P. Developments in bio-hydrogen production from algae: A review. *Res. J. Appl. Sci. Eng. Technol.* **2016**, *12*, 968–982. [CrossRef]

112. Akhlaghi, N.; Najafpour-Darzi, G. A comprehensive review on biological hydrogen production. *Int. J. Hydrogen Energy* **2020**, *45*, 22492–22512. [CrossRef]

113. Yu, J.; Takahashi, P. Biophotolysis-based hydrogen production by cyanobacteria and green microalgae. *Commun. Curr. Res. Educ. Top. Trends Appl. Microbiol.* **2007**, *1*, 79–89.

114. Kayfeci, M.; Keçebaş, A.; Bayat, M. Hydrogen production. In *Solar Hydrogen Production*; Academic Press: Cambridge, MA, USA, 2019; pp. 45–83.
115. Kossalbayev, B.D.; Tomo, T.; Zayadan, B.K.; Sadvakasova, A.K.; Bolatkhan, K.; Alwasel, S.; Allakhverdiev, S.I. Determination of the potential of cyanobacterial strains for hydrogen production. *Int. J. Hydrogen Energy* **2020**, *45*, 2627–2639. [CrossRef]
116. Najafpour, G.D.; Shahavi, M.H.; Neshat, S.A. Assessment of biological hydrogen production processes: A review. In *IOP Conference Series: Earth and Environmental Science*; IOP Publishing: Bristol, UK, 2016; Volume 36, p. 012068. [CrossRef]
117. Kothari, R.; Tyagi, V.V.; Pathak, A. Renewable and sustainable energy reviews. *Renew. Sustain. Energ. Rev.* **2010**, *14*, 1744–1751.
118. Benemann, J.R. Feasibility analysis of photobiological hydrogen production. *Int. J. Hydrogen Energy* **1997**, *22*, 979–987. [CrossRef]
119. Sen, U.; Shakdwipee, M.; Banerjee, R. Status of biological hydrogen production. *J. Sci. Ind. Res.* **2008**, *67*, 980–993.
120. Sarangi, P.K.; Nanda, S. Biohydrogen production through dark fermentation. *Chem. Eng. Technol.* **2020**, *43*, 601–612. [CrossRef]
121. Saratale, G.D.; Chen, S.D.; Lo, Y.C.; Saratale, R.G.; Chang, J.S. Outlook of biohydrogen production from lignocellulosic feedstock using dark fermentation–a review. *J. Sci. Ind. Res.* **2008**, *67*, 11.
122. Lee, D.J.; Show, K.Y.; Su, A. Dark fermentation on biohydrogen production: Pure culture. *Bioresour. Technol.* **2011**, *102*, 8393–8402. [CrossRef] [PubMed]
123. Srivastava, N.; Srivastava, M.; Malhotra, B.D.; Gupta, V.K.; Ramteke, P.W.; Silva, R.N.; Shukla, P.; Dubey, K.K.; Mishra, P.K. Nanoengineered cellulosic biohydrogen production via dark fermentation: A novel approach. *Biotechnol. Adv.* **2019**, *37*, 107384. [CrossRef] [PubMed]
124. Tapia-Venegas, E.; Ramirez-Morales, J.E.; Silva-Illanes, F.; Toledo-Alarcón, J.; Paillet, F.; Escudie, R.; Lay, C.H.; Chu, C.Y.; Leu, H.J.; Marone, A. Biohydrogen production by dark fermentation: Scaling-up and technologies integration for a sustainable system. *Rev. Environ. Sci. Bio/Technol.* **2015**, *14*, 761–785. [CrossRef]
125. Hay, J.X.W.; Wu, T.Y.; Juan, J.C.; Jahim, J.M. Biohydrogen production through photo fermentation or dark fermentation using waste as a substrate: Overview, economics, and future prospects of hydrogen usage. *Biofuels Bioprod. Biorefining* **2013**, *7*, 334–352. [CrossRef]
126. Khanna, N.; Das, D. Biohydrogen production by dark fermentation. *Wiley Interdiscip. Rev. Energy Environ.* **2013**, *2*, 401–421. [CrossRef]
127. Argun, H.; Kargi, F. Bio-hydrogen production by different operational modes of dark and photo-fermentation: An overview. *Int. J. Hydrogen Energy* **2011**, *36*, 7443–7459. [CrossRef]
128. Budiman, P.M.; Wu, T.Y. Role of chemicals addition in affecting biohydrogen production through photofermentation. *Energy Convers. Manag.* **2018**, *165*, 509–527. [CrossRef]
129. Hitam, C.N.C.; Jalil, A.A. A review on biohydrogen production through photo-fermentation of lignocellulosic biomass. *Biomass Convers. Biorefinery* **2020**, 1–19. [CrossRef]
130. Sampath, P.; Reddy, K.R.; Reddy, C.V.; Shetti, N.P.; Kulkarni, R.V.; Raghu, A.V. Biohydrogen production from organic waste–a review. *Chem. Eng. Technol.* **2020**, *43*, 1240–1248. [CrossRef]
131. Adessi, A.; De Philippis, R. Hydrogen production: Photofermentation. In *Microbial Technologies in Advanced Biofuels Production*; Springer: Boston, MA, USA, 2012; pp. 53–75.
132. Androga, D.D.; Özgür, E.; Eroglu, I.; Gündüz, U.; Yücel, M. Photofermentative hydrogen production in outdoor conditions. In *Hydrogen Energy—Challenges and Perspectives*; In Tech, 2012; pp. 77–120.
133. Keskin, T.; Abo-Hashesh, M.; Hallenbeck, P.C. Photofermentative hydrogen production from wastes. *Bioresour. Technol.* **2011**, *102*, 8557–8568. [CrossRef]
134. Tian, H.; Li, J.; Yan, M.; Tong, Y.W.; Wang, C.H.; Wang, X. Organic waste to biohydrogen: A critical review from technological development and environmental impact analysis perspective. *Appl. Energy* **2019**, *256*, 113961. [CrossRef]
135. Osman, A.I.; Deka, T.J.; Baruah, D.C.; Rooney, D.W. Critical challenges in biohydrogen production processes from the organic feedstocks. *Biomass Convers. Biorefin.* **2020**, 1–19. [CrossRef]
136. Bermudez, J.M.; Fidalgo, B. Production of bio-syngas and bio-hydrogen via gasification. In *Handbook of Biofuels Production*, 2nd ed.; Elsevier: Amsterdam, The Netherlands, 2016; pp. 431–494.
137. Reddy, S.N.; Nanda, S.; Dalai, A.K.; Kozinski, J.A. Supercritical water gasification of biomass for hydrogen production. *Int. J. Hydrogen Energy* **2014**, *39*, 6912–6926. [CrossRef]
138. Arregi, A.; Amutio, M.; Lopez, G.; Bilbao, J.; Olazar, M. Evaluation of thermochemical routes for hydrogen production from biomass: A review. *Energy Convers. Manag.* **2018**, *165*, 696–719. [CrossRef]
139. Dou, B.; Zhang, H.; Song, Y.; Zhao, L.; Jiang, B.; He, M.; Ruan, C.; Chen, H.; Xu, Y. Hydrogen production from the thermochemical conversion of biomass: Issues and challenges. *Sustain. Energy Fuels* **2019**, *3*, 314–342. [CrossRef]
140. Ding, C.; Yang, K.L.; He, J. Biological and fermentative production of hydrogen. In *Handbook of Biofuels Production*, 2nd ed.; Elsevier: Amsterdam, The Netherlands, 2016; pp. 303–333.
141. Dincer, I. *Comprehensive Energy Systems*; Elsevier: Amsterdam, The Netherlands, 2018; Volume 3, pp. 1–40.
142. Bolatkhan, K.; Kossalbayev, B.D.; Zayadan, B.K.; Tomo, T.; Veziroglu, T.N.; Allakhverdiev, S.I. Hydrogen production from phototrophic microorganisms: Reality and perspectives. *Int. J. Hydrogen Energy* **2019**, *44*, 5799–5811. [CrossRef]
143. Melitos, G.; Voulkopoulos, X.; Zabaniotou, A. Waste to Sustainable Biohydrogen Production Via Photo-Fermentation and Biophotolysis—A Systematic Review. *Renew. Energy Environ. Sustain.* **2021**, *6*, 45. [CrossRef]